Volcanic Plumes

Volcanic Plumes

Impacts on the Atmosphere and Insights into Volcanic Processes

Special Issue Editors

Pasquale Sellitto
Giuseppe Salerno
Andrew McGonigle

MDPI • Basel • Beijing • Wuhan • Barcelona • Belgrade

MDPI

Special Issue Editors

Pasquale Sellitto
Laboratoire de Météorologie Dynamique
France

Giuseppe Salerno
Istituto Nazionale di Geofisica e Vulcanologia
Italy

Andrew McGonigle
Department of Geography, University of Sheffield
UK

Editorial Office
MDPI
St. Alban-Anlage 66
4052 Basel, Switzerland

This is a reprint of articles from the Special Issue published online in the open access journal *Geosciences* (ISSN 2076-3263) from 2017 to 2018 (available at: https://www.mdpi.com/journal/geosciences/special_issues/volcanic_processes)

For citation purposes, cite each article independently as indicated on the article page online and as indicated below:

LastName, A.A.; LastName, B.B.; LastName, C.C. Article Title. *Journal Name* **Year**, *Article Number*, Page Range.

ISBN 978-3-03897-628-8 (Pbk)
ISBN 978-3-03897-629-5 (PDF)

Credit: Mt Etna eruption on 8th of September 2011 from the town of Fleri(east flank of Mt. Etna). Boris Behncke, Istituto Nazionale di Geofisica e Vulcanologia, Osservatorio Etneo.

Contents

About the Special Issue Editors

Pasquale Sellitto is an Associate Professor at Laboratoire Interuniversitaire des Systèmes Atmospheriques, Université Paris-Est, France. Formed as a physicist, he received his PhD in Geo-Information and Remote Sensing at Tor Vergata University, Rome. He was Visiting Scientist at NASA-GSFC, Greenbelt, MD, USA, and worked as an Assistant Professor at Ecole Normale Supérieure, Paris, France, and as an Atmospheric Scientist at Rutherford Appleton Laboratory, Chilton-Oxford, UK. He is an expert in atmospheric spectroscopy and radiative transfer, remote sensing and modelling of the atmosphere and climate forcing, with applications to boundary layer to upper-tropospheric–lower-stratospheric sources, like volcanic activity and Monsoon transport of Asian pollution. He has published over 30 papers. He won "La Recherche" prize 2014 (section: "Environment"), awarded by "La Recherche" magazine, and the "Italian Physical Society–Associazione Geofisica Italiana 2006" prize. He is Principal Investigator and co-Principal Investigator of several national (French) and international projects. He is Principal Investigator of the Etna Plume Lab research cluster.

Giuseppe Salerno is researcher at INGV Catania, PhD in Volcanlogy at the University of Cambridge (UK). His research focus on volcanic degassing and integration of geophysical data for exploring eruptive mechanism process and the time scales over which these processes occur. He is responsible person of the gas geochemistry team of INGV Catania, and he has been very closely involved in the monitoring and emergency response to Etnean and Stromboli and El Salvador eruptions. He has also carried out several international research campaigns in North, Central, and South America, Iceland, USA and Antartica. Dr Salerno has published 51 ISI publications and 4 book chapters.

Andrew McGonigle is a Reader in volcanology at the University of Sheffield. He was previously a NERC Independent Research Fellow at the University of Cambridge, following on from degrees at the Universities of St. Andrews and Oxford. He is a Laureat of the Rolex Awards for Enterprise for his work on developing new remote sensing tools for volcano monitoring.

Preface to "Volcanic Plumes: Impacts on the Atmosphere and Insights into Volcanic Processes"

Volcanoes release a mixture of gas and particles into the atmosphere. These ejecta not only have fundamental implications on the style and timing of eruptions but may also have significant effects on the global climate and environment. Volcanic gases and particles may alter the chemical composition of the tropo-stratosphere, perturbing the Earth's radiation budget and climate system over a range of temporal and spatial scales. Furthermore, volcanic plumes can affect the air quality, pose hazards to aviation and human health, and damage ecosystems. The chemical compositions and emission rates of volcanic plumes can be observed via a range of direct sampling and remote sensing instrumentation in order to gain insights into subterranean processes and to monitor plume cloud dispersion in the atmosphere. Over the last decades, thanks to technological advances, major progress has been made in the understanding of volcanic plumes. New instruments have enabled the widening of the global volcanic gas inventory, and novel data analytic procedures have advanced the understanding of eruptive mechanisms and the impact of gas in atmospheric chemistry and physics as well as allowing the refinement and tuning of models.

Hence, it is an appropriate time to produce a book of recent research in volcanic plumes by world-class scientists in the fields of the Geosciences. Though several excellent texts exist that cover some of the subjects tackled in this book, none explore volcanic plumes in a synoptic approach from volcanology to atmospheric sciences and from observation to modeling. The content of the book includes selected chapters covering four main sections of research. The first two explore the controls of volcanic degassing process and ash emissions in eruptive mechanisms together with modelling of physical dispersal and chemical and microphysical evolution of plumes in the atmosphere. Sections three and four pave the way for exciting developments in volcano monitoring instrumentation from both ground- and space-based platforms and improved data analytic procedures.

Our goal for this book was to produce a concise, well oriented book for both undergraduate students and researchers in Geosciences who wish to gain further insight into the subject. We hope that the wide scientific coverage of the book will provide the reader with a good overview of the state-of-the-art research across the breadth of this field. The contributing authors and reviewers are sincerely acknowledged by the editors, as well as the effort by MDPI staff for their technical coordination in producing this book.

Pasquale Sellitto, Giuseppe Salerno, Andrew McGonigle
Special Issue Editors

geosciences

MDPI

Editorial
Volcanic Plumes: Impacts on the Atmosphere and Insights into Volcanic Processes

Andrew J. S. McGonigle [1,2,3,*], **Pasquale Sellitto** [4,5] **and Giuseppe G. Salerno** [6]

1   Department of Geography, University of Sheffield, Winter Street, Sheffield S10 2TN, UK
2   School of Geosciences, The University of Sydney, Sydney NSW2006, Australia
3   Istituto Nazionale di Geofisica e Vulcanologia, Sezione di Palermo, Via Ugo La Malfa 153,
     90146 Palermo, Italy
4   Laboratoire de Météorologie Dynamique, Institut Pierre Simon Laplace, École Normale Supérieure, PSL
     Research University, École Polytechnique, Université Paris-Saclay, Sorbonne Université, UPMC Université
     Paris 6, CNRS, 75005 Paris, France; psellitto@lmd.ens.fr
5   Remote Sensing Group, UK Research and Innovation, Science and Technology Facilities Council, Rutherford
     Appleton Laboratory, Harwell, Oxford OX11 0QX, UK
6   Istituto Nazionale di Geofisica e Vulcanologia, Osservatorio Etneo, I95123 Catania, Italy;
     giuseppe.salerno@ingv.it
*   Correspondence: a.mcgonigle@sheffield.ac.uk; Tel.: +44-114-222-7961

Received: 22 April 2018; Accepted: 24 April 2018; Published: 30 April 2018

Keywords: volcanic plumes; volcanic gases; volcanic geochemistry; atmospheric remote sensing; radiative forcing; atmospheric chemistry

Here we introduce a Special Issue of Geosciences focused on the scientific research field of 'Volcanic Plumes: Impacts on the atmosphere and insights into volcanic processes'. We would like to firstly thank those who have participated in this endeavour, in particular the authors who have chosen to submit their research outputs to this collection of articles, as well as the reviewers who have devoted their time, in a manner which has been invaluable in improving all the papers that appear here. This research theme provides a truly inter-system perspective on the dynamics of planet Earth, spanning the geosphere and the atmosphere and covering processes, which occur over a wide variety of timescales, and have significant impacts on human beings and the biosphere, not least via volcanogenic climate change. The title was deliberately chosen to encompass to the wide variety of scientific research occurring within this field and we are very pleased to see that the range of articles appearing here does span that of current enquiry regarding volcanic plumes.

Firstly, there are articles focused on the impacts on the atmosphere of volcanic plumes. In particular, Roberts [1] outlines an intriguing study into the rather recently discovered ozone destroying chemical processes, which occur within tropospheric volcanic plumes. By presenting results from measurements of ozone in the plume of a low halogen emitting volcano (Kīlauea), and combining these data with those from higher halogen emitters (Etna and Mt. Redoubt), as well as model simulations, new insights are offered into the role of halogen species in driving these chemical processes. On this theme, sun photometric observations of aerosol optical depth are also reported by Sellitto et al. [2], concerning the plume released from Pacaya volcano, Guatemala. Constraining the microphysics of aerosol particles in volcanic plumes is important to enable better understanding of the radiative properties of these emissions, and hence their potential role in modulation of radiative transfer and climatic dynamics. Another work, which is focused on measuring from the ground the properties of volcanic gas plumes is that of Pfeiffer et al. [3], who deployed a variety of spectroscopic and in-situ sampling tools, under challenging environmental conditions, to constrain the gas chemistry, emission rate and aerosol properties of the plume arising from the Bárðarbunga fissure eruption in Iceland. This eruption was the greatest in Iceland in the last 200 years and one of the most polluting volcanic events in centuries. Capture of data such as these is very important in terms of parameterising plume dispersal models, with a view to better mitigating the impacts on human beings of eruption clouds.

The second major focus of the volume title is on the use of plume data to inform our understanding of the subterranean processes, which drive volcanic activity at the surface. Here two articles appeared, the first of which was by Terray et al. [4]. In this paper, the authors report on the disequilibria between the radioactive species: ^{210}Pb, ^{210}Bi and ^{210}Po in the plume of Mt. Etna. Such data have been used for some decades in attempts to constrain underground magma dynamics and degassing kinetics. Here a novel basaltic degassing model is put forward, based on a Monte-Carlo simulation, which the authors argue delivers a better fit to the observed data than those adopted previously. Linkage between models and observed gas data is also the theme of the next article, concerning magmatic-degassing dynamics, by Pering and McGonigle [5]. Here the authors use high time resolution remotely sensed degassing data (from ultraviolet cameras; see below) in tandem with mathematical models to provide an overarching model classification for puffing and strombolian degassing modes in basaltic volcanism for the first time.

Beyond the above topics, there were a number of articles, which focused on the development of hardware and software/algorithmic protocols for remotely sensing the properties of volcanic plumes. Within this realm there were three pieces focused on aerial/satellite observation platforms, and a further four concerning ground based configurations. In terms of the first approach, which is relevant for aviation in the event of large eruptions, or constraining the climatic impacts of eruptions, there was a contribution from Guermazi et al. [6] regarding thermal infrared methodologies. In particular, the authors focus on better constraining the concurrent impact of both sulphur dioxide and secondary sulphate aerosols on the signals received in the sensors' measurement bands. This radiative interference between sulphur dioxide and secondary sulfate aerosols has not been investigated before, hence this study paves the way towards more precise quantitative observation of these components of volcanic plumes. In addition, Licciardi et al. [7] report on a study concerning hyperspectral imaging in volcanology, aimed at unmixing the relative spectral effects of the ground covering and plume composition on the radiation signal received by the sensor. Here a nonlinear approach, based on machine learning, is adopted, which is in contrast to the linear techniques applied hitherto with a view to better resolving these relative effects, and it hence informs a more robust model interpretation of captured volcano-hyperspectral data. Furthermore, Corradini et al. [8] attempted a down-scaling of satellite-based observations to characterise proximal parameters of the volcanic activity, such as the start date and duration of eruptions, plume discharge rates and plume heights. This is of particular relevance to the anticipated increased future use of space-based platforms in volcanic monitoring.

Finally, there are a number of pieces focused on the ground based remote sensing techniques applied to volcanic plume sensing, an approach of particular importance in volcano monitoring, where high time resolution is helpful and the pre-eruptive plumes are often rather too weak to be resolved from space. An excellent overview of this domain is provided by Platt et al. [9], who review the significant development of this field in recent decades, which has led to a large number of volcanoes, spanning the globe, now being the focus of routine remotely sensed gas observations, enabling the observer to remain stationed at a safe distance from the source. This article provides an overview of a range of imaging and spectroscopic approaches, which have been applied in this arena. McGonigle et al. [10] focus on a particular modality within this genre: ultraviolet imaging of volcanic plumes, which has emerged over the last decade to provide plume emission rate data with unprecedented time resolutions of order 1 s. The associated hardware and software protocols are covered as well as the significant novel scientific possibilities that this approach now enables, in particular the capacity to relate high time resolution gas flux data with volcano geophysical data for the first time to bridge between two previously rather separate branches of volcanology.

Gliß et al. [11] push the theme of UV imaging further by reporting on open access Python code aimed at handling all the processing steps, which are required to generate volcanic sulphur dioxide gas fluxes from raw acquired camera data. In particular, this provides calibration, plume

Geosciences **2018**, *8*, 158

speed determination and light dilution correction functionality, with the aim of expediting the wider uptake of this methodology amongst the international volcanology community. The final article, by Santoro et al. [12] concerns active remote sensing of carbon dioxide emissions from Mt. Etna using a LiDAR based system. Increases in the emission of this species can be a signature of forthcoming volcanic eruptions, hence this approach has very great potential in hazard assessment. The technique can also be applied at considerable distances from the source, providing significant safety benefits relative to the traditionally applied proximal measurements of this species at/near active vents.

It seems an apposite moment to publish this Special Issue, given the very significant developments, which have occurred in the area of volcanic plumes in the last decade or so. It is also hoped that the wide scientific coverage of the articles presented here will provide the reader with a good overview of the state of the art across the breadth of this field. These articles additionally pave the way for the exciting developments that are likely to follow in the following decade based on anticipated improvements in volcano monitoring instrumentation, deployed from both ground and space based platforms, in addition to improved data analytic procedures. With additional improvements in models concerning both underground gas behaviour as well as the physical dispersal and chemical and microphysical evolution of plumes in the atmosphere, we look forward to the further development of this field in the decade to come.

Author Contributions: The authors contributed equally to the writing of this Editorial.

Conflicts of Interest: The authors declare no conflict of interest.

References

1. Roberts, T.J. Ozone Depletion in Tropospheric Volcanic Plumes: From Halogen-Poor to Halogen-Rich Emissions. *Geosciences* **2018**, *8*, 68. [CrossRef]
2. Sellitto, P.; Spampinato, L.; Salerno, G.G.; La Spina, A. Aerosol Optical Properties of Pacaya Volcano Plume Measured with a Portable Sun-Photometer. *Geosciences* **2018**, *8*, 36. [CrossRef]
3. Pfeffer, M.A.; Bergsson, B.; Barsotti, S.; Stefánsdóttir, G.; Galle, B.; Arellano, S.; Conde, V.; Donovan, A.; Ilyinskaya, E.; Burton, M.; et al. Ground-Based Measurements of the 2014–2015 Holuhraun Volcanic Cloud (Iceland). *Geosciences* **2018**, *8*, 29. [CrossRef]
4. Terray, L.; Gauthier, P.-J.; Salerno, G.; Caltabiano, T.; La Spina, A.; Sellitto, P.; Briole, P. A New Degassing Model to Infer Magma Dynamics from Radioactive Disequilibria in Volcanic Plumes. *Geosciences* **2018**, *8*, 27. [CrossRef]
5. Pering, T.D.; McGonigle, A.J.S. Combining Spherical-Cap and Taylor Bubble Fluid Dynamics with Plume Measurements to Characterize Basaltic Degassing. *Geosciences* **2018**, *8*, 42. [CrossRef]
6. Guermazi, H.; Sellitto, P.; Serbaji, M.M.; Legras, B.; Rekhiss, F. Assessment of the Combined Sensitivity of Nadir TIR Satellite Observations to Volcanic SO_2 and Sulphate Aerosols after a Moderate Stratospheric Eruption. *Geosciences* **2017**, *7*, 84. [CrossRef]
7. Licciardi, G.A.; Sellitto, P.; Piscini, A.; Chanussot, J. Nonlinear Spectral Unmixing for the Characterisation of Volcanic Surface Deposit and Airborne Plumes from Remote Sensing Imagery. *Geosciences* **2017**, *7*, 46. [CrossRef]
8. Corradini, S.; Guerrieri, L.; Lombardo, V.; Merucci, L.; Musacchio, M.; Prestifilippo, M.; Scollo, S.; Silvestri, M.; Spata, G.; Stelitano, D. Proximal Monitoring of the 2011–2015 Etna Lava Fountains Using MSG-SEVIRI Data. *Geosciences* **2018**, *8*, 140. [CrossRef]
9. Platt, U.; Bobrowski, N.; Butz, A. Ground-Based Remote Sensing and Imaging of Volcanic Gases and Quantitative Determination of Multi-Species Emission Fluxes. *Geosciences* **2018**, *8*, 44. [CrossRef]
10. McGonigle, A.J.S.; Pering, T.D.; Wilkes, T.C.; Tamburello, G.; D'Aleo, R.; Bitetto, M.; Aiuppa, A.; Willmott, J.R. Ultraviolet Imaging of Volcanic Plumes: A New Paradigm in Volcanology. *Geosciences* **2017**, *7*, 68. [CrossRef]

11. Gliß, J.; Stebel, K.; Kylling, A.; Dinger, A.S.; Sihler, H.; Sudbø, A. Pyplis-A Python Software Toolbox for the Analysis of SO_2 Camera Images for Emission Rate Retrievals from Point Sources. *Geosciences* **2017**, *7*, 134. [CrossRef]
12. Santoro, S.; Parracino, S.; Fiorani, L.; D'Aleo, R.; Di Ferdinando, E.; Giudice, G.; Maio, G.; Nuvoli, M.; Aiuppa, A. Volcanic Plume CO_2 Flux Measurements at Mount Etna by Mobile Differential Absorption Lidar. *Geosciences* **2017**, *7*, 9. [CrossRef]

geosciences

MDPI

Review

Ultraviolet Imaging of Volcanic Plumes: A New Paradigm in Volcanology

Andrew J. S. McGonigle [1,2,3,]*, Tom D. Pering [1], Thomas C. Wilkes [1], Giancarlo Tamburello [4], Roberto D'Aleo [5], Marcello Bitetto [5], Alessandro Aiuppa [2,5] and Jon R. Willmott [6]

[1] Department of Geography, University of Sheffield, Sheffield S10 2TN, UK; t.pering@sheffield.ac.uk (T.D.P.); tcwilkes1@sheffield.ac.uk (T.C.W.)
[2] Istituto Nazionale di Geofisica e Vulcanologia, Sezione di Palermo, Via Ugo La Malfa 153, 90146 Palermo, Italy; aiuppa@unipa.it
[3] School of Geosciences, The University of Sydney, Sydney NSW 2006, Australia
[4] Istituto Nazionale di Geofisica e Vulcanologia, Sezione di Bologna, Via Donato Creti, 12, 40100 Bologna, Italy; giancarlo.tamburello@ingv.it
[5] DiSTeM, Università di Palermo, via Archirafi, 22, 90123 Palermo, Italy; roberto.daleo01@unipa.it (R.D.); marcellobitetto@gmail.com (M.B.)
[6] Department of Electronic & Electrical Engineering, University of Sheffield, Sheffield S1 4DE, UK; j.r.willmott@sheffield.ac.uk
* Correspondence: a.mcgonigle@sheffield.ac.uk; Tel.: +44-114-222-7961

Academic Editors: Pasquale Sellitto, Giuseppe Salerno and Jesús Martínez Frías
Received: 1 April 2017; Accepted: 1 August 2017; Published: 8 August 2017

Abstract: Ultraviolet imaging has been applied in volcanology over the last ten years or so. This provides considerably higher temporal and spatial resolution volcanic gas emission rate data than available previously, enabling the volcanology community to investigate a range of far faster plume degassing processes than achievable hitherto. To date, this has covered rapid oscillations in passive degassing through conduits and lava lakes, as well as puffing and explosions, facilitating exciting connections to be made for the first time between previously rather separate sub-disciplines of volcanology. Firstly, there has been corroboration between geophysical and degassing datasets at ≈1 Hz, expediting more holistic investigations of volcanic source-process behaviour. Secondly, there has been the combination of surface observations of gas release with fluid dynamic models (numerical, mathematical, and laboratory) for gas flow in conduits, in attempts to link subterranean driving flow processes to surface activity types. There has also been considerable research and development concerning the technique itself, covering error analysis and most recently the adaptation of smartphone sensors for this application, to deliver gas fluxes at a significantly lower instrumental price point than possible previously. At this decadal juncture in the application of UV imaging in volcanology, this article provides an overview of what has been achieved to date as well as a forward look to possible future research directions.

Keywords: ultraviolet cameras; volcanic plumes; interdisciplinary volcanology

1. Introduction

Volcanic activity is observed in a number of primary ways: firstly, by measurements of geophysical signatures, e.g., seismic, thermal, and acoustic; and secondly, through observations of gases released from summit craters, flanks, or fumaroles [1]; petrology also plays a key role here in respect of magma geochemistry. However, historically, the degassing data have been considered somewhat secondary to those from geophysics, in particular seismic data, largely because of limitations in the applied instrumentation. However, during the last two decades, there has been a major renaissance in volcanic gas monitoring, arising from the implementation of exciting new ground-based technologies for measuring the gases released in volcanic plumes. These approaches have been of utility in increasing our understanding of the underground processes that drive surface activity, as well as in routine volcano monitoring operations.

These recently applied techniques fall into two categories: firstly, those that concern the chemical composition of the gases, e.g., Fourier Transform Infrared (FTIR) spectroscopy [2] and MultiGAS units [3]; and secondly, those that capture emission rates or fluxes, for example correlation spectrometers (COSPECs), differential optical absorption spectrometers (DOAS units), and ultraviolet (UV) cameras. The emission rate data have been largely focused on sulphur dioxide (SO_2), which is straightforward to remotely sense in volcanic plumes due to its strong UV absorption bands and low ambient concentrations. There have also been exciting recent developments concerning laser LIght Detection And Ranging (LIDAR) remote sensing of carbon dioxide (CO_2) emissions, (e.g., [4,5]) from volcanoes.

UV remote sensing of SO_2 emissions has been conducted since the 1970s, initially with COSPEC units developed for monitoring smokestack emissions from coal burning power stations, leading to the generation of a number of valuable long-term datasets [6,7]. Since the turn of the century, these units have been replaced with low cost USB-coupled linear array spectrometers, costing only a few thousand dollars, an order of magnitude less than COSPEC [8,9]. Data analysis to deliver SO_2 column amounts is achieved using DOAS routines, and the units have been applied from mobile platforms, e.g., on cars and airplanes, whilst traversing beneath a plume, as well as in fixed position deployments involving scanning optics [10,11]. These scanning spectrometers are now in routine operation on numerous volcanoes worldwide [12,13].

Notwithstanding the benefits of the above technology, and its service within the volcanology community, the flux data are limited in time resolution to a datum every 100 s or so, due to the requirement to physically scan or traverse the plume, which effectively provides time-integrated assessments of emissions on this timescale. This is too slow to resolve many rapid gas-driven volcanic processes, e.g., puffing and strombolian explosions, such that the acquired data cannot be used to investigate the driving underground fluid dynamics in these cases. Indeed, the only way to scrutinise these more rapid phenomena was via geophysical data, which are acquired at frequencies of at least 1 Hz, leading to potentially a somewhat indirect proxy understanding. This prompted several research groups (e.g., [14,15]) to pioneer UV imaging approaches, which provide image snapshots of the plume gas column amounts every second or so, from which gas fluxes can be generated at the same time resolution. In this article, we cover the technological aspects of the application of UV imagery within volcanology, followed by an overview of the present and potential future scientific possibilities that this approach brings to the field.

2. Ultraviolet Camera Instrumentation

The UV camera's operation is based on imaging gas plumes, which arise from volcanic craters, vents, or fumarole fields, with a bandpass filter mounted to the fore of the unit, centered around 310 nm, where SO_2 absorbs incident radiation. Typically, imagery at a wavelength of around 330 nm is also acquired, where there is no SO_2 absorption, to factor out broadband aerosol-related issues, which apply to both wavebands. This can be achieved using two co-aligned cameras, or a single camera, and a filter wheel. Below is a brief overview of the measurement approach, which is detailed further in Kantzas et al. [16], for the two cameras, two filter setup.

Firstly, optimal exposure settings are determined for each camera, based on the skylight illumination intensity, to maximize signal-to-noise and avoid saturation whilst viewing the sky. The next step is to measure dark images, at these exposure times, in order to account for the camera response when light is blocked from entering the fore-optics. Following this, background sky images are acquired for each camera by imaging a region of sky adjacent to the plume, e.g., containing no gas absorption. At this stage, the cameras are pointed at the plume and the measurement sequence begins.

Following Beer's law, these images are processed to provide the uncalibrated apparent absorption, *AA*, for each pixel via the following relationship:

$$AA = -\log_{10}\left[\frac{IP_A - ID_A}{IB_A - ID_A} \Big/ \frac{IP_{NA} - ID_{NA}}{IB_{NA} - ID_{NA}}\right]. \quad (1)$$

Here, *IP* is the intensity whilst viewing the plume, *IB* is the background sky intensity, and *ID* is the dark intensity for the pixel in question, where the subscripts pertain to the camera filter wavelengths where there is (*A*) and is not (*NA*) absorption from the SO_2, e.g., in the region of 310 nm and 330 nm, respectively. Following the determination of the apparent absorption images, calibration is required. This can be achieved with quartz cells containing known column amounts of SO_2. In this case, *AA* values are determined for each cell and averaged over a section in the centre of the image. These data are plotted on a scatter plot of axes: cell column amount vs. apparent absorption. The slope of the best-fit line is then extracted, acting as the calibration factor, which all volcanic plume image pixel *AA* values are then multiplied by. An alternative approach to calibration is to use a co-aligned spectrometer to determine a column amount value corresponding to a small section of the image to enable scaling to calibrated concentration values across the whole image. Once the calibrated images are generated, a cross-section line through the plume is defined, and all column amounts are integrated along this to determine the so-called integrated column amount. The plume speed is then found, often by determining an integrated column amount time series from cross-sections drawn through the plume at two different distances above the crater. These series are then cross-correlated to determine the temporal lag between them, from which the transit speed can be found [17,18]. Alternately, more sophisticated motion-tracking algorithms have also been applied [19], as has the correlation of temporally successive spatial series/longitudinal profiles of the plume to better exploit the available two-dimensional (2D) information in determining plume velocities [20]. Multiplication of the transport speed by one of the integrated column amount time series then leads to the generation of the flux time series.

Errors in flux computations are thought to be in the region of 20–30% [21] for individual camera measurements. Furthermore, in a detailed inter-comparison of the performance characteristics of multiple camera units in establishing SO_2 emissions, a one standard deviation precision of 20% was established for the ensemble of tested units [22]. Errors arise from the scattering of radiation between the sensor and the plume, e.g., light dilution, as well as scattering within the plume itself [23,24]. There are also uncertainties arising from cell calibration [25], as well as from light transit through the filters at different incident angles [26], which can cause the peak transmission wavelength to alter across the image. A further point relates to the requirement to image the plume at two wavelengths. Where this involves a single camera and a filter wheel, there will be a short time delay between the filter acquisitions. This will result in slight offsets between the plume locations in each case, due to the advection of the gases in the atmosphere, which can create issues in the retrieval. The use of dual cameras may also be problematic, as the retrieval is predicated on pixel-to-pixel correspondence between the imagery in both bands. In practice, this can be complicated by parallax effects: for deployments close to the plume, these are thermal and vibrational effects causing misalignments, as well as slight imperfections in the applied lenses, and non-identical optical settings for the cameras, e.g., in terms of the different applied filters. For this reason, the images can be shifted relative to one another in software in order to achieve the best possible spatial matching.

One approach that could mitigate against radiative transfer-related errors is a Fabry–Perot configuration [27,28]. Such optical devices allow light transmission at regularly spaced wavelengths, blocking the intervening radiation. In the context of volcanic SO_2 measurements, the devices are tuned such that the interval between the peaks of this transmission spectrum corresponds to that between the peaks in the comb-like absorption spectrum of sulphur dioxide around 310 nm. The devices can be set to sample radiation at these maxima in absorption, as well as the radiance in intervening wavelengths, and by comparing the two outputs, gas column amounts can be derived. To date, most of the UV imaging systems applied in volcanic SO_2 measurements have been based on commercially available UV cameras, with price points of thousands of USD. Recently, however, low-cost sensors, designed primarily for the smartphone market, have been adapted for this application, such that a usable UV sensitivity of these units has been demonstrated [29], as well as adequate signal-to-noise characteristics for the SO_2 monitoring application [30] (Figure 1).

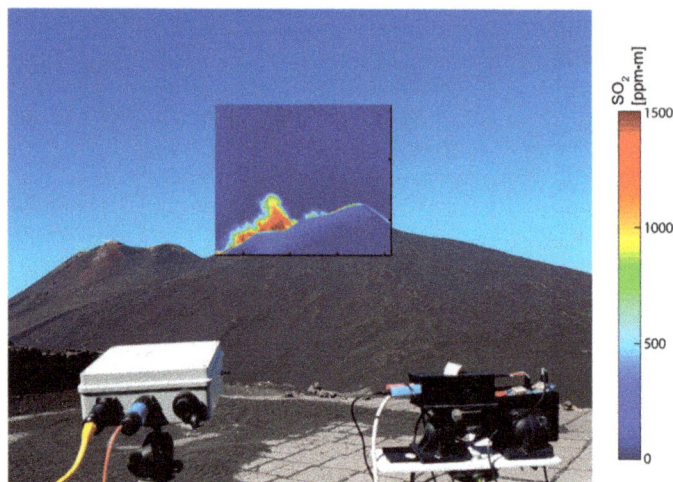

Figure 1. Deployment of inexpensive smartphone sensor-based ultraviolet (UV) camera instrumentation (**right**) in tandem with more traditionally applied scientific grade cameras (**left**) on Mt. Etna. A false colour gas column amount inset image is included in the graphic, with scale to right, for the cheaper units, which were based on modified Raspberry Pi cameras (Raspberry Pi Foundation, Cambridge, UK). For further detail, see [29,30].

3. Improving the Spatio-Temporal Resolution of Volcanic Degassing

The cameras have now been deployed on a significant number of volcanoes worldwide, due in part to the convenience of being able to set up and operate from fixed positions during discrete field campaigns (e.g., [31,32]). To date, the targets covered by permanent network installations have been rather few, e.g., Etna and Stromboli in Italy and Kīlauea in Hawaii [33–37], potentially as a consequence of the requirement to image the plume, e.g., without cloud cover between the camera and summit area. There is, of course, meteorological cloud cover at the top of volcanoes, which can occlude observations. Herein lies one advantage of conventional spectroscopic gas flux assessments, in that imaging is not a requirement for this class of observation.

The cameras provide the possibility of resolving spatio-temporal degassing characteristics in unprecedented details. For instance, spatial information was typically only available heretofore from volcanoes with multiple craters, by the rare occurrence of walking traverse observations made very close to the source [38]. By gathering spatial information, the cameras implicitly provide scope for the resolution of gas fluxes from heterogenous sources, as exploited on Vulcano island (Italy), to measure gas fluxes from individual fumaroles [39]. This capability has also been exploited in respect of multiple vent scenarios, e.g., Fuego (Guatemala) [20] and Mt. Etna, where a shifting of degassing from one vent to another was observed in tandem with a transference of eruptive activity between craters [34].

In terms of temporal information, the UV cameras have enabled us to capture rapid trends in passive and explosive degassing. In particular, fluctuations in passive degassing on timescales of 10 s to 1000 s of seconds have been resolved using UV cameras [40], building on earlier observations of this phenomenon using a non-imaging dual spectrometer approach involving units with cylindrical lenses and quasi-horizontal fields of view [41,42]. Based on observations on Mt. Etna, Mayon (Philippines), and Erebus (Antarctica), using contemporaneous Multi-GAS observations and/or ancillary visible/near-infrared (IR) cameras, these fluctuations appear to also be manifested in the degassing of CO_2 and water vapour emissions e.g., [43,44]. This behaviour has been observed in both conduit degassing scenarios (e.g., Mt. Etna) as well as from lava lakes. In terms of conduit degassing, arguments have been put forward that this behaviour arises from the arrangement of rising bubbles into layers of elevated gas concentrations, leading to periodic enhancements in passive non-overpressurised bubble bursting at the surface [40].

The situation with lava lakes is intriguing, in that rather different degassing trends have been observed from each of the volcanoes targeted to date with high time resolution gas flux observations, e.g., Villarrica, Chile [45], Kīlauea [46], and Erebus [42]. This potentially points to a wide variety of gas flow processes occurring across these systems, which range significantly, both in magmatic viscosities as well as in gas flux magnitudes. In particular, 'gas pistoning' is evident in the Kīlauea data, involving pronounced spikes in degassing, followed by a gradual waning in emissions, on timescales of tens of minutes, potentially caused by a gas accumulation and release mechanism. In contrast, the Erebus volcano demonstrates stable periodic degassing behaviour, present in both the acquired gas flux and gas composition time series [42], which is thought to arise from a stable bidirectional flow in the conduit, such that gas rich magma batches periodically rise, degas, then sink down again into the conduit. In Villarrica, gas flux time series data revealed no stable periodicity in degassing. This is thought to be precluded by turbulent mixing in the lava lake arising from continuous inflow of magma from the conduit [45].

One major application of the UV cameras has been to measure gas masses released during discrete explosions. Whilst this has been achieved spectroscopically with high temporal resolution differential optical absorption spectroscopy observations [41], and even with a correlation spectrometer [47], it is far easier to resolve these emissions with the cameras' imaging capacity. The eruptions where SO_2 masses have been constrained with UV imagery have been ash-poor, strombolian, or weakly vulcanian events. Whilst the UV imaging of ash rich plumes has been acheived, yielding interesting insights into ash phase plume dynamics [48], the reduction in optical thickness caused by ash in these cases rules out the retrieval of SO_2 emissions. Interestingly, these explosive UV camera studies typically point toward the non-explosive release of gas as being the dominant means by which these volcanoes release volatiles to the atmosphere [21,49–54], especially for basaltic open conduit cases, such as Etna and Stromboli in Italy, where gas bubbles are free to move through the melt. Indeed, in the Stromboli case, degassing was partitioned as 77% passive gas release (e.g., from spherical bubbles), 16% from puffing, e.g., from cap bubbles, and with only 7% from explosions, e.g., from gas slugs (Taylor bubbles) [49]. This study, incidentally, also constituted the first direct measurement of puffing gas masses from a volcano, pointing to the real benefits of the camera technology in terms of its high spatial resolution and sufficiently good sensitivity to capture these subtle degassing features.

In these reports, the subdivision of fluxing between the degassing classes appears to be most strongly tipped towards explosive release (although it is still often dominated by passive degassing) in the scenarios where eruptions are more vulcanian in nature [21,52–54], e.g., the Santiaguito (Guatemala), Asama (Japan), Semeru (Indonesia), and Fuego volcanoes. In particular, Smekens et al. [52] suggest, in respect of the Semeru observations, that accumulation and pressurisation beneath a viscous plug are in operation before breach and explosive release. This assertion is intuitive and follows on from that put forth following pioneering observations on the Karymsky volcano (Russia) by Fischer et al. [47], where a predominantly explosive gas release was reported based on correlation spectrometer observations made long before the advent of UV imaging.

4. Combination of UV Camera Degassing Data with Geophysical Data and Conduit Fluid Dynamics

The above studies point to the absolute necessity of models to facilitate the interpretation of the acquired data. In this respect, the high time resolution of the UV cameras has an enormous benefit for volcanology. Specifically, the cameras can resolve gas release processes, which are caused by a variety of subterranean fluid dynamic mechanisms, which have been the subject of considerable prior numerical, mathematical, and laboratory modeling efforts within both the volcanic and engineering research communities. This is especially so in the case of strombolian explosions, which are thought to arise from the bursting of conduit filling gas slugs at the surface [55]. Here, at the very exciting current frontier in volcanology, UV camera data are enabling the first substantive bridging between the volcanic gas flux measurement and volcanic conduit fluid dynamic modeling communities.

In particular, in the first study of this nature, explosions on Stromboli were investigated, involving a tail or coda in emissions following the events [36]. The authors interpreted the activity, with the aid of computational fluid dynamic models, as arising from the fissioning of smaller "daughter" bubbles from the bases of the rising slugs (Figure 2). Furthermore, during a study of rapid (0.25 Hz) strombolian activity on Mt. Etna, explosive data were plotted on a scatter plot of repose time following the event vs. event mass [51]. An absence of large mass, long repose time data were noted, which was interpreted as being due to coalescence of adjacent rising slugs, e.g., leading to a longer repose interval before the arrival of the next distinct slug, and constituting the first direct empirical evidence of slug interaction in volcanic conduits. A follow-on report, based on thermal observations of puffing on Stromboli, also affirms the potential importance of this process in respect of cap bubbles [56].

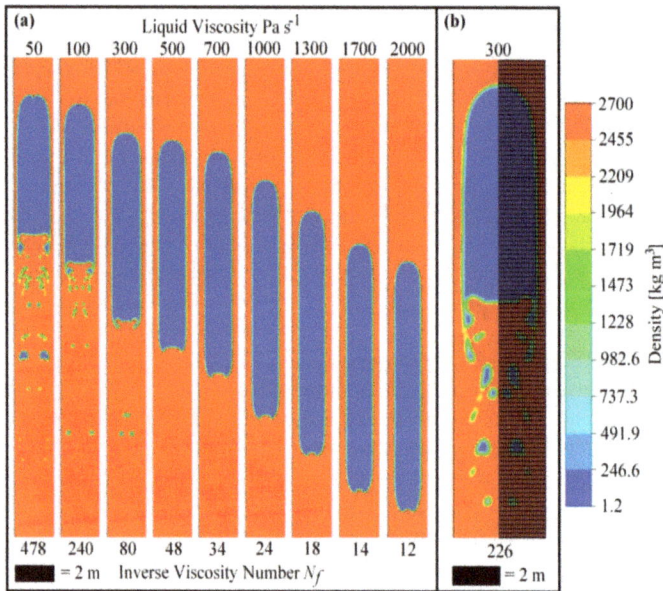

Figure 2. Computational fluid dynamic modeling of rising gas slugs on Stromboli, illustrating the fissuring of daughter bubbles from the slug base. This has been linked to codas in UV camera gas flux time series following strombolian explosions, illustrating the potential of combining models with high time resolution field degassing data to unravel the subterranean drivers of surface activity; (**a**) shows simulations for a range of fluid dynamical conditions e.g., in terms of liquid viscosity and inverse viscosity number, and (**b**) shows a zoom of behaviour for one such parameterisation; see main text and [36] for more detail.

Another important benefit of the UV camera data is that the high time resolution gas fluxes can be compared with contemporaneous geophysical data, with far less aliasing than necessary previously; e.g., the latter data are acquired at frequencies ≥ 1 Hz. This is highly significant in that many geophysical manifestations on volcanoes arise from gas-based processes, e.g., seismic signals are caused by the ascent of gas slugs in conduits, and thermal and infrasonic signatures are generated from the surficial bursting of these bubbles. Until now, we have had to rely purely on geophysical means to understand the rapid degassing processes on volcanoes, e.g., strombolian explosions and puffing. Not only do the UV camera data provide a more direct means of understanding these phenomena, but they also enable the possibility of making a far more direct comparison with geophysical series, which could lead the way towards a better interpretation of geophysical observations on volcanoes, and a more holistic understanding of volcanic behaviour.

The first report of high time resolution (≈ 1 Hz) degassing data being corroborated with geophysical data concerned explosions on the Stromboli volcano [41]. This study was performed using spectroscopy, rather than UV imaging, but, in common with a more recent UV camera study on this target [49], it revealed linear correlations between the magnitudes of the recorded degassing, thermal, and very long period (VLP) seismic signatures for the events. This fits with the conventionally held, although previously rather hypothetical, view that VLP signals on Stromboli arise from volumetric changes associated with the ascent of gas masses, e.g., slugs in the conduit, such that the larger the rising gas mass, the greater the VLP signal. Further related investigations have been performed on the Asama [53] and Fuego [54] volcanoes, where proportionality between VLP signals and released SO_2 masses was also observed. The relative scaling of seismic moment with released SO_2 mass, however, does not equate for all targets, which may be related to an absence of gas ratio information, e.g., the total released gas masses are not being considered. However, it is also highly likely that volcano-specific features, e.g., the conduit geometry, magma rheology, and the precise mechanism of VLP generation, which is likely to vary between the targets, will affect the degree to which the degassing processes are coupled into seismic energy in each case.

Ultraviolet imaging degassing fluxes have also been linked to tremor, a class of seismicity associated with pressure fluctuations in degassing magmas. Here, a number of studies have noted a relationship between these time series for the Etna, Fuego, and Kīlauea volcanoes, including conduit and lava lake degassing scenarios [20,40,44,46]. In one case, this has involved novel signal processing techniques based on wavelet analysis to isolate commonality in periodicity in pairs of geophysical datastreams [57]. These experimental outcomes are as would be expected, given that tremor is anticipated to increase with elevated bubble concentrations in magmas, or the more rapid flow of bubbles in the conduit, scenarios which would both correspond to periods of elevated gas release. In two of the studies, intriguing lag relationships were identified [20,44]. In particular, in Nadeau et al. [20], a trend of increasing temporal lags between the seismic and gas flux time series was observed in the period following explosions, implying perhaps that the source of seismic energy was becoming progressively deeper within the conduit due to the rheological stiffening of the magma downwards from the top of the column. In Pering et al. [44], bursts in CO_2 outgassing were reported (derived from the UV cameras in tandem with Multi-GAS [3] units) which preceded spikes in seismicity, raising intriguing questions regarding the causal processes that might link these phenomena in this case.

Thermal observations have also been correlated with UV camera data, with the most detailed study to date in this area concerning explosions on Stromboli [49], which builds on an earlier more limited treatment on this target [41]. Both resulting articles demonstrated a linear relationship between these two parameters, although in the more recent one, two populations arose, corresponding to events, which were ash-free/with ash, respectively. This is as would be expected, e.g., the ashier eruptions will be thermally brighter due to the larger quantity of radiating solid ejecta. In addition, UV camera/high time resolution spectroscopic instrumentation-based attempts have been made to compare SO_2 fluxes with acoustic data in respect of explosive activity, resulting in reports of either no [41] or somewhat limited [58] correlations being apparent between these time series. A significant breakthrough in this regard has come with the application of linear acoustic processing of the acquired infrasonic signals to infer gas masses. This methodology, which involves consideration of the entire waveform, and not merely the initial pressure conditions, led to 1:1 linear correlations with contemporaneously acquired UV camera-derived gas masses [35].

5. Future Directions

Ultraviolet camera technology provides an unprecedented opportunity to investigate volcanic degassing behaviour with far better spatial and temporal resolution than possible previously. This has expedited the linkage of gas fluxes with geophysical data and conduit fluid dynamic models in ways that were impossible previously, opening the path to a number of potentially very fruitful future research directions.

In particular, the comparison of high time resolution UV camera gas fluxes with modeled emission rates from laboratory and computational fluid dynamics is an area of study in its absolute infancy. By simulating degassing behaviour from a series of underground gas flow mechanisms, and comparing the results against field data to determine best matches, e.g., using correlative approaches, new avenues will be opened in terms of being able to understand how subterranean processes drive volcanism. We recently combined these approaches for the first time [38], illustrating the exciting scientific potential contained therein, in a study of strombolian activity on the Strombolian volcano. There is now much work remaining to be done in expanding this methodology to unravel degassing dynamics across a wide spectrum of activity styles and volcanic targets. This will encourage further developments in the models themselves, which have been somewhat focused on slug dynamics to date (e.g., [55]), to give greater attention to other potential fluid processes in volcanoes, e.g., bubbly, cap, annular, and churn flow mechanisms.

The linking of UV camera degassing data to geophysical data on timescales of ≈ 1 Hz has led to the establishment of correlations between degassing data and volcano-seismic and acoustic signals. This could enable the calibration of acquired acoustic and seismic signals to infer gas masses, thereby helping to overcome one of the key limitations of the UV camera approach, namely inoperability at night-time and when the plume cannot be imaged due to cloud cover [35,53]. Hence, whilst the UV cameras deliver the most direct estimates of the degassing output from volcanoes, these calibrated geophysical proxies could be used to straddle windows where the cameras cannot be used, enabling continuous monitoring to take place. Another dimension of corroboration between the geophysical and UV camera degassing signals will be to better understand how underground degassing processes are responsible for generating geophysical manifestations on volcanoes. A key aim here would be to attempt to identify ≈ 1 Hz multi-parametric (e.g., geophysical and degassing) signatures, which precede eruptions, to establish precursory templates. This will build on pioneering studies, which have already illustrated the profound scientific insights achievable by blending these disparate data, albeit on the basis of previously available coarse time resolution gas flux data (e.g., [59]). This linkage of ground based gas fluxes, models, and geophysical data will also mirror similar integrated initiatives to combine these approaches on the basis of satellite observations, which are enabling significant breakthroughs in volcanology [60].

Notwithstanding the capacity of camera observations to advance science, the uptake of this technology in routine monitoring, as opposed to discrete field campaign operation, has been somewhat limited in comparison with spectrometers employing DOAS retrieval algorithms. There are a number of reasons for this; for example, the requirement to image the plume does exclude observations under conditions where cloud obscures the gases. The lack of hyper-spectral data, unless the units are spectroscopically, rather than cell, calibrated, also provides less opportunity to mitigate against radiative transfer errors. One factor which might assist the further dissemination of the camera technology is the recent demonstration that order of magnitude cheaper smartphone sensor-based systems can be used in volcanic SO_2 monitoring [29,30] (Figure 3). These units are so light that they could also be straightforwardly mounted on inexpensive drones available in the consumer electronics market for aerial observations to broaden the prior reach of drone-based volcanic gas surveillance [61,62]. There is, furthermore, the prospect of augmenting UV SO_2 imaging with thermal IR camera systems, which are capable of measuring SO_2 (and ash) release by day and night, hence overcoming the limitation of the former spectral region in being reliant upon solar radiation [63].

Recently, we have briefly investigated the possible use of multi-band imaging to study plume gas ratios. This work was performed on the Stromboli volcano during 22–24 July 2015. Here, conventional UV SO_2 observations were made in tandem with near-IR ($\approx$900 nm) imaging, in the latter case following the technique of Girona et al. [43], who posited that visible pixel brightness is quantitatively related to plume H_2O abundance due to plume scattering from water-rich aerosol droplets. We thereby obtained uncalibrated H_2O/SO_2 ratio images, demonstrating higher ratios for the fumarolic discharges vs. crater degassing, as would be expected. We furthermore measured the ratios associated with twenty explosions, and found that the H_2O/SO_2 values were reduced systematically by some 50% relative to those during passive degassing, which follows the ratio decrease during such events noted in a prior Fourier transform infrared spectroscopy-based study on this volcano [64]. These results are very tentative, and in particular do not consider environmental effects which will significantly change the partitioning of the water between vapour and aerosol phases, with implications for the H_2O retrievals, particularly when plume temperature changes, e.g., during explosions. Nevertheless, one can ascertain clear differences between fumarolic vs. crater degassing and the explosive vs. passive gas discharge signature in this case. Considerably more follow-on work would be required in order to establish whether calibrated and therefore quantitatively meaningful outputs could be achieved from this measurement approach.

Figure 3. Smartphone sensor (e.g., Rasperry Pi camera)-based UV imaging deployments on the Masaya volcano, Nicaragua, during June 2017. Measurement of the plume taken from outside the crater (**left**); measurements looking down at the lava lake surface (**right**).

A further possible research focus relates to the capacity of the cameras to image rates of volcanic gas–ambient air mixing, which are known to exert a key control on plume bromine monoxide (BrO) chemistry [65,66]. To date, the atmospheric chemistry modeling approaches applied to investigate this phenomenon have been based on assumed Gaussian-type plume dispersion scenarios. Here, the capacity of the cameras to empirically capture the heterogeneity in plume dilution in the 'real world' could be of utility in helping to advance this area of science. This could, for example, build on prior work using imaging DOAS instrumentation to provide profile information of the SO_2 and BrO column amounts in volcanic plumes [67]. The advantage of the cameras, however, in this case would be the ability to capture data faster than these alternate devices, as there is no requirement for the scanning of the spectrometer field of view across the image scene.

Finally, there are fresh emerging strands in respect of attempting to better constrain and reduce errors associated with the measurements. In particular, recent work has been invested in better determining the plume transport direction in order to decrease uncertainty [68]. This is related to the fact that when viewing plumes obliquely, the overestimate in the associated integrated column amount is only offset by the underestimation in apparent plume speed in the case of slightly non-parallel views; for more extreme viewing scenarios appreciable error can result.

6. Conclusions

Ultraviolet imaging has been applied in volcanology over the last ten years or so, leading to step change improvements in our ability to resolve volcanic gas emissions, both in the temporal and spatial domain, with a user-friendly single point measurement configuration. This has led, in particular, to the capture of rapid gas flux trends associated with explosions, puffing, and passive degassing in a way that would have been simply impossible in the past. A number of groups have now developed UV camera instrumentation [22,69] and operating software [70], leading to constraints on gas release budgets and studies into a variety of degassing-driven processes on a wide range of volcanoes, worldwide. In addition, the cameras have the potential to lead to significant scientific breakthroughs by combining the acquired high temporal resolution degassing data with contemporaneous geophysical datasteams and models for underground gas flows. In recent years, we have seen the first steps towards realising these objectives, and evidence of the significant scientific value of these blended approaches. This article looks back at what has been achieved to date, and the considerable promise of UV plume imaging in volcanology going forward.

Acknowledgments: A.J.S.M. acknowledges a Leverhulme Trust Research Fellowship (RF-2016-580), the Rolex Awards for Enterprise and a Google Faculty Research Award (r/136958-11-1). T.D.P. acknowledges the support of a NERC studentship (NE/K500914/1), the University of Sheffield, and ESRC Impact Acceleration funding. T.C.W. acknowledges scholarship funding from the University of Sheffield. A.A., M.B., R.D. and G.T. acknowledge support from the European Research Council starting independent research grant (agreement 305377). We are most grateful to anonymous reviewers for the time they have taken to provide thoughtful and meaningful comments on an earlier version of this article.

Author Contributions: A.J.S.M., T.D.P. and T.C.W. wrote the paper. A.J.S.M., T.D.P., T.C.W., G.T., M.B., R.D., A.A. and J.R.W. were involved in designing the instrumentation, and performing the fieldwork and data analysis reported on here.

Conflicts of Interest: The authors declare no conflict of interest. The founding sponsors had no role in the design of the study; in the collection, analyses, or interpretation of data; in the writing of the manuscript, and in the decision to publish the results.

References

1. Fischer, T.P.; Chiodini, G. Volcanic, Magmatic and Hydrothermal Gases. In *The Encyclopedia of Volcanoes*, 2nd ed.; Sigurdsson, H., Houghton, B., McNutt, S.R., Rymer, H., Stix, J., Eds.; Academic Press: London, UK, 2015; pp. 779–798.

2. Notsu, K.; Mori, T.; Igarashi, G.; Tohjima, Y.; Wakita, H. Infrared spectral radiometer: A new tool for remote measurement of SO_2 of volcanic gas. *Geochem. J.* **1993**, *27*, 361–366. [CrossRef]

3. Shinohara, H. A new technique to estimate volcanic gas composition: Plume measurements with a portable multi-sensor system. *J. Volcanol. Geotherm. Res.* **2005**, *143*, 319–333. [CrossRef]

4. Santoro, S.; Parracino, S.; Fioriani, L; D'Aleo, R.; Di Ferdinando, E.; Giudice, G.; Maio, G.; Nuvoli, M.; Aiuppa, A. Volcanic plume CO_2 measurements at Mount Etna by mobile differential absorption Lidar. *Geosciences* **2017**, *7*, 9. [CrossRef]

5. Queißer, M.; Granieri, D.; Burton, M. A new frontier in CO_2 flux measurements using a highly portable DIAL laser system. *Sci. Rep.* **2016**, *6*, 33834. [CrossRef] [PubMed]

6. Stoiber, R.E.; Malinconico, L.L.; Williams, S.N. Use of the correlation spectrometer at volcanoes. In *Forecasting Volcanic Events*; Tazieff, H., Sabroux, J.C., Eds.; Elsevier: Amsterdam, The Netherlands, 1983; pp. 425–444.

7.   Sutton, A.J.; Elias, T.; Gerlach, T.M.; Stokes, J.B. Implications for eruptive processes as indicated by sulfur dioxide emissions from Kilauea Volcano, Hawaii, 1979–1997. *J. Volcanol. Geotherm. Res.* **2001**, *108*, 283–302. [CrossRef]

8.   McGonigle, A.J.S.; Oppenheimer, C.; Galle, B.; Mather, T.A.; Pyle, D.M. Walking traverse and scanning DOAS measurements of volcanic gas emission rates. *Geophys. Res. Lett.* **2002**, *29*, 461–464. [CrossRef]

9.   Galle, B.; Oppenheimer, C.; Geyer, A.; McGonigle, A.J.S.; Edmonds, M.; Horrocks, L.A. A miniaturised UV spectrometer for remote sensing of SO_2 fluxes: A new tool for volcano surveillance. *J. Volcanol. Geotherm. Res.* **2003**, *119*, 241–254. [CrossRef]

10. Edmonds, M.; Herd, R.A.; Galle, B.; Oppenheimer, C.M. Automated, high time-resolution measurements of SO_2 flux at Soufrière Hills Volcano, Montserrat. *Bull. Volcanol.* **2003**, *65*, 578–586. [CrossRef]

11. McGonigle, A.J.S.; Oppenheimer, C.; Hayes, A.R.; Galle, B.; Edmonds, M.; Caltabiano, T.; Salerno, G.; Burton, M.; Mather, T.A. Sulphur dioxide flux measurements at Mount Etna, Vulcano and Stromboli measured with an automated scanning static ultraviolet spectrometer. *J. Geophys. Res.* **2003**, *108*, 2455. [CrossRef]

12. Salerno, G.G.; Burton, M.R.; Oppenheimer, C.; Caltabiano, T.; Randazzo, D.; Bruno, N.; Longo, V. Three-years of SO_2 flux measurements of Mt. Etna using an automated UV scanner array: Comparison with conventional traverses and uncertainties in flux retrieval. *J. Volcanol. Geotherm. Res.* **2009**, *183*, 76–83. [CrossRef]

13. Galle, B.; Johansson, M.; Rivera, C.; Zhang, Y.; Kihlman, M.; Kern, C.; Lehmann, T.; Platt, U.; Arellano, S.; Hidalgo, S. Network for Observation of Volcanic and Atmospheric Change (NOVAC)—A global network of volcanic gas monitoring: Network layout and instrument description. *J. Geophys. Res.* **2010**, *115*, D05304. [CrossRef]

14. Mori, T.; Burton, M.R. The SO_2 camera: A simple, fast and cheap method for ground-based imaging of SO_2 in volcanic plumes. *Geophys. Res. Lett.* **2006**, *33*. [CrossRef]

15. Bluth, G.; Shannon, J.; Watson, I.M.; Prata, A.J.; Realmuto, V. Development of an ultra-violet digital camera for volcanic SO_2 imaging. *J. Volcanol. Geotherm. Res.* **2007**, *161*, 47–56. [CrossRef]

16. Kantzas, E.P.; McGonigle, A.J.S.; Tamburello, G.; Aiuppa, A.; Bryant, R.G. Protocols for UV camera volcanic SO_2 measurements. *J. Volcanol. Geotherm. Res.* **2010**, *194*, 55–60. [CrossRef]

17. Williams-Jones, G.; Horton, K.A.; Elias, T.; Garbeil, H.; Mouginis-Mark, P.J.; Sutton, A.J.; Harris, A.J.L. Accurately measuring volcanic plume velocity with multiple UV spectrometers. *Bull. Volcanol.* **2006**, *68*, 328–332. [CrossRef]

18. McGonigle, A.J.S.; Hilton, D.R.; Fischer, T.P.; Oppenheimer, C. Plume velocity determination for volcanic SO_2 flux measurements. *Geophys. Res. Lett.* **2005**, *32*, L11302. [CrossRef]

19. Peters, N.; Hoffmann, A.; Barnie, T.; Herzog, M.; Oppenheimer, C. Use of motion estimation algorithms for improved flux measurements using SO_2 cameras. *J. Volcanol. Geotherm. Res.* **2015**, *300*, 58–69. [CrossRef]

20. Nadeau, P.A.; Palma, J.L.; Waite, G.P. Linking volcanic tremor, degassing, and eruption dynamics with SO_2 imaging. *Geophys. Res. Lett.* **2011**, *38*, L013404. [CrossRef]

21. Holland, A.S.P.; Watson, I.M.; Phillips, J.C.; Caricchi, L.; Dalton, M.P. Degassing processes during lava dome growth: Insights from Santiaguito Lava Dome, Guatemala. *J. Volcanol. Geotherm. Res.* **2011**, *202*, 153–166. [CrossRef]

22. Kern, C.; Lübcke, P.; Bobrowski, N.; Campion, R.; Mori, T.; Smekens, J.-F.; Stebel, K.; Tamburello, G.; Burton, M.R.; Platt, U.; et al. Intercomparison of SO_2 camera systems for imaging volcanic gas plumes. *J. Volcanol. Geotherm. Res.* **2015**, *300*, 22–36. [CrossRef]

23. Campion, R.A.; Delgado-Granados, H.; Mori, T. Image-based correction of the light dilution effect for SO_2 camera measurements. *J. Volcanol. Geotherm. Res.* **2015**, *300*, 48–57. [CrossRef]

24. Kern, C.; Werner, C.; Elias, T.; Sutton, A.J.; Lübcke, P. Applying UV cameras for SO_2 detection to distant or optically thick volcanic plumes. *J. Volcanol. Geotherm. Res.* **2013**, *262*, 80–89. [CrossRef]

25. Lübcke, P.; Bobrowski, N.; Illing, S.; Kern, C.; Alvarez Nieves, J.M.; Vogel, L.; Zielcke, J.; Delgado Granados, H.; Platt, U. On the absolute calibration of SO_2 cameras. *Atmos. Meas. Tech.* **2013**, *6*, 677–696. [CrossRef]

26. Kern, C.; Kick, F.; Lübcke, P.; Vogel, L.; Wöhrbach, M.; Platt, U. Theoretical description of functionality, applications, and limitations of SO_2 cameras for the remote sensing of volcanic plumes. *Atmos. Meas. Tech.* **2010**, *3*, 733–749. [CrossRef]

27. Kuhn, J.; Bobrowski, N.; Lübcke, P.; Vogel, L.; Platt, U. A Fabry-Perot interferometer-based camera for two-dimensional mapping of SO_2 distributions. *Atmos. Meas. Tech.* **2014**, *7*, 3705–3715. [CrossRef]

28. Platt, U.; Lübcke, P.; Kuhn, J.; Bobrowski, N.; Prata, F.; Burton, M.; Kern, C. Quantitative imaging of volcanic plumes—Results, needs, and future trends. *J. Volcanol. Geotherm. Res.* **2015**, *300*, 7–21. [CrossRef]

29. Wilkes, T.C.; McGonigle, A.J.S.; Pering, T.D.; Taggart, A.J.; White, B.S.; Bryant, R.G.; Willmott, J.R. Ultraviolet Imaging with Low Cost Smartphone Sensors: Development and Application of a Raspberry Pi-Based UV Camera. *Sensors* **2016**, *16*, 1649. [CrossRef] [PubMed]

30. Wilkes, T.C.; Pering, T.D.; McGonigle, A.J.S.; Tamburello, G.; Willmott, J.R. A low cost smartphone sensor-based UV camera for volcanic SO_2 emission measurements. *Remote Sens.* **2017**, *9*, 27. [CrossRef]

31. Campion, R.; Martinez-Cruz, M.; Lecocq, T.; Caudron, C.; Pacheco, J.; Pinardi, G.; Hermans, C.; Carn, S.; Bernard, A. Space- and ground-based measurements of sulphur dioxide emissions from Turrialba Volcano (Costa Rica). *Bull. Volcanol.* **2012**, *74*, 1757–1770. [CrossRef]

32. Stebel, K.; Amigo, A.; Thomas, H.E.; Prata, A.J. First estimates of fumarolic SO_2 fluxes from Putana volcano, Chile, using an ultraviolet imaging camera. *J. Volcanol. Geotherm. Res.* **2015**, *300*, 112–120. [CrossRef]

33. Kern, C.; Sutton, J.; Elias, T.; Lee, L.; Kamibayashi, K.; Antolik, L. An automated SO_2 camera system for continuous, real-time monitoring of gas emissions from Kīlauea Volcano's summit Overlook Crater. *J. Volcanol. Geotherm. Res.* **2015**, *300*, 81–94. [CrossRef]

34. D'Aleo, R.; Bitetto, M.; Delle Donne, D.; Tamburello, G.; Battaglia, A.; Coltelli, M.; Patanè, D.; Prestifilippo, M.; Sciotto, M.; Aiuppa, A. Spatially resolved SO_2 flux emissions from Mt Etna. *Geophys. Res. Lett.* **2016**, *43*, 7511–7519. [CrossRef] [PubMed]

35. Delle Donne, D.; Ripepe, M.; Lacanna, G.; Tamburello, G.; Bitetto, M.; Aiuppa, A. Gas mass derived by infrasound and UV cameras: Implications for mass flow rate. *J. Volcanol. Geotherm. Res.* **2016**, *325*, 169–178. [CrossRef]

36. Pering, T.D.; McGonigle, A.J.S.; James, M.R.; Tamburello, G.; Aiuppa, A.; Delle Donne, D.; Ripepe, M. Conduit dynamics and post explosion degassing on Stromboli: A combined UV camera and numerical modelling treatment. *Geophys. Res. Lett.* **2016**, *43*, 5009–5016. [CrossRef] [PubMed]

37. Burton, M.R.; Salerno, G.G.; D'Auria, L.; Caltabiano, T.; Murè, F.; Maugeri, R. SO_2 flux monitoring at Stromboli with the new permanent INGV SO_2 camera system: A comparison with the FLAME network and seismological data. *J. Volcanol. Geotherm. Res.* **2015**, *300*, 95–102. [CrossRef]

38. Aiuppa, A.; Giudice, G.; Gurrieri, S.; Liuzzo, M.; Burton, M.; Caltabiano, T.; McGonigle, A.J.S.; Salerno, G.; Shinohara, H.; Valenza, M. Total volatile flux from Mount Etna. *Geophys. Res. Lett.* **2008**, *35*, L24302. [CrossRef]

39. Tamburello, G.; Kantzas, E.P.; McGonigle, A.J.S.; Aiuppa, A.; Giudice, G. UV camera measurements of fumarole field degassing (La Fossa crater, Vulcano Island). *J. Volcanol. Geotherm. Res.* **2011**, *199*, 47–52. [CrossRef]

40. Tamburello, G.; Aiuppa, A.; McGonigle, A.J.S.; Allard, P.; Cannata, A.; Giudice, G.; Kantzas, E.P.; Pering, T.D. Periodic volcanic degassing behavior: The Mount Etna example. *Geophys. Res. Lett.* **2013**, *40*, 4818–4822. [CrossRef]

41. McGonigle, A.J.S.; Aiuppa, A.; Ripepe, M.; Kantzas, E.P.; Tamburello, G. Spectroscopic capture of 1 Hz volcanic SO_2 fluxes and integration with volcano geophysical data. *Geophys. Res. Lett.* **2009**, *36*, L21309. [CrossRef]

42. Boichu, M.; Oppenheimer, C.; Tsanev, V.; Kyle, P.R. High temporal resolution SO_2 flux measurements at Erebus volcano, Antarctica. *J. Volcanol. Geotherm. Res.* **2010**, *190*, 325–336. [CrossRef]

43. Girona, T.; Costa, F.; Taisne, B.; Aggangan, B.; Ildefonso, S. Fractal degassing from Erebus and Mayon volcanoes revealed by a new method to monitor H_2O emission cycles. *J. Geophys. Res. Solid Earth* **2015**, *120*, 2988–3002. [CrossRef]

44. Pering, T.D.; Tamburello, G.; McGonigle, A.J.S.; Aiuppa, A.; Cannata, A.; Giudice, G.; Patanè, D. High time resolution fluctuations in volcanic carbon dioxide degassing from Mount Etna. *J. Volcanol. Geotherm. Res.* **2014**, *270*, 115–121. [CrossRef]

45. Moussallam, Y.; Bani, P.; Curtis, A.; Barnie, T.; Moussallam, M.; Peters, N.; Schipper, C.I.; Aiuppa, A.; Giudice, G.; Amigo, Á.; et al. Sustaining persistent lava lakes: Observations from high-resolution gas measurements at Villarrica volcano, Chile. *Earth Planet. Sci. Lett.* **2016**, *454*, 237–247. [CrossRef]

46. Nadeau, P.A.; Werner, C.; Waite, G.P.; Carn, S.A.; Brewer, I.D.; Elias, T.; Sutton, A.J.; Kern, C. Using SO_2 camera imagery to examine degassing and gas accumulation at Kīlauea volcano. *J. Volcanol. Geotherm. Res.* **2015**, *300*, 103–111. [CrossRef]

47. Fischer, T.P.; Roggensack, K.; Kyle, P.R. Open and almost shut case for explosive eruptions: Vent processes determined by SO_2 emission rates at Karymsky volcano, Kamchatka. *Geology* **2002**, *30*, 1059–1062. [CrossRef]

48. Yamamoto, H.; Watson, I.M.; Phillips, J.C.; Bluth, G.J. Rise dynamics and relative ash distribution in vulcanian eruption plumes at Santiaguito Volcano, Guatemala, revealed using an ultraviolet imaging camera. *Geophys. Res. Lett.* **2008**, *35*, L08314. [CrossRef]

49. Tamburello, G.; Aiuppa, A.; Kantzas, E.P.; McGonigle, A.J.S.; Ripepe, M. Passive vs. active degassing modes at an open-vent volcano (Stromboli, Italy). *Earth Planet. Sci. Lett.* **2012**, *359*, 106–116. [CrossRef]

50. Mori, T.; Burton, M. Quantification of the gas mass emitted during single explosions on Stromboli with the SO_2 imaging camera. *J. Volcanol. Geotherm. Res.* **2009**, *188*, 395–400. [CrossRef]

51. Pering, T.D.; Tamburello, G.; McGonigle, A.J.S.; Aiuppa, A.; James, M.R.; Lane, S.J.; Sciotto, M.; Cannata, A.; Patanè, D. Dynamics of mild strombolian activity on Mt. Etna. *J. Volcanol. Geotherm. Res.* **2015**, *300*, 103–111. [CrossRef]

52. Smekens, J.-F.; Clarke, A.B.; Burton, M.R.; Harijoko, A.; Wibowo, H. SO_2 emissions at Semeru volcano, Indonesia: Characterization and quantification of persistent periodic explosive activity. *J. Volcanol. Geotherm. Res.* **2015**, *300*, 121–128. [CrossRef]

53. Kazahaya, R.; Mori, T.; Takeo, M.; Ohminato, T.; Urabe, T.; Maeda, Y. Relation between single very-long-period pulses and volcanic gas emissions at Mt. Asama, Japan. *Geophys. Res. Lett.* **2011**, *38*, L11307. [CrossRef]

54. Waite, G.P.; Nadeau, P.A.; Lyons, J.J. Variability in eruption style and associated very long period events at Fuego volcano, Guatemala. *J. Geophys. Res. Solid Earth* **2013**, *118*, 1526–1533. [CrossRef]

55. James, M.R.; Lane, S.J.; Wilson, L.; Corder, S.B. Degassing at low magma-viscosity volcanoes: Quantifying the transition between passive bubble-burst and Strombolian eruption. *J. Volcanol. Geotherm. Res.* **2009**, *180*, 81–88. [CrossRef]

56. Gaudin, D.; Taddeucci, J.; Scarlato, P.; Harris, A.; Bombrun, M.; Del Bello, E.; Ricci, T. Characteristics of puffing activity revealed by ground-based, thermal infrared imaging: The example of Stromboli volcano (Italy). *Bull. Volcanol.* **2007**, *79*, 24. [CrossRef]

57. Pering, T.D.; Tamburello, G.; McGonigle, A.J.S.; Hanna, E.; Aiuppa, A. Correlation of oscillatory behaviour in Matlab using wavelets. *Comput. Geosci.* **2014**, *70*, 206–212. [CrossRef]

58. Dalton, M.P.; Waite, G.P.; Watson, I.M.; Nadeau, P.A. Multiparameter quantification of gas release during weak Strombolian eruptions at Pacaya Volcano, Guatemala. *Geophys. Res. Lett.* **2010**, *37*, L09303. [CrossRef]

59. Fischer, T.P.; Morrissey, M.M.; Calvache, M.L.V.; Gómez, D.M.; Torres, R.C.; Stix, J.; Williams, S.N. Correlations between SO_2 flux and long-period seismicity at Galeras volcano. *Nature* **1994**, *368*, 135–137. [CrossRef]

60. McCormick Kilbride, B.; Edmonds, M.; Biggs, J. Observing eruptions of gas-rich compressible magmas from space. *Nat. Commun.* **2016**, *7*, 13744. [CrossRef] [PubMed]

61. McGonigle, A.J.S.; Aiuppa, A.; Guidice, G.; Tamburello, G.; Hodson, A.J.; Gurrieri, S. Unmanned aerial vehicle measurements of volcanic carbon dioxide fluxes. *Geophys. Res. Lett.* **2008**, *35*, L06303. [CrossRef]

62. Xi, X.; Johnson, M.S.; Jeong, S.; Fladeland, M.; Pieri, D.; Diaz, J.A.; Bland, G.L. Constraining the sulfur dioxide degassing flux from Turrialba volcano, Costa Rica using unmanned aerial system measurements. *J. Volcanol. Geotherm. Res.* **2016**, *325*, 110–118. [CrossRef]

63. Lopez, T.; Thomas, H.; Prata, A.J.; Amigo, A.; Fee, D.; Moriano, D. Volcanic Plume Characteristics Determined Using an Infrared Imaging Camera. *J. Volcanol. Geotherm. Res.* **2015**, *300*, 148–166. [CrossRef]

64. Burton, M.; Allard, P.; Muré, F.; La Spina, A. Magmatic gas composition reveals the source depth of slug-driven strombolian explosive activity. *Science* **2007**, *317*, 227. [CrossRef] [PubMed]

65. Von Glasow, R. Atmospheric chemistry in volcanic plumes. *Proc. Natl. Acad. Sci. USA* **2010**, *107*, 6594–6599. [CrossRef] [PubMed]

66. Roberts, T.J.; Martin, R.S.; Jourdain, L. Reactive bromine chemistry in Mount Etna's volcanic plume: The influence of total Br, high-temperature processing, aerosol loading and plume-air mixing. *Atmos. Chem. Phys.* **2014**, *14*, 11201–11219. [CrossRef]

67. Loubain, I.; Bobrowski, N.; Rouwet, D.; Inguaggiato, S.; Platt, U. Imaging DOAS for volcanological applications. *Bull. Volcanol.* **2009**, *71*, 753–765. [CrossRef]
68. Klein, A.; Lübcke, P.; Bobrowski, N.; Kuhn, J.; Platt, U. Plume propagation direction determination with SO_2 cameras. *Atmos. Meas. Tech.* **2017**, *10*, 979–987. [CrossRef]
69. Burton, M.R.; Prata, F.; Platt, U. Volcanological applications of SO_2 cameras. *J. Volcanol. Geotherm. Res.* **2015**, *300*, 2–6. [CrossRef]
70. Tamburello, G.; Kantzas, E.P.; McGonigle, A.J.S.; Aiuppa, A. Vulcamera: A program for measuring volcanic SO_2 using UV cameras. *Ann. Geophys.* **2011**, *54*, 219–221. [CrossRef]

geosciences

MDPI

Review

Volcanic Plume Impact on the Atmosphere and Climate: O- and S-Isotope Insight into Sulfate Aerosol Formation

Erwan Martin

Sorbonne Université, CNRS-INSU, Institut des Sciences de la Terre Paris, ISTeP UMR 7193, F-75005 Paris, France; erwan.martin@sorbonne-universite.fr; Tel.: +33-144-274-181

Received: 4 May 2018; Accepted: 26 May 2018; Published: 31 May 2018

Abstract: The impact of volcanic eruptions on the climate has been studied over the last decades and the role played by sulfate aerosols appears to be major. S-bearing volcanic gases are oxidized in the atmosphere into sulfate aerosols that disturb the radiative balance on earth at regional to global scales. This paper discusses the use of the oxygen and sulfur multi-isotope systematics on volcanic sulfates to understand their formation and fate in more or less diluted volcanic plumes. The study of volcanic aerosols collected from air sampling and ash deposits at different distances from the volcanic systems (from volcanic vents to the Earth poles) is discussed. It appears possible to distinguish between the different S-bearing oxidation pathways to generate volcanic sulfate aerosols whether the oxidation occurs in magmatic, tropospheric, or stratospheric conditions. This multi-isotopic approach represents an additional constraint on atmospheric and climatic models and it shows how sulfates from volcanic deposits could represent a large and under-exploited archive that, over time, have recorded atmospheric conditions on human to geological timescales.

Keywords: volcanic sulfate aerosols; oxygen and sulfur multi-isotopes; atmospheric chemistry

1. Introduction

Globally on Earth, each year, volcanoes release an average of 10–20 Mt of sulfur-bearing gases, and occasionally much more during super eruptions (e.g., [1,2]). Atmospheric sulfur plays a paramount role in the terrestrial radiative balance. Consequently, determining its source and understanding its physico-chemical transformations and fate in the atmosphere appear crucial in predicting its impact on the atmosphere and climate. Indeed, when released into the atmosphere S-bearing gases (mainly SO_2: $\leq 25\%$; H_2S: $\leq 10\%$; and COS and CS_2: $\leq 0.01\%$ of the volume of the emitted gases; [3–5] and references therein) are oxidized, being thus transformed into sulfate aerosols that directly and indirectly lead to a negative radiative forcing. First, sulfate aerosols directly reflect part of the solar radiations, thus decreasing the amount of sun energy reaching the Earth's surface. On the other hand, sulfate aerosols absorb part of the incoming solar radiation in the IR wavelengths and this results in a warming of the aerosol-bearing atmospheric layer associated to a temperature decrease between it and the Earth's surface. Additionally, volcanic aerosols play the role of cloud nucleation leading to more cloud formation, which reinforces the albedo and therefore causes the Earth's surface to cool (e.g., [6,7]). Such "volcanic winters", that can be more or less severe and global depending on the strength and location of the eruption, have been observed after several major eruptions such as those at Mt Agung in 1963, El Chichón in 1983, and Pinatubo in 1991 (e.g., [8–11]).

Atmospheric modeling aims at predicting climate evolution on Earth. However, including the sulfur volcanic aerosol formation processes would increase the models' accuracy regarding the volcanic forcing climate. This paper discusses the use of oxygen and sulfur multi-isotopes in improving our understanding of sulfate aerosol formation, fate, and sources. The $^{18}O/^{16}O$ and $^{17}O/^{16}O$ as well as

^{34}S/^{32}S, ^{33}S/^{32}S, and ^{36}S/^{32}S are measured and expressed as δ^{18}O and δ^{17}O as well as δ^{34}S, δ^{33}S and, δ^{36}S respectively (Equation (1)). During any process, the isotopes fractionate, changing the isotopic ratios, depending on the mass differences between heavy and light isotopes. For instance, the mass difference between ^{18}O and ^{16}O is twice as much as between ^{17}O and ^{16}O, leading to the relation ^{17}O/^{16}O ~0.5 * ^{18}O/^{16}O, commonly expressed as δ^{17}O = $(\delta^{18}$O + 1$)^{0.524}$ − 1. This very widespread rule suffers some exceptions, where isotopic fractionation does not depend on the mass differences between isotopes, it is referred to as Mass Independent Fractionation (MIF) (e.g., [12–15]). The difference to the isotopic mass dependent fractionation relation is quantified by Δ^{17}O, Δ^{33}S and Δ^{36}S (Equations (2)–(4)). Therefore, isotopic mass dependent fractionation results in a $\Delta = 0‰$ whilst MIF has $\Delta \neq 0‰$.

$$\delta A = (R_A/R_{st}) - 1; \text{ for instance } \delta^{18}O = (^{18}O/^{16}O_{(sample)}/^{18}O/^{16}O_{(standard)}) - 1 \tag{1}$$

$$\Delta^{17}O = \delta^{17}O - [(\delta^{18}O + 1)^{0.524} - 1] \tag{2}$$

$$\Delta^{33}S = \delta^{33}S - [(\delta^{34}S + 1)^{0.515} - 1] \tag{3}$$

$$\Delta^{36}S = \delta^{36}S - [(\delta^{34}S + 1)^{1.89} - 1] \tag{4}$$

considering that: δA: isotopic composition of a sample (e.g., $\delta^{18}O$); R_A: isotope ratio of the measured sample (e.g., $^{18}O/^{16}O$); Rst: isotope ratio of a standard. The mass dependent coefficients 0.524, 0.515, and 1.89 are from references [16,17].

In the atmosphere, sulfate aerosols are generated by oxidation of S-bearing gases through different possible reactions or channels (Ch1, Ch2, Ch3, and Ch4; e.g., [18]). The multi O-isotopic composition of the different atmospheric oxidants is relatively well-known such that the resulting sulfate composition can be estimated for each oxidation channel (Figure 1). Therefore, coupling O- and S-isotopic composition with atmospheric chemistry allows the oxidation channel(s) from which they formed to be retraced. This not only helps to elucidate the sulfate aerosol formation in the atmosphere but also permits us to probe into the composition of the atmosphere and more specifically its oxidant capacity during specific volcanic eruptions.

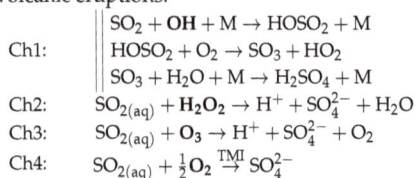

Ch1:
$$SO_2 + OH + M \rightarrow HOSO_2 + M$$
$$HOSO_2 + O_2 \rightarrow SO_3 + HO_2$$
$$SO_3 + H_2O + M \rightarrow H_2SO_4 + M$$

Ch2: $\quad SO_{2(aq)} + H_2O_2 \rightarrow H^+ + SO_4^{2-} + H_2O$

Ch3: $\quad SO_{2(aq)} + O_3 \rightarrow H^+ + SO_4^{2-} + O_2$

Ch4: $\quad SO_{2(aq)} + \frac{1}{2}O_2 \xrightarrow{TMI} SO_4^{2-}$

where M: any inert molecule that removes excess energy without participating in the reaction; TMI: transition metal ion; for (Ch4); $SO_{2(aq)}$ is in aqueous phase and can be present as $SO_2 \cdot H_2O$, HSO_3^-, or SO_3^{2-} (for pH from 2 to 7, which is most likely the case in volcanic plumes, HSO_3^- dominates). Ch1, Ch2, Ch3 and Ch4 are the same as in Figure 1.

This isotopic approach opens new perspectives when considering volcanic deposits formed by different eruptions all around the world during the whole history of Earth. As volcanic activity has always been present, sulfates from volcanic deposits represent an underexploited archive that, over time, has recorded atmospheric conditions on human to geological timescales.

This paper reviews the different methods used for sulfate aerosol sampling, as well as for O- and S-isotopic measurements, the final purpose being a discussion of the mechanisms of formation of the volcanic sulfate during the different kind of volcanic eruptions, from passive degassing to large caldera-forming eruptions (super-eruptions).

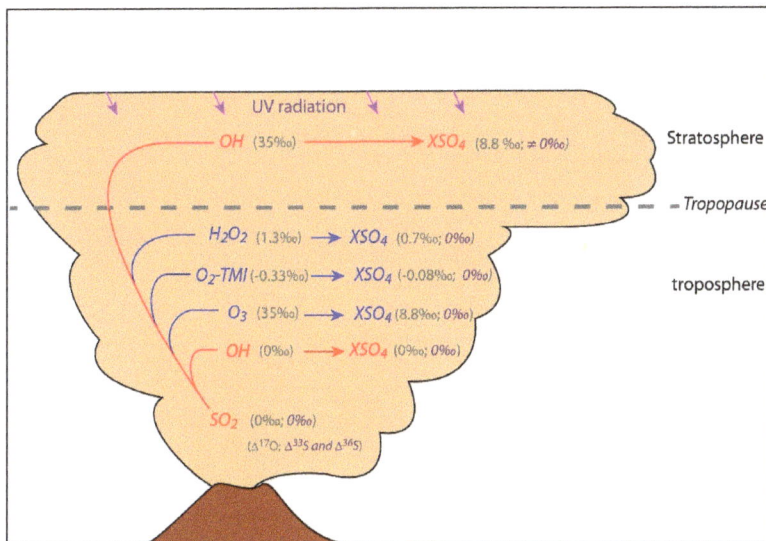

Figure 1. Volcanic SO_2 main oxidation channels in the troposphere and stratosphere (gas phase reactions in red and aqueous reactions in blue). X can be two monovalents or one bivalent cation (e.g., H^+, K^+, Na^+, or Ca^{2+}). The average isotopic values of volcanic SO_2, atmospheric oxidants and expected sulfates are indicated in grey values for $\Delta^{17}O$ and purple for $\Delta^{33}S$ and $\Delta^{36}S$. The detailed oxidation reactions are discussed in the text (Ch1, Ch2, Ch3 and Ch4). Note that sulfate oxygen atoms come partially from the atmospheric oxidants, at 25% from OH in Ch1, 50% from H_2O_2 in Ch2, 25% from O_3 in for Ch3, and 25–50% from O_2 in Ch4 [19]. Using these proportions makes a theoretical estimation of sulfate $\Delta^{17}O$ [20] possible. $\Delta^{33}S$ and $\Delta^{36}S \neq 0$ are generated mainly by SO_2 photolysis and photoexcitation by UV radiation [12,21,22], process that can most likely take place in the stratosphere above the O_3 layer, where UV radiations are not filtered. Note that in the troposphere $\Delta^{33}S$ and $\Delta^{36}S = 0$ is expected but some mass dependent processes can generate small isotopic anomalies as discussed in the text (Section 3.2) and in Figure 3.

2. Methods

Several laboratories use a variety of methods to extract sulfate from natural samples and to analyze their isotopic composition. Some methods are still in development, nonetheless the present paper describes the methods classically used nowadays.

2.1. Volcanic SO_2 and H_2S Sampling

Gas sampling at volcanic vents have been used for the last 40 years [23] and is based on the fact that acidic volcanic gases, including SO_2 and H_2S can be trapped in alkaline solutions. The most quantitative method appears to be the Giggenbach bottle system [24], but it requires approaching the volcanic vent, making the sampling and isotopic characterization of volcanic S-bearing gases difficult. The Giggenbach bottle system consists of pumping the volcanic gases, which are collected into a bottle containing an alkaline solution. The acid gases are dissolved in the solution and the low solubility gases trapped in the headspace of the bottle. The development of such a method allows the simultaneous and separate collection of SO_2 and H_2S [25,26]. Indeed, both gases are trapped in the alkaline solution, but the presence of Zn-acetate for instance leads to the immediate precipitation of H_2S into ZnS, allowing the physical separation of sulfur from SO_2 and H_2S. This method is rather quantitative, which is necessary especially when isotopic analyses and mass balances are considered. Another method is based on passive alkaline traps, and consists simply of plastic beakers containing alkaline solution close the volcanic vents [27]. This method permits the volcanic gases to be collected over a longer period of time, allowing a better estimation of the average volcanic gas fluxes, but it does not allow the separation of different gases (SO_2 vs. H_2S for instance). Furthermore, during the absorption of S-bearing gases, some isotopic fractionation may occur, which may lead to a significant bias for isotopic studies [28]. Finally, filters soaked in alkaline solution are also used, but no separation of SO_2 and H_2S is possible either and some isotopic fraction occurs when the SO_2 flux is too high [29,30].

2.2. Sulfate Aerosol Sampling

Air sampling and ash leaching are the two classic methods for collecting volcanic sulfates. The first method consists of pumping the air near volcanic vents; the particles (aerosols) are then gathered on filters, usually polytetrafluoroethylene (PTFE) filter packs (e.g., [29]). Pumps with a flow rate of about 30 L min^{-1} are generally used, as they are able to collect—in a few hours—enough sulfate aerosols for isotopic measurements [30,31]. Filters are then leached with deionized water in order to extract and concentrate sulfate aerosols into solution. Special care must be taken as, even if the reason is not yet fully understood, there is some variability in the $\delta^{18}O$ measured in sulfates collected on filters [29,30]; this could be linked to the collecting method as it cannot be explained by any magmatic processes. However, this variability is not observed for $\Delta^{17}O$ and S-isotopes. The other method for sulfate sampling consists of directly leaching volcanic ash. Indeed, a few tens to thousands of ppm of sulfate can be present in volcanic ash deposits (e.g., [32–34]). With the exception of barite ($BaSO_4$), sulfates are highly soluble in water, making their preservation in volcanic ash problematic or impossible in humid regions as well as in old volcanic deposits. Nevertheless, in arid to semi-arid environments, sulfates are at least partially preserved in volcanic ash layers up to tens of Ma [33,35]. Although the scavenging of sulfate is possible, isotopic exchange between water and sulfate is rather negligible in such environments [36]. Therefore, even if sulfate is partially removed by leaching or weathering from a volcanic deposit, its overall isotopic composition is preserved. In some cases, as in a sedimentary basin (an alkaline lake for instance), volcanic deposits can undertake the formation of authigenic sulfates that have a different isotopic composition compared to volcanic sulfates. Furthermore, long-term atmospheric deposition can also contribute to the sulfate from volcanic deposits, but it is rather negligible compared to the volcanic and authigenic sulfates [33]. The overall isotopic composition of sulfate from these volcanic deposits can possibly be progressively modified by dilution of the initial volcanic sulfate. However, using isotopic mixing models, it is possible to estimate the proportion of volcanic sulfate left in such deposits [33,35]. Deionized water is used for ash leaching and diluted HCl is added when carbonates were formed and could have trapped some sulfate. For efficient leaching, the preferred water/ash mass ratio is of about 1/20 [37].

2.3. Sulfate Chemical Composition and Preparation

Even if the volcanic sulfate aerosol mineralogy is very complex and not totally understood, it appears that at volcanic vents, dominating sulfates are commonly K-Na-sulfates [38,39]. Considering the high volatility of Na and K, the formation of these sulfate aerosol particles can be the result of the reaction between sulfuric acid and volatilized alkali-chlorides, the condensation of alkali-sulfate directly volatilized out of the magma or emitted by the hydrothermal system [40–43].

On the other hand, further away from the vent ($\geq$1–2 km), where the plume is more diluted, it seems that Ca-sulfates dominate [38,39]. They are found as particles but also as well-formed crystals and coating on other particles such as volcanic ash. This testifies that Ca-sulfates are less likely to have been produced by mechanical aerosol formation like in volcanic conduits or vents. However, considering the low volatility of calcium, Ca-sulfates are most likely formed by the alteration of volcanic glass in the plume itself. This is also consistent with the fact that in volcanic deposits sampled far from the volcanic centers (between 5 and 25 km [30]; $\geq$500 km [33,35]), Ca-sulfates seem to be the dominant sulfate species. Ayris et al. [44] show that at high temperatures the high Ca^{2+} diffusivity in ash particles leads to $CaSO_4$ formation in the volcanic conduit, which during the cooling down of the plume and its mixing with air, can be hydrated and generate gypsum as observed in volcanic ash deposits. However, in their experiments, K-Na-sulfates are not generated at high temperature by diffusion-driven mechanism. Therefore, this mechanism does not seem to explain the dominance of alkali-sulfate at volcanic vents.

In order to be able to measure the multi-isotopic composition of the collected sulfates, the latter is reacted with salts, which results in precipitates such as $BaSO_4$, Ag_2SO_4, or Ag_2S.

- $BaSO_4$: For about 30 years (e.g., [36,45]) sulfates were transformed into highly insoluble $BaSO_4$ by adding $BaCl_2$ in the leachate obtained from different kind of samples such as ash or filters. However, when precipitated from a multi-anion solution, barite ($BaSO_4$) can occlude impurities such as nitrate, which can introduce an analytical bias in the measurement of O-isotopes [46]. For this reason, it is of paramount importance to purify the collected sulfate and make sure that no nitrate remains in the leachate. In order to purify barite, Bao et al. [47] developed the DDARP (diethylenetriaminepentaacetic acid dissolution and re-precipitation) method, while more recently, Le Gendre et al. [48] worked out a Resin Method for Sulfate Extraction and Purification (RMSEP), that passes the leachate solution through an anionic exchange resin, allowing the purification and concentration of sulfate. Then, by adding $BaCl_2$, pure $BaSO_4$ is precipitated.
- Ag_2SO_4: Via anionic exchange resins, sulfate from the leachate is converted into Na_2SO_4. Then Ag_2SO_4 is produced by passing the Na_2SO_4 through another resin conditioned into Ag^+ [49].
- Ag_2S: Sulfate from leachate or $BaSO_4$ or Na_2SO_4 is converted into H_2S (gas) via acid attack [50] in order to react with $AgNO_3$ to finally precipitate and the sulfur as Ag_2S [51].

2.4. Oxygen Isotope Measurements

Oxygen isotope ratios were conventionally determined via graphite-reduction techniques that generate CO_2 from the oxygen extracted from the sulfates [52,53]. However, due to an isobaric interference from ^{13}C (mass 45: $^{13}C^{16}O^{16}O$ or $^{12}C^{16}O^{17}O$), only $\delta^{18}O$ can be determined via these methods. More recently, Bao and Thiemens [54] developed a laser fluorination method, in order to precisely and simultaneously measure $^{18}O/^{16}O$ and $^{17}O/^{16}O$ from O_2 extracted from sulfate samples. It is noteworthy that this method typically requires >20–30 μmol of sulfate (>4–7 mg of $BaSO_4$), so that enough material remains available for duplicate or triplicate analyses. Using a CO_2 laser, 2–3 mg of $BaSO_4$ reacted with BrF_5 at a high temperature, leading to O_2 being released, which is then extracted and purified by successive cryogenic traps in an extraction line. Finally, it is trapped onto a molecular sieve and then sent to the mass spectrometer. However, O_2 extraction from $BaSO_4$ by fluorination is never total (only 30–45% yield) [33,54], which induces an isotopic fractionation. Typical fractionation of $\delta^{18}O$ is of about +8‰, yet this fractionation is mass dependent, such that it has no effect on the $\Delta^{17}O$ value. Overall, the method uncertainties (sulfate transformation into $BaSO_4$ + O_2 extraction line + mass spectrometer) are: $\delta^{18}O \pm 0.5$‰ and $\Delta^{17}O \pm 0.1$‰ (2σ).

Whilst the laser fluorination method requires a minimum of 8–10 μmol of sulfate for an O-isotopes analysis, samples as small as 0.2 μmol of sulfate can be analyzed by pyrohydrolysis of Ag_2SO_4 using an elemental analyzer [55,56]. This method is ideally adapted for analyzing small quantities of sulfate aerosols. However, the uncertainties ($\delta^{18}O \pm 2$‰ and $\Delta^{17}O \pm 0.2$–0.3‰ (2σ)) are greater than with the laser fluorination method.

2.5. Sulfur Isotope Measurements

While $^{34}S/^{32}S$ in sulfates can be easily and directly analyzed using an elemental analyzer, multi S-isotope measurements ($^{34}S/^{32}S$, $^{33}S/^{32}S$ and $^{36}S/^{32}S$) require that Ag_2S precipitate is converted into SF_6 [51]. SF_6 is obtained by fluorination (via F_2) of Ag_2S; it is then purified by gas phase chromatography, and subsequently concentrated in a cryogenic trap before being injected into the mass spectrometer. Recently, Au Yang et al. [57] developed a method allowing the analysis of less than

0.1 µmol of SF_6. Overall, the method uncertainties (sulfate transformation into Ag_2S + SF_6 extraction line + mass spectrometer) are: $\delta^{34}S \pm 0.2‰$; $\Delta^{33}S \pm 0.02‰$ $\Delta^{36}S \pm 0.09‰$ (2σ).

3. Volcanic Sulfate Formation

Sulfate aerosols can possibly be generated at high temperature in the magma itself in very oxidizing conditions or more likely in the volcanic conduit during the gas ascent. Indeed, as discussed above, they mostly consist of K-Na-sulfates and can result from the reaction between sulfuric acid and volatilized alkali-chlorides, from the condensation of alkali-sulfate directly volatilized out of the magma or emitted by the hydrothermal systems [40–43]. These sulfates are referred to as primary sulfates. Their proportion in volcanic plumes during passive magma degassing is usually low with a sulfate/SO_2 ratio, usually <1% near the volcanic vents [25,58,59]. The S-bearing gases that are carried by the colder (dense or diluted) volcanic plumes and clouds produce secondary sulfate aerosols via different possible oxidation pathways in the atmosphere [20] and also possibly by chemical and photochemical reactions on mineral and dust surfaces [60]. As discussed above, these secondary sulfates consist most likely of Ca-sulfates due to interaction between sulfuric acid and volcanic glass. Below is detailed the possible mechanisms responsible for the formation of primary volcanic sulfates at high temperatures and the secondary volcanic sulfates in the troposphere, stratosphere, or during super-volcanic events.

3.1. Isotopic Composition of S-Bearing Gases

Due to obvious difficulties regarding a quantitative sampling close to volcanic vents, the previous determination of isotopic composition of volcanic S-bearing gases collected close to open degassing vents were only achieved at volcanic systems, where only minor eruptions occur [25–27,29,61–68]. In such open degassing systems, it appears that the $\delta^{34}S$ of the bulk volcanic S-bearing gases (mainly SO_2 + H_2S) are rather similar to their magmatic sources, as expected since sulfur is almost thoroughly degassed (>90%) from degassing magmas [25]. However, Menyailov et al. [26] were able to measure lower $\delta^{34}S$ for H_2S than for SO_2 by ~4‰ at 700–800 °C, ~12‰ at 500 °C and ≥16‰ at temperature ≤100 °C, which is consistent with the tendency previously observed [66] and the isotopic fractionation expected between the two S-bearing gases [69,70]. While the bulk S-bearing gas has a similar composition to its magmatic source, the latter strongly depends on the oxygen fugacity of the magma, with light isotopic ratios for low oxygen fugacity (MORB-like magmas: $\delta^{34}S$ ~−1‰ to 0‰; e.g., [71]) and heavier ratios when the magmatic conditions are more oxidizing (arc lavas: $\delta^{34}S$ ~ 5‰; e.g., [72]). In Figure 2, the range of −1 to 6‰ is reported as representative of the bulk volcanic S-bearing gases. No measures of multi-S and -O isotopes ratios have been done on the S-bearing gases. However, as it has been observed for $\delta^{34}S$, the $\Delta^{33}S$, $\Delta^{36}S$, $\delta^{18}O$, and $\Delta^{17}O$ of S-bearing gases is expected to be similar to the magmatic sources and the mantle values that are close to 0‰ for $\Delta^{33}S$, $\Delta^{36}S$, and $\Delta^{17}O$ and 5.5 to 7‰ for $\delta^{18}O$ depending on the geodynamic context ([17,71–74]; Figure 2).

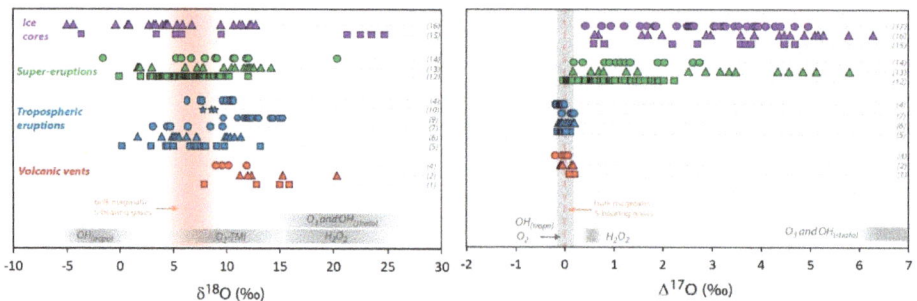

Figure 2. *Cont.*

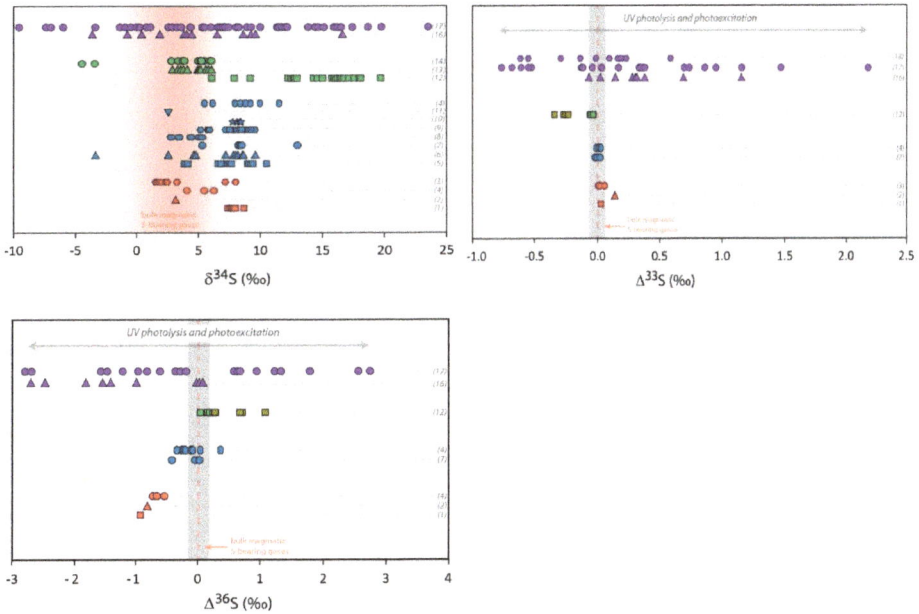

Figure 2. Multi O- and S-isotopic composition measured in volcanic sulfates. Red symbols represent primary sulfate samples collected at volcanic vents (aerosols or sulfate extracted from ashes). Blue symbols are sulfate samples extracted from volcanic deposits from tropospheric eruptions gathered at distances up to 100 km from the vent. Green symbols represent volcanic sulfates collected in deposits of super-eruptions at distances between 500 km and 5000 km from the volcanic systems. Pale green symbols correspond to $\Delta^{33}S$ and $\Delta^{36}S$ recalculated from the values represented by the dark green symbols, assuming dilution by non-volcanic sulfates [33]. Purple symbols represent sulfate samples from Antarctica ice cores. Mass dependent (non-MIF) compositions are emphasized by grey areas (taking into account the analytical uncertainties in 2σ) and bulk magmatic S-bearing gases composition by orange areas or dotted lines (see the Section 3.1 isotopic composition of S-bearing gases for further discussion). The theoretical $\delta^{18}O$ and $\Delta^{17}O$ composition of secondary sulfate generated in the atmosphere by different oxidation pathways are reported in grey areas. The different oxidation pathways via OH, H_2O_2, O_3, and O_2-TMI are detailed in the introduction of this paper and in Figure 1. The effect of UV photolysis and photoexcitation of SO_2 on the sulfate $\Delta^{33}S$ and $\Delta^{36}S$ are shown by grey arrows. References of the figure: sulfate aerosols (1): [31]; (2): [30]; (3): [25] and references therein; sulfate extracted from volcanic ash (4): [30]; (5): [20]; (6): [35]; (7): [75]; (8): [76]; (9): [32]; (10): [77]; (11): [34]; (12): [33]; (13): [35]; (14): Unpublished data; (15): [78]; (16): [79–81]; (17): [82]. For the (16) and (17) dataset and due to significant background corrections, only samples with high volcanic fraction (>65%) are considered here (see [82] for further discussion); (18): [83]. Note that for stratospheric aerosols collected in ice cores, even if only high volcanic fraction samples are considered, the uncertainties are typically of about: $\Delta^{17}O \pm 0.5‰$, $\Delta^{33}S \pm 0.1‰$ and $\Delta^{36}S \pm 0.8‰$ (in 2σ; see Figure 3).

3.2. High Temperature Primary Sulfates

In order to study the primary volcanic sulfates, the sampling of aerosols must be performed as close as possible to the volcanic vents (up to a few hundred meters). Figure 2 shows a compilation of the isotopic composition of such primary sulfates. The variation range for O-isotopes is from 7‰ to 20‰ in $\delta^{18}O$ and it becomes very narrow for $\Delta^{17}O$ whose values are close to 0‰ (from −0.18 to +0.2‰). S-isotopes are more homogeneous with only 8‰ variation in $\delta^{34}S$ (from 1 to 9‰), $\Delta^{33}S$

very close to 0‰ or slightly positive (from 0.01 to 0.14‰) and $\Delta^{36}S$ significantly negative values from −0.5 to −0.9‰.

Overall, the aerosols $\delta^{34}S$ tend to be higher than the bulk S-bearing gases, and when measured simultaneously the $\delta^{34}S$ of bulk S-bearing gases is systematically a few permil lower than in sulfate [26], in agreement with the expected isotopic fractionation [84]. The $\Delta^{33}S$ of ~0‰ is consistent with a direct oxidation of S-bearing gases at high temperature where, like in magmatic conditions, a mass dependent isotopic fractionation is expected (e.g., [85]). In contrast, the negative $\Delta^{36}S$ values are consistent with non-MIF processes as small non-zero $\Delta^{33}S$ (between 0 and 0.14‰; Figure 2) and $\Delta^{36}S$ (between −0.5 and −0.9; Figure 2) can be generated by mass dependent processes. The observed $\Delta^{33}S/\Delta^{36}S$ ratio of about −8 is indeed in the same range as the mass dependent fractionation line that has a $\Delta^{33}S/\Delta^{36}S$ ratio between −5 and −10 [51,86,87] (Figure 3).

The measured $\delta^{18}O$ values are systematically above the magmatic composition, which can be interpreted in terms of isotopic fractionation during distillation/condensation processes in the volcanic conduit and/or to sulfur oxidants that have systematically higher $\delta^{18}O$ than the magma. Such oxidants could be atmospheric oxidants like O_2 or H_2O_2, which could be responsible for these high $\delta^{18}O$ as they can have $\delta^{18}O$ higher than 20‰ (e.g., [20] and references therein; Figure 2). However, it is noteworthy that when sulfate aerosols are collected on filters, their $\delta^{18}O$ seems to increase with the sulfate concentration [30,31]. This seems to show that some isotopic fractionation may occur on the filter during the sampling. In locations where such measurements have been done (Stromboli), LeGendre [30] also analyzed sulfate collected on volcanic ash and the $\delta^{18}O$ is much more reproducible and corresponds to the lowest range of what was measured on filters (around 10‰; Figure 2). These values are closer than the magmatic values, but still slightly higher (Figure 2); therefore, we cannot rule out that atmospheric oxidants play a significant role in the high temperature chemistry at volcanic vents and therefore on the primary volcanic sulfate formation.

The $\Delta^{17}O$ values are all very close to zero, indicating that the S-bearing gases oxidant is mass dependent, which is consistent with magmatic conditions, where all the compounds (volcanic gases, silicates minerals and glass) are mass dependent. If, as suggested above, some atmospheric oxidants play a significant role in the formation of the primary sulfate, these oxidants should have a $\Delta^{17}O$ close to 0‰, which is only consistent with O_2 ($\Delta^{17}O = -0.33‰$) and not H_2O_2. Indeed, even if $\Delta^{17}O$ of atmospheric O_2 is significantly negative, when combined with magmatic oxygen atoms in the resulting sulfate, the overall $\Delta^{17}O$ should be very close to $0 \pm 0.1‰$ (Figure 1). Overall, we could expect sulfate $\delta^{18}O$ to be magmatic-like or slightly higher if atmospheric O_2 played a significant role and did not isotopically exchange with magmatic compounds.

It is very challenging to assess processes that occur at high temperatures, but some studies try to do that via equilibrium thermodynamic models [88–91]. They show that, considering volcanic gas mixtures with atmospheric air during the plume cooling, SO_3 can be generated from S-bearing gases and subsequently form H_2SO_4 by co-condensation along with water. Combustion experiment studies also demonstrate that ash particles and their iron oxides are excellent catalyzers to the SO_2 to SO_3 conversion and that the resulting sulfate is isotopically mass dependent [92,93]. The models also predict the formation of H, OH, and OH_2 radicals that are initially coming from thermal dissociation of H_2O, which produces H and OH radicals reacting with atmospheric O_2 to generate OH and OH_2 [88]. The incorporation of atmospheric OH is also possible, but probably quickly depleted in the high temperature plume as the highly concentrated SO_2 oxidation (via OH) would rapidly consume it.

3.3. Tropospheric Secondary Sulfates

Secondary sulfates are generated by oxidation of sulfur precursors in the atmosphere, either in the gas or condensed phases. When the volcanic plume does not reach the stratosphere, and remains in the troposphere, it is possible to address the formation of secondary sulfates in a more or less diluted volcanic tropospheric plume or cloud, by collecting sulfates from volcanic ashes deposited at distances even up to a few hundred kilometers from the volcanic vent [20]. Such sulfates measured all over the world show compositions ranging from 0 to 15‰ for both $\delta^{18}O$ and $\delta^{34}S$; their $\Delta^{17}O$, $\Delta^{33}S$ and $\Delta^{36}S$ values are very homogeneous and very close to 0‰ as well (Figure 2). In detail, $\Delta^{36}S/\Delta^{33}S$ of about −9.4 is similar to what is observed close to volcanic vents and close to the mass dependent fractionation line (Figure 3).

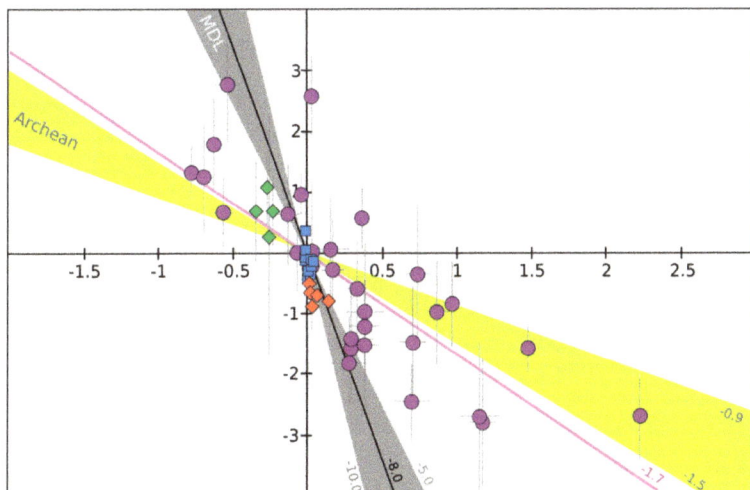

Figure 3. $\Delta^{33}S$ versus $\Delta^{36}S$ relationship for volcanic sulfates collected close the vents (in red), from volcanic deposits of tropospheric eruptions (in blue), super-eruptions (in green), and from ice cores recording stratospheric eruptions (in purple). All the dataset is the same as in the Figure 2. MDL: mass dependent fractionation line which has a slope between −10.0 and −5.0 (−6.58 in average) and Archean (yellow zone), corresponds to reference array defined by Archean rock samples, and have a slope between −1.5 and −0.9. For more details, the reader is invited to read the recent review on the subject by Ono [94]. Despite the analytical uncertainties, specifically on $\Delta^{36}S$ from ice core sulfates samples (see the error bars in 2σ), which induce possible large uncertainties on the $\Delta^{33}S/\Delta^{36}S$ ratios (lines) defined by volcanic samples, some general conclusion can still be addressed. Volcanic sulfates produced in volcanic vent at high temperatures and in the troposphere define a line with a slope of about −8.0 ($R^2 = 0.61$), which is comparable to the MDL. However, volcanic sulfate from stratospheric and super-eruptions define a line with a slope of about −1.7 ($R^2 = 0.65$), which is clearly not comparable to the MDL but very close to the Archean array.

Models of tropospheric chemistry show that SO_2 must be overwhelmingly oxidized by H_2O_2 when pH < 6, which is expected to be the case in volcanic plume. However, as all non-volcanic tropospheric sulfates have a $\Delta^{17}O$ of about 0.7‰ (e.g., references [30,92,95–97]), this should also be the case for volcanic tropospheric sulfates (Figure 1). The fact that $\Delta^{17}O$ of tropospheric volcanic sulfate is close to 0‰, clearly shows that H_2O_2 is not the main oxidant. Based on the fact that H_2O_2 concentration in the troposphere (H_2O_2 concentration = $[H_2O_2]$ < 0.5 DU; Dobson unit; 1 DU = 2.68 × 10^{16} molecule cm^{-2}) is much lower than the concentration of volcanic SO_2 in the volcanic plumes ($[SO_2]$ = 10–100 DU; e.g., [98,99]), Martin et al. [20] proposed that H_2O_2 reacts and is rapidly consumed in the tropospheric column. The second main oxidants are OH (Ch1) or O_2 (Ch4) that are non-MIF (or slightly negative for O_2). Consequently, they logically produce sulfate aerosols with $\Delta^{17}O$ very close to 0‰ (Figure 1). In tropospheric volcanic plumes, when tropospheric humidity is high (water condensation in the plume), heterogeneous aqueous oxidation reaction via O_2-TMI is expected to be faster than homogeneous gas phase reaction via OH. However, in relatively low humidity conditions (no condensing plume), the oxidation via OH should dominate [100]. This is in agreement with observations made at Kilauea (Hawaii) where diurnal variations in the sulfate fraction ($[SO_4^{2-}]/([SO_4^{2-}] + [SO_2])$) have been observed in the volcanic plume, indicating that photochemically produced oxidants play a major role in the SO_2 oxidation [101]. OH and H_2O_2 are photochemically produced oxidants but, as discussed above, oxidation via H_2O_2 cannot explain the non-MIF isotopic composition of volcanic sulfate produced in the troposphere. Therefore, oxidation via OH seems to play the main role in relatively dry condition plumes.

Finally, collecting tropospheric secondary sulfates from volcanic ashes could be biased by the fact that primary sulfates adsorbed on ash particles are still present, which dilutes the secondary sulfate isotopic signature. Unfortunately, based only on petrographic observations and on O-isotopic signatures of these sulfates, distinguishing a primary vs. secondary origin (Figure 2) is a strenuous task. However, as measured at volcanic vents, the sulfate/SO_2 ratio is usually <1% [25,58,59], indicating that further away from the vent, after oxidation of some SO_2 in the volcanic plume, secondary sulfate should rapidly (in a few hours) dominate the primary ones. Therefore, in ash deposit collected at more than ~30–50 km from the volcanic vent, secondary sulfate should dominate. However, secondary sulfate could also be generated in volcanic plumes by other processes such as S-bearing gases oxidation by OH radicals produced on ash particles (e.g., [60,102–105]). Additionally, another way of producing secondary sulfate in the volcanic plume can be the SO_2 reaction with halogen species (HOBr or HOCl) dissolved in the aqueous phase [106,107]. Unfortunately, yet very little is known on the conditions prevailing inside the dense and hot volcanic plumes. For instance, in such an environment, the halogen chemical behaviors are poorly constrained so it is difficult to estimate the role played in volcanic plumes by these oxidation channels.

In a tropospheric volcanic plume, the estimation of the conversion of SO_2 into sulfate aerosols is a key parameter in order to have accurate volcanic S-spices fluxes, which is crucial in volcano monitoring (see [5] for a review). The SO_2 oxidation rate depends on the oxidation pathways, which as discussed above, depend on the SO_2 concentration, the relative humidity, the time of the day and the season of the year. In tropospheric volcanic plumes, the estimations of SO_2 oxidation rates usually range from ~10^{-7} s^{-1} to ~10^{-4} s^{-1} ([101] and references therein). It is noteworthy that these estimations can be biased by the content of primary sulfate, loss of SO_2 by non-oxidative processes, and the presence of aerosols other than sulfates. These parameters are not easy to quantify and are often not taken into account, leading to large uncertainties on the SO_2 oxidation rate. Kroll et al. [101] measured in real-time the SO_2 and sulfate concentrations in the Kilauea volcanic plume and inferred SO_2 oxidation rates up to 2.5×10^{-6} s^{-1} at noon and an average over 24 h of 5.3×10^{-7} s^{-1}, which is in the lower range of all the previous estimations. Values of about 10^{-7} s^{-1} are also inferred from atmospheric chemistry models that do not include halogen chemistry, which is comparable to the Kilauea case, where halogen emissions are rather low (Galeazzo, personal communication).

3.4. Stratospheric Secondary Sulfates

During major explosive volcanic eruptions, the volcanic plume can reach the stratosphere. There, S-bearing gases can be oxidized and form sulfate aerosols that, due to strong stratospheric winds, are able to travel long distances, to be potentially dispersed globally and finally to slowly sediment on the Earth's surface. It is striking that, due to atmospheric circulation, a volcanic plume emitted in the tropics is easily spread out all over the globe, while at a higher latitude it remains in the same hemisphere. Furthermore, and for the same reasons, the concentration of volcanic sulfate aerosols is unlikely to be homogeneous on the deposition area.

This has been very well demonstrated for the 1991 Pinatubo eruption (e.g., [108]). These stratospheric volcanic sulfate aerosols can be ideally sampled in polar ice cores. Indeed, for the last hundred thousand years, ice accumulation at the poles has progressively recorded and archived atmosphere compositions. Thanks to its chemical stability (very low volatility and reactivity with other compounds), and taking into account some snow redistribution by the wind, the sulfate concentration in the ice has been used for decades to identify volcanic events (references [109–113] among others).

In Antarctica, O- and S-isotope compositions of these sulfates have been studied [78–81]. It is noteworthy that background correction is necessary as concentration up to 100 ppb of mostly biogenic sulfate can be present. There, $\delta^{18}O$ ranges from $-5‰$ to $+25‰$, and $\delta^{34}S$: from $-5‰$ to $+20‰$ (Figure 2). Furthermore, S- and O-MIF signatures ($-0.8‰ < \Delta^{33}S < 2.2‰$; $-2.7‰ < \Delta^{36}S < 2.8‰$ and $\Delta^{17}O$ up 6.5‰; Figure 2) were recorded in the sulfates from more than ten stratospheric eruptions (e.g., Pinatubo, Agung, Krakatoa, Kuwae and different unknown eruptions). Note that the average $\Delta^{33}S/\Delta^{36}S$ ratio from this dataset is about -1.7, which is clearly different from the mass dependent fractionation line ratio (-5 to -10; see [94] for a review; Figure 3).

The most accepted process for generating such S-MIF signatures consists of the photolysis and photoexcitation of SO_2 by UV radiations, in an oxygen-poor atmosphere ([12,21,22,114–116] and [94] for a review). On the early Earth, before the great oxidation event (2.4–2.2 Ga), there was no atmospheric oxygen and consequently, no ozone layer. So, due to the lack of this shield, UV radiation reached the surface of the Earth, such that sulfur photolysis and photoexcitation could take place through the whole atmospheric column, resulting in possibly large S-MIF. Since then, the ozone layer has partially protected the Earth's surface from UV radiation. Consequently, the only place where such processes can take place (and generate large S-MIF signatures) is in the stratosphere above or in the upper part of the ozone layer (≥ 25 km). Volcanic sulfate aerosols generated in the stratosphere are thus expected to have S-MIF signatures [114,117]. Therefore, volcanic sulfate aerosols having large S-MIF signatures clearly testify to their stratospheric origin.

In ice-cores, an evolution of $\Delta^{33}S$ from negative to positive and $\Delta^{36}S$ from positive to negative is recorded during a single volcanic event [80,82]. This can be explained by formation in the stratosphere of ^{33}S enriched and ^{36}S depleted sulfate aerosols at first and then, by mass balance, the resulting SO_2 pool generates sulfates that are ^{33}S depleted and ^{36}S enriched [82]. However, in the stratosphere, the S-bearing gases oxidation after an eruption takes a few months, while the ice volcanic sulfate records are on a timescale of years. This requires a physical separation in space and time of the SO_2 and generated sulfates right from the beginning of the SO_2 oxidation in the stratosphere and maintained separated for a few years while traveling in the stratosphere and while depositing in formation-time order in the ice.

The large O-MIF signature ($\Delta^{17}O$ up to 6.5‰) observed in the same stratospheric sulfates [78,79] can only be explained by the oxidation of S-bearing gases by OH radicals that have high $\Delta^{17}O$. Indeed, as OH results from O_3 photo-dissociation, they carry the same isotopic anomaly as O_3 (Figure 1). Due to isotopic exchange with non-MIF water, tropospheric OH is O-mass dependent ($\Delta^{17}O = 0‰$; Figure 1). The "cold trap" effect played by the tropopause prevents significant water transfers between troposphere and stratosphere. This lack of exchanges results in the low water content in the stratosphere (2–6 ppmv which is 3–4 orders of magnitude lower than in the troposphere) [118]. These relatively anhydrous stratospheric conditions favor homogeneous (gas phase) oxidation of SO_2; in that case, the main oxidant of volcanic sulfur is OH (Figure 1). It is noteworthy that during major eruptions, the water-rich volcanic plumes reach the stratosphere where they could provide significant amounts of water (at least locally), such that heterogeneous gas phase oxidation by O_3 could potentially play a significant role as well. Unfortunately, taken alone, O-isotopes do not allow precise quantification of the relative role played by O_3 or OH in the stratospheric oxidation of sulfur. If the ozone plays a role in such diluted stratospheric volcanic plumes, it means that the conditions are rather basic and not acidic as in denser plumes. In the case where acidic conditions are still preserved, then the SO_2 oxidation via O_2-TMI should prevail, which should lead to a decrease of the sulfate $\Delta^{17}O$. This could explain the fact that some sulfates generated during large eruptions and collected in ice-core do not have high $\Delta^{17}O$ signatures [78]. However, these low values seem to correlate with very large eruptions that inject more than 100 Mt of SO_2 into the stratosphere. Savarino et al. [78] proposed that in such conditions, OH radicals could be rapidly exhausted and that oxidation via $O(^3P)$, which quickly reacts with O_2 that has a $\Delta^{17}O$ very close to 0‰, could be the main mechanism taking place in the stratosphere during these exceptional volcanic eruptions.

3.5. Ash Layer from Large Caldera-Forming Eruptions (Super-Eruptions)

Super-eruptions correspond to volcanic eruptions that emit more than 1000 km^3 of fragmental material (mainly ash) and form large calderas. Sulfates extracted from these deposits display a wide range of isotopic compositions: $\delta^{18}O$ and $\delta^{34}S$ generally spread from 0‰ to 15‰, $\Delta^{17}O$ from 0‰ up to 6‰, while $\Delta^{33}S$ and $\Delta^{36}S$ remain very close to 0‰ (Figure 2). The study of super-eruption deposits raises some questions that are still under debate [33,35]. During super-eruptions, such as those from Yellowstone (Lava Creek Tuff, 0.64 Ma; Huckleberry Ridge Tuff, 2.1 Ma) and the Long Valley eruption (Bishop tuff, 0.76 Ma), ashes have been deposited at distances up to 5000 km from their source calderas. As discussed above, at such distances, the fraction of primary sulfates should be negligible compared to secondary sulfates. Even if these old ash layers (hundreds of ka to tens of Ma), have been diluted by non-MIF sedimentary sulfates, O- and S-MIF signatures could still have been preserved. Indeed, Martin and Bindeman [33] have shown that up to 25% of O- and S-MIF volcanic sulfates are still preserved in some Yellowstone and Bishop Tuff volcanic deposits. Taking into account the effects of dilution, the authors were able to recalculate the $\Delta^{33}S$ and $\Delta^{36}S$ of the initial volcanic sulfates, which have an unambiguous S-MIF signature ($\Delta^{33}S$ down to -0.4‰ and $\Delta^{36}S$ up to 1.2‰; Figure 2). Therefore, even in deposits located within a radius of a few thousands of kilometers from the caldera, S-MIF sulfates are observed, which indicates that a significant amount of stratospheric sulfates is still present in super-eruption deposits. It is noteworthy that Bao et al. [119] proposed that the sulfate O-MIF signature from super-eruption deposits could be accounted by tropospheric oxidation via O_3. This is possible in basic environments (pH > 6), however it is inconsistent with the acidic (low pH) character of dense volcanic plumes. Nonetheless, such conclusions raise new questions.

(A) The tropospheric origin of sulfates: If the O-MIF (up to 6‰) sulfates are generated in the troposphere, O_3 must have played an important role (Figure 1), which seems unlikely considering acidic conditions of volcanic plumes. Furthermore, if S-MIF is also generated in the troposphere during super-eruptions, it has to be considered that the ozone layer was depleted, at least locally, such that a significant fraction of UV radiation was able to reach the troposphere, thus making SO_2 photoexcitation and photolysis possible. It is noteworthy that such a process is compatible with the UV-B flux variations possibly linked to the large igneous provinces emplacements ([120,121] and reference therein).

(B) The stratospheric origin of sulfates: If sulfates are generated by O_3 (aqueous phase oxidation) in the stratosphere, does it get at least locally hydrated? Such a scenario was observed after the 1982 El Chichon eruption [122,123], but the injection of SO_2 in the stratosphere could also lead to the drying out of the stratosphere [124] which makes it difficult to give a general answer to this question. Indeed, the formation of sulfates in the stratosphere by OH oxidation would account for both O- and S-MIF signatures. In turn, this scenario implies that the watering of the stratosphere must be negligible, otherwise OH should react with H_2O and dilute its positive $\Delta^{17}O$ signature. It must be noted that in very high SO_2 concentration conditions, OH can react significantly with SO_2 before isotopic exchange with water (Galeazzo, personal communication). The fact that all the $\Delta^{17}O$ measured is lower than 8.8‰, as expected by OH oxidation (Figure 1), can obviously be explained by non-MIF sediment sulfate dilution in the volcanic deposit, but it could also be explained by another oxidation pathway in the stratosphere. For large stratospheric eruptions, it has been suggested that, due to OH depletion, oxidation via O(^{3}P) could generate non-MIF sulfate [78]. In any case, the stratospheric origin is unable to explain why such an amount of sulfate (up to a few hundreds of ppm) can be found in an area of a few hundreds to thousands of kilometers around the caldera while it should have been spread out all around the world (or at least hemispherically for high latitude eruptions) as it is observed for stratospheric eruptions.

(C) The mixed origin of sulfates: This raises questions about the physico-chemical properties of the tropopause during such volcanic events. After the 1982 El Chichon and the 1991 Pinatubo eruptions for instance the temperature of the low stratosphere increased up to 1.5 °C [9]. If a significant temperature increase had occurred at the tropopause, its "cold trap" effect would have been partially reduced, thus facilitating chemical fluxes (O_3 and H_2O for instance) between the troposphere and stratosphere. Even if the impact was global, the apogee of such an effect could have been restricted to a few hundreds to thousands of kilometers in the atmospheric column located above the volcanic systems.

4. Summary and Conclusions

Oxygen and sulfur isotopes provide good constraints on the formation and fate of volcanic sulfate aerosols and ultimately the real nature of the physico-chemical interactions between the atmosphere and volcanic eruptions. Multi O- and S-isotope analyses of sulfates collected by air sampling or volcanic ash leaching allow us to decipher the oxidation pathways of S-bearing gases in the atmosphere. They also discriminate between primary vs. secondary aerosols and give information about the place where they were generated (troposphere vs. stratosphere).

Some aspects need to be improved:

- High temperature chemistry (including sulfate aerosol formation) is assessed by equilibrium thermodynamic models, but at volcanic vents the plume cools down and dilutes very quickly, therefore modeling using a kinetic approach should be more appropriate.
- The isotopic approach alone can hardly differentiate between different possible mass-dependent processes responsible for sulfate formation. In volcanic plumes, sulfates can be generated by oxidation channels such as OH or O_2-TMI oxidation, but the exploration of other possible oxidation processes is required. The role played by halogens and OH radicals generated from ash particles in the SO_2 oxidation needs to be quantified as in volcanic plumes they may play a more preponderant role than expected. This would improve our understanding of sulfate aerosol formation in a relatively particle-dense plume or cloud.
- In the stratosphere, the low $\Delta^{17}O$ sulfates are not totally understood yet. Oxidation channels other than oxidation via OH need to be explored. Furthermore, the fact that in ice-cores the evolution of $\Delta^{33}S$ (from negative to positive) and $\Delta^{36}S$ (from positive to negative) is recorded during a single volcanic event on a year timescale, while the SO_2 oxidation in the stratosphere takes a few months after an eruption, is still unexplained.
- The impact of super-eruptions on the atmosphere and more specifically on the tropopause needs to be explored in more detail in order to better constrain the potential chemical fluxes between the troposphere and the stratosphere during such an event. This would have an impact on the sulfate aerosol formation and on the atmospheric and climatic impact of super-eruptions in general.

In dry environments, sulfates can be preserved in volcanic deposits for millions of years, thus they can be considered as a reliable archive having recorded the impact on the atmosphere of ancient volcanic eruptions. Furthermore, volcanoes are widespread all over the world, and they have erupted more or less regularly over the Earth's history. Consequently, volcanic sulfates can also be used as proper proxies for the oxidant capacity of the atmosphere on a geological timescale. This would represent an additional constraint on the climatic models for the past but also the future periods of times. From this point of view, the study of volcanic sulfates could provide an open window on past and future climatic changes. Additionally, a better understanding of sulfate aerosol formation and fate in the atmosphere would provide a new parameter in our understanding of atmospheric chemistry evolution. A similar approach performed on anthropogenic S-bearing emissions would significantly improve our knowledge on present-day atmospheric chemistry. This appears to be a fundamental parameter for climatic modeling and for a better prediction of forthcoming climatic changes.

Acknowledgments: The author expresses his sincere thanks the "Emergence program" from the city of Paris, the Agence National de la recherche (ANR) via the contracts 14-CE33-0009-02-FOFAMIFS and 16-CE31-0010-PaleOX, the Sorbonne Universités and the KIC-Climate PhD programs for their support. He is also grateful to E. LeGendre, H. Martin, S. Bekki, I. Bindeman, T. Galeazzo, A. Aroskay, E. Gautier and L. Whiteley for their informal reviews, discussions and proof-reading of the manuscript. The author thanks the anonymous reviewers for their comments that certainly improved the manuscript.

Conflicts of Interest: The author declares no conflict of interest.

References

1. Andres, R.J.; Kasgnoc, A.D. A time-averaged inventory of subaerial volcanic sulfur emissions. *J. Geophys. Res. Atmos.* **1998**, *103*, 25251–25261. [CrossRef]
2. Graf, H.-F.; Feichter, J.; Langmann, B. Volcanic sulfur emissions: Estimates of source strength and its contribution to the global sulfate distribution. *J. Geophys. Res. Atmos.* **1997**, *102*, 10727–10738. [CrossRef]
3. Textor, C.; Graf, H.-F.; Timmreck, C.; Robock, A. Emissions from volcanoes. In *Emissions of Atmospheric Trace Compounds*; Advances in Global Change Research; Springer: Dordrecht, The Netherlands, 2004; pp. 269–303, ISBN 978-90-481-6605-3.
4. Delmelle, P.; Stix, J. Volcanic gases. In *Encyclopedia of Volcanoes*; Sigurdsson, H., Houghton, B., McNutt, S.R., Rymer, H., Stix, J., Eds.; Elsevier: New York, NY, USA, 2000; pp. 803–815.
5. Oppenheimer, C.; Scaillet, B.; Martin, R.S. Sulfur Degassing From Volcanoes: Source Conditions, Surveillance, Plume Chemistry and Earth System Impacts. *Rev. Mineral. Geochem.* **2011**, *73*, 363–421. [CrossRef]
6. Haywood, J.; Boucher, O. Estimates of the direct and indirect radiative forcing due to tropospheric aerosols: A review. *Rev. Geophys.* **2000**, *38*, 513–543. [CrossRef]
7. Robock, A. Volcanic eruption and climate. *Rev. Geophys.* **2000**, *38*, 191–219. [CrossRef]
8. Hansen, J.E.; Wang, W.-C.; Lacis, A.A. Mount Agung Eruption Provides Test of a Global Climatic Perturbation. *Science* **1978**, *199*, 1065–1068. [CrossRef] [PubMed]
9. Parker, D.E.; Wilson, H.; Jones, P.D.; Christy, J.R.; Folland, C.K. The impact of mount pinatubo on world-wide temperatures. *Int. J. Climatol.* **1996**, *16*, 487–497. [CrossRef]
10. Parker, D.E.; Brownscombe, J.L. Stratospheric warming following the El Chichón volcanic eruption. *Nature* **1983**, *301*, 406–408. [CrossRef]
11. Rampino, M.R.; Self, S. Historic Eruptions of Tambora (1815), Krakatau (1883), and Agung (1963), their Stratospheric Aerosols, and Climatic Impact. *Quat. Res.* **1982**, *18*, 127–143. [CrossRef]
12. Farquhar, J.; Savarino, J.; Airieau, S.; Thiemens, M.H. Observation of wavelength-sensitive mass-independent sulfur isotope effects during SO_2 photolysis: Implications for the early atmosphere. *Geophys. Res.* **2001**, *106*, 32829–32839. [CrossRef]
13. Farquhar, J.; Bao, H.; Thiemens, M. Atmospheric Influence of Earth's Earliest Sulfur Cycle. *Science* **2000**, *289*, 756–758. [CrossRef] [PubMed]
14. Thiemens, M.H. History and applications of mass-independent isotope effects. *Annu. Rev. Earth Planet. Sci.* **2006**, *34*, 217–262. [CrossRef]
15. Thiemens, M.H.; Heidenreich, J.E. The Mass-Independent Fractionation of Oxygen: A Novel Isotope Effect and Its Possible Cosmochemical Implications. *Science* **1983**, *219*, 1073–1075. [CrossRef] [PubMed]
16. Farquhar, J.; Wing, B.A. Multiple sulfur isotopes and the evolution of the atmosphere. *Earth Planet. Sci. Lett.* **2003**, *213*, 1–13. [CrossRef]
17. Rumble, D.; Miller, M.F.; Franchi, I.A.; Greenwood, R.C. Oxygen three-isotope fractionation lines in terrestrial silicate minerals: An inter-laboratory comparison of hydrothermal quartz and eclogitic garnet. *Geochim. Cosmochim. Acta* **2007**, *71*, 3592–3600. [CrossRef]
18. Seinfeld, J.H.; Pandis, S.N. *Atmospheric Chemistry and Physics: From Air Pollution to Climate Change*, 2nd ed.; John Wiley & Sons: New York, NY, USA, 2006.
19. Savarino, J.; Lee, C.C.W.; Thiemens, M.H. Laboratory oxygen isotopic study of sulfur (IV) oxidation: Origin of the mass-independent oxygen isotopic anomaly in atmospheric sulfates and sulfate mineral deposits on Earth. *J. Geophys. Res. Atmos.* **2000**, *105*, 29079–29088. [CrossRef]
20. Martin, E.; Bekki, S.; Ninin, C.; Bindeman, I. Volcanic sulfate aerosol formation in the troposphere. *J. Geophys. Res. Atmos.* **2014**, *119*, 12:660–12:673. [CrossRef]
21. Ono, S.; Whitehill, A.R.; Lyons, J.R. Contribution of isotopologue self-shielding to sulfur massindependent fractionation during sulfur dioxide photolysis. *Ournal Geophys. Res. Atmos.* **2013**, *118*, 2444–2454. [CrossRef]
22. Whitehill, A.R.; Xie, C.; Hu, X.; Xie, D.; Guo, H.; Ono, S. Vibronic origin of sulfur mass-independent isotope effect in photoexcitation of SO_2 and the implications to the early earth's atmosphere. *Proc. Natl. Acad. Sci. USA* **2013**, *110*, 17697–17702. [CrossRef] [PubMed]
23. Giggenbach, W.F. A simple method for the collection and analysis of volcanic gas samples. *Bull. Volcanol.* **1975**, *39*, 132–145. [CrossRef]

24. Giggenbach, W.F.; Goguel, R.L. *Methods for the Collection and Analysis of Geothermal and Volcanic Water and Gas Samples*; Chemistry Division, Department of Scientific and Industrial Research: Petone, New Zealand, 1989; p. 53.

25. De Moor, J.M.; Fischer, T.P.; Sharp, Z.D.; King, P.L.; Wilke, M.; Botcharnikov, R.E.; Cottrell, E.; Zelenski, M.; Marty, B.; Klimm, K.; et al. Sulfur degassing at Erta Ale (Ethiopia) and Masaya (Nicaragua) volcanoes: Implications for degassing processes and oxygen fugacities of basaltic systems: Sulfur Degassing at Basaltic Volcanoes. *Geochem. Geophys. Geosyst.* **2013**, *14*, 4076–4108. [CrossRef]

26. Menyailov, I.A.; Nikitina, L.P.; Shapar, V.N.; Pilipenko, V.P. Temperature increase and chemical change of fumarolic gases at Momotombo Volcano, Nicaragua, in 1982–1985: Are these indicators of a possible eruption? *J. Geophys. Res. Solid Earth* **1986**, *91*, 12199–12214. [CrossRef]

27. Goff, F.; Janik, C.J.; Delgado, H.; Werner, C.; Counce, D.; Stimac, J.A.; Siebe, C.; Love, S.P.; Williams, S.N.; Fischer, T.; et al. Geochemical surveillance of magmatic volatiles at Popocatépetl volcano, Mexico. *GSA Bull.* **1998**, *110*, 695–710. [CrossRef]

28. Ohba, T.; Nogami, K.; Hirabayashi, J.-I.; Mori, T. Isotopic fractionation of SO_2 and H_2S gases during the absorption by KOH solution, with the application to volcanic gas monitoring at Miyakejima Island, Japan. *Geochem. J.* **2008**, *42*, 119–131. [CrossRef]

29. Mather, T.A.; Pyle, D.M.; Heaton, T.H.E. Investigation of the use of filter packs to measure the sulphur isotopic composition of volcanic sulphur dioxide and the sulphur and oxygen isotopic composition of volcanic sulphate aerosol. *Atmos. Environ.* **2008**, *42*, 4611–4618. [CrossRef]

30. Le Gendre, E. *Étude des Anomalies Isotopiques de L'oxygène Etdusoufre Dans les Sulfates D'origine Volcanique et Anthropique*; UPMC: Pittsburgh, PA, USA, 2016.

31. Mather, T.A.; McCabe, J.R.; Rai, V.K.; Thiemens, M.H.; Pyle, D.M.; Heaton, T.H.E.; Sloane, H.J.; Fern, G.R. Oxygen and sulfur isotopic composition of volcanic sulfate aerosol at the point of emission. *J. Geophys. Res. Atmos.* **2006**, *111*, D18205. [CrossRef]

32. Armienta, M.A.; De la Cruz-Reyna, S.; Soler, A.; Cruz, O.; Ceniceros, N.; Aguayo, A. Chemistry of ash-leachates to monitor volcanic activity: An application to Popocatépetl volcano, central Mexico. *Appl. Geochem.* **2010**, *25*, 1198–1205. [CrossRef]

33. Martin, E.; Bindeman, I. Mass-independent isotopic signatures of volcanic sulfate from three supereruption ash deposits in Lake Tecopa, California. *Earth Planet. Sci. Lett.* **2009**, *282*, 102–114. [CrossRef]

34. Risacher, F.; Alonso, H. Geochemistry of ash leachates from the 1993 Lascar eruption, northern Chile. Implication for recycling of ancient evaporites. *J. Volcanol. Geotherm. Res.* **2001**, *109*, 319–337. [CrossRef]

35. Bao, H.; Thiemens, M.H.; Loope, D.B.; Yuan, X.-L. Sulfate oxygen-17 anomaly in an Oligocene ash bed in mid-North America: Was it the dry fogs? *Geophys. Res. Lett.* **2003**, *30*, 11–14. [CrossRef]

36. Van Stempvoort, D.R.; Krouse, H.R. Controls of d^{18}O in sulfate: Review of experimental data and application to specific environments. In *Environmental Geochemistry of Sulfide Oxidation*; Alpers, C.N., Blowes, D.W., Eds.; American Chemical Society: Washington, DC, USA, 1994; Volume 550.

37. Witham, C.S.; Oppenheimer, C.; Horwell, C.J. Volcanic ash-leachates: A review and recommendations for sampling methods. *J. Volcanol. Geotherm. Res.* **2005**, *141*, 299–326. [CrossRef]

38. Botter, C. *Utomated Single-Particle SEM/EDS Analysis of Volcanic Aerosols from Stromboli, Aeolian Arc, Italy*; Mineralogical Characterization of the PM10 Fraction; University of Fribourg: Fribourg, Switzerland, 2011.

39. Botter, C.; Meier, M.; Wiedenmann, D.; Grobety, B.; Ricci, T. Single Particle Analysis of Volcanic Aerosols from Stromboli, Italy. In Proceedings of the Cities on Volcanoes 6 (COV6), Tenerife, Spain, 31 May–4 June 2010.

40. Toutain, J.; Quisefit, J.; Briole, P.; Aloupogiannis, P.; Blanc, P.; Robaye, G. Mineralogy and chemistry of solid aerosols emitted from Mount Etna. *Geochem. J.* **1995**, *29*, 163–173. [CrossRef]

41. Stoiber, R.E.; Rose, W.I. The Geochemistry of Central American Volcanic Gas Condensates. *GSA Bull.* **1970**, *81*, 2891–2912. [CrossRef]

42. Symonds, R.B.; Reed, M.H.; Rose, W.I. Origin, speciation, and fluxes of trace-element gases at Augustine volcano, Alaska: Insights into magma degassing and fumarolic processes. *Geochim. Cosmochim. Acta* **1992**, *56*, 633–657. [CrossRef]

43. Quisefit, J.P.; Bergametti, G.; Tedesco, D.; Pinart, J.; Colin, J.L. Origin of particulate potassium in Mt Etna emissions before and during the 1983 eruption. *J. Volcanol. Geotherm. Res.* **1988**, *35*, 111–119. [CrossRef]

44. Ayris, P.M.; Lee, A.F.; Wilson, K.; Kueppers, U.; Dingwell, D.B.; Delmelle, P. SO$_2$ sequestration in large volcanic eruptions: High-temperature scavenging by tephra. *Geochim. Cosmochim. Acta* **2013**, *110*, 58–69. [CrossRef]

45. Van Stempvoort, D.R.; Reardon, E.J.; Fritz, P. Fractionation of sulfur and oxygen isotopes in sulfate by soil sorption. *Geochim. Cosmochim. Acta* **1990**, *54*, 2817–2826. [CrossRef]

46. Michalski, G.; Kasem, M.; Rech, J.A.; Adieu, S.; Showers, W.S.; Genna, B.; Thiemens, M. Uncertainties in the oxygen isotopic composition of barium sulfate induced by coprecipitation of nitrate. *Rapid Commun. Mass Spectrom.* **2008**, *22*, 2971–2976. [CrossRef] [PubMed]

47. Bao, H. Purifying barite for oxygen isotope measurement by dissolution and reprecipitation in a chelating solution. *Anal. Chem.* **2006**, *78*, 304–309. [CrossRef] [PubMed]

48. Le Gendre, E.; Martin, E.; Villemant, B.; Cartigny, P.; Assayag, N. A simple and reliable anion-exchange resin method for sulfate extraction and purification suitable for multiple O- and S-isotope measurements: Anion-exchange method for multiple O- and S-isotope analysis. *Rapid Commun. Mass Spectrom.* **2017**, *31*, 137–144. [CrossRef] [PubMed]

49. Schauer, A.J.; Kunasek, S.A.; Sofen, E.D.; Erbland, J.; Savarino, J.; Johnson, B.W.; Amos, H.M.; Shaheen, R.; Abaunza, M.; Jackson, T.L.; et al. Oxygen isotope exchange with quartz during pyrolysis of silver sulfate and silver nitrate: Oxygen isotope exchange during pyrolysis of Ag$_2$SO$_4$ and AgNO$_3$. *Rapid Commun. Mass Spectrom.* **2012**, *26*, 2151–2157. [CrossRef] [PubMed]

50. Pepkowitz, L.; Shirley, E. microdetection of sulfur. *Anal. Chem.* **1951**, *23*, 1709–1710. [CrossRef]

51. Farquhar, J.; Peters, M.; Johnston, D.T.; Strauss, H.; Masterson, A.; Wiechert, U.; Kaufman, A.J. Isotopic evidence for Mesoarchaean anoxia and changing atmospheric sulphur chemistry. *Nature* **2007**, *449*, 706–709. [CrossRef] [PubMed]

52. Rafter, T.A. Oxygen isotopic composition of sulfates. Part I. A method for the extraction of oxygen and its quantitative conversion to carbon dioxide for isotope radiation measurements. *N. Z. J. Sci.* **1967**, *10*, 493–510.

53. Mizutani, Y. An improvement in the carbon-reduction method for the oxygen isotopic analysis of sulphates. *Geochem. J.* **1971**, *5*, 69–77. [CrossRef]

54. Bao, H.; Thiemens, M.H. Generation of O$_2$ from BaSO$_4$ using a CO$_2$-laser fluorination system for simultaneous analysis of d^{18}O and d^{17}O. *Anal. Chem.* **2000**, *72*, 4029–4032. [CrossRef] [PubMed]

55. Geng, L.; Schauer, A.J.; Kunasek, S.A.; Sofen, E.D.; Erbland, J.; Savarino, J.; Allman, D.J.; Sletten, R.S.; Alexander, B. Analysis of oxygen-17 excess of nitrate and sulfate at sub-micromole levels using the pyrolysis method: Analysis of oxygen-17 excess of nitrate and sulfate. *Rapid Commun. Mass Spectrom.* **2013**, *27*, 2411–2419. [CrossRef] [PubMed]

56. Savarino, J.; Alexander, B.; Darmohusodo, V.; Thiemens, M.H. Sulfur and Oxygen Isotope Analysis of Sulfate at Micromole Levels Using a Pyrolysis Technique in a Continuous Flow System. *Anal. Chem.* **2001**, *73*, 4457–4462. [CrossRef] [PubMed]

57. Au Yang, D.; Landais, G.; Assayag, N.; Widory, D.; Cartigny, P. Improved analysis of micro- and nanomole-scale sulfur multi-isotope compositions by gas source isotope ratio mass spectrometry: Improved analysis of S isotopes at micro/nanomole levels by IRMS. *Rapid Commun. Mass Spectrom.* **2016**, *30*, 897–907. [CrossRef] [PubMed]

58. Martin, R.S.; Sawyer, G.M.; Spampinato, L.; Salerno, G.G.; Ramirez, C.; Ilyinskaya, E.; Witt, M.L.I.; Mather, T.A.; Watson, I.M.; Phillips, J.C.; et al. A total volatile inventory for Masaya Volcano, Nicaragua. *J. Geophys. Res.* **2010**, *115*. [CrossRef]

59. Allen, A.G.; Oppenheimer, C.; Ferm, M.; Baxter, P.J.; Horrocks, L.A.; Galle, B.; McGonigle, A.J.S.; Duffell, H.J. Primary sulfate aerosol and associated emissions from Masaya Volcano, Nicaragua. *J. Geophys. Res. Atmos.* **2002**, *107*, 4682. [CrossRef]

60. Cwiertny, D.M.; Young, M.A.; Grassian, V.H. Chemistry and Photochemistry of Mineral Dust Aerosol. *Annu. Rev. Phys. Chem.* **2008**, *59*, 27–51. [CrossRef] [PubMed]

61. Sakai, H.; Casadevall, T.J.; Moore, J.G. Chemistry and isotope ratios of sulfur in basalts and volcanic gases at Kilauea volcano, Hawaii. *Geochim. Cosmochim. Acta* **1982**, *46*, 729–738. [CrossRef]

62. Allard, P. 13C/12C and 34S/32S ratios in magmatic gases from ridge volcanism in Afar. *Nature* **1979**, *282*, 56–58. [CrossRef]

63.   Sakai, H.; Gunnlaugsson, E.; Tòmasson, J.; Rouse, J.E. Sulfur isotope systematics in icelandic geothermal systems and influence of seawater circulation at Reykjanes. *Geochim. Cosmochim. Acta* **1980**, *44*, 1223–1231. [CrossRef]

64.   Goff, F.; McMurtry, G.M. Tritium and stable isotopes of magmatic waters. *J. Volcanol. Geotherm. Res.* **2000**, *97*, 347–396. [CrossRef]

65.   Taran, Y.; Gavilanes, J.C.; Cortés, A. Chemical and isotopic composition of fumarolic gases and the SO_2 flux from Volcán de Colima, México, between the 1994 and 1998 eruptions. *J. Volcanol. Geotherm. Res.* **2002**, *117*, 105–119. [CrossRef]

66.   Sakai, H.; Matsubaya, O. Stable isotopic studies of japanese geothermal systems. *Geothermics* **1977**, *5*, 97–124. [CrossRef]

67.   Giggenbach, W. The chemical and isotopic composition of gas discharges from New Zealand andesitic volcanoes. *Bull. Volcanol.* **1982**, *45*, 253–255. [CrossRef]

68.   Williams, S.N.; Sturchio, N.C.; Calvache, V.M.L.; Mendez, F.R.; Londoño, C.A.; García, P.N. Sulfur dioxide from Nevado del Ruiz volcano, Colombia: Total flux and isotopic constraints on its origin. *J. Volcanol. Geotherm. Res.* **1990**, *42*, 53–68. [CrossRef]

69.   Richet, P.; Bottinga, Y.; Javoy, M. A Review of Hydrogen, Carbon, Nitrogen, Oxygen, Sulphur, and Chlorine Stable Isotope Fractionation Among Gaseous Molecules. *Annu. Rev. Earth Planet. Sci.* **1977**, *5*, 65–110. [CrossRef]

70.   Ohmoto, H.; Rye, R.O. Isotope of sulfur and carbon. In *Geochemestry of Hydrothermal Deposits*; Barnes, H.L., Ed.; John Wiley & Sons: Hoboken, NJ, USA, 1979; pp. 509–567.

71.   Labidi, J.; Cartigny, P.; Birck, J.L.; Assayag, N.; Bourrand, J.J. Determination of multiple sulfur isotopes in glasses: A reappraisal of the MORB $\delta^{34}S$. *Chem. Geol.* **2012**, *334*, 189–198. [CrossRef]

72.   De Hoog, J.C.M.; Taylor, B.E.; van Bergen, M.J. Sulfur isotope systematics of basaltic lavas from Indonesia: Implications for the sulfur cycle in subduction zones. *Earth Planet. Sci. Lett.* **2001**, *189*, 237–252. [CrossRef]

73.   Martin, E.; Bindeman, I.; Grove, T. The origin of high-Mg magmas in Mt Shasta and Medicine Lake volcanoes, Cascade Arc (California): Higher and lower than mantle oxygen isotope signatures attributed to current and past subduction. *Contrib. Miner. Petrol.* **2011**, *162*, 945–960. [CrossRef]

74.   Eiler, J. Oxygen isotope variations of basaltic lavas and upper mantle rocks. In *Reviews in Mineralogy and Geochemistry*; Walley, J.W., Cole, D.R., Eds.; The Mineralogical Society of America: Washington, DC, USA, 2001; Volume 43, pp. 319–364.

75.   Bindeman, I.N.; Eiler, J.M.; Wing, B.A.; Farquhar, J. Rare sulfur and triple oxygen isotope geochemistry of volcanogenic sulfate aerosols. *Geochim. Cosmochim. Acta* **2007**, *71*, 2326–2343. [CrossRef]

76.   De Moor, J.M.; Fischer, T.P.; Hilton, D.R.; Hauri, E.; Jaffe, L.A.; Camacho, J.T. Degassing at Anatahan volcano during the May 2003 eruption: Implications from petrology, ash leachates, and SO_2 emissions. *J. Volcanol. Geotherm. Res.* **2005**, *146*, 117–138. [CrossRef]

77.   Rye, R.O.; Luhr, J.F.; Wasserman, M.D. Sulfur and oxygen isotopic systematics of the 1982 eruptions of El Chichón Volcano, Chiapas, Mexico. *J. Volcanol. Geotherm. Res.* **1984**, *23*, 109–123. [CrossRef]

78.   Savarino, J.; Bekki, S.; Cole-Dai, J.H.; Thiemens, M.H. Evidence from sulfate mass independent oxygen isotopic compositions of dramatic changes in atmospheric oxidation following massive volcanic eruptions. *J. Geophys. Res. Atmos.* **2003**, *108*. [CrossRef]

79.   Baroni, M. *Etude des Anomalies Isotopiques du Soufre et de L'oxygène Dans le Sulfate D'origine Volcanique Enregistré dans les Archives Glaciaires*; Univeristé Joseph Fourier: Grenoble, France, 2006.

80.   Baroni, M.; Thiemens, M.H.; Delmas, R.J.; Savarino, J. Mass-Independent Sulfur Isotopic Compositions in Stratospheric Volcanic Eruptions. *Science* **2007**, *315*, 84–87. [CrossRef] [PubMed]

81.   Baroni, M.; Savarino, J.; Cole-Dai, J.; Rai, V.K.; Thiemens, M.H. Anomalous sulfur isotope compositions of volcanic sulfate over the last millennium in Antarctic ice cores. *J. Geophys. Res. Atmos.* **2008**, *113*, D20112. [CrossRef]

82.   Gautier, E. *Empreinte Isotopique et Histoire du Volcanisme Stratosphérique des 2600 Dernières Années, Enregistrées à Dome C, Antarctique*; Université Grenoble Alpes: Grenoble, France, 2015.

83.   Cole-Dai, J.; Ferris, D.; Lanciki, A.; Savarino, J.; Baroni, M.; Thiemens, M.H. Cold decade (AD 1810–1819) caused by Tambora (1815) and another (1809) stratospheric volcanic eruption. *Geophys. Res. Lett.* **2009**, *36*. [CrossRef]

84. Miyoshi, T.; Sakai, H.; Chiba, H. Experimental study of sulfur isotope fractionation factors between sulfate and sulfide in high temperature melts. *Geochem. J.* **1984**, *18*, 75–84. [CrossRef]

85. Hoefs, J. *Stable Isotope Geochemistry*; Springer: Berlin/Heidelberg, Germany, 2015.

86. Farquhar, J.; Johnston, D.T.; Wing, B.A. Implications of conservation of mass effects on mass-dependent isotope fractionations: Influence of network structure on sulfur isotope phase space of dissimilatory sulfate reduction. *Geochim. Cosmochim. Acta* **2007**, *71*, 5862–5875. [CrossRef]

87. Ono, S.; Wing, B.; Johnston, D.; Farquhar, J.; Rumble, D. Mass-dependent fractionation of quadruple stable sulfur isotope system as a new tracer of sulfur biogeochemical cycles. *Geochim. Cosmochim. Acta* **2006**, *70*, 2238–2252. [CrossRef]

88. Martin, R.S.; Mather, T.A.; Pyle, D.M. High-temperature mixtures of magmatic and atmospheric gases. *Geochem. Geophys. Geosyst.* **2006**, *7*, Q04006. [CrossRef]

89. Roberts, T.J.; Braban, C.F.; Martin, R.S.; Oppenheimer, C.; Adams, J.W.; Cox, R.A.; Jones, R.L.; Griffiths, P.T. Modelling reactive halogen formation and ozone depletion in volcanic plumes. *Chem. Geol.* **2009**, *263*, 151–163. [CrossRef]

90. Roberts, T.J.; Martin, R.S.; Jourdain, L. Reactive bromine chemistry in Mount Etna's volcanic plume: The influence of total Br, high-temperature processing, aerosol loading and plume–air mixing. *Atmos. Chem. Phys.* **2014**, *14*, 11201–11219. [CrossRef]

91. Gerlach, T.M. Volcanic sources of tropospheric ozone-depleting trace gases. *Geochem. Geophys. Geosyst.* **2004**, *5*. [CrossRef]

92. Lee, C.C.-W.; Savarino, J.H.; Cachier, H.; Thiemens, M.H. Sulfur (32S, 33S, 34S, 36S) and oxygen (16O,17O,18O) isotopic ratios of primary sulfate produced from combustion processes. *Tellus B* **2002**, *54*, 193–200. [CrossRef]

93. Belo, L.P.; Elliott, L.K.; Stanger, R.J.; Spörl, R.; Shah, K.V.; Maier, J.; Wall, T.F. High-Temperature Conversion of SO_2 to SO_3: Homogeneous Experiments and Catalytic Effect of Fly Ash from Air and Oxy-fuel Firing. *Energy Fuels* **2014**, *28*, 7243–7251. [CrossRef]

94. Ono, S. Photochemistry of Sulfur Dioxide and the Origin of Mass-Independent Isotope Fractionation in Earth's Atmosphere. *Annu. Rev. Earth Planet. Sci.* **2017**, *45*, 301–329. [CrossRef]

95. Jenkins, K.A.; Bao, H. Multiple oxygen and sulfur isotope compositions of atmospheric sulfate in Baton Rouge, LA, USA. *Atmos. Environ.* **2006**, *40*, 4528–4537. [CrossRef]

96. Lee, C.C.-W.; Thiemens, M.H. The $\delta^{17}O$ and $\delta^{18}O$ measurements of atmospheric sulfate from a coastal and high alpine region: A mass-independent isotopic anomaly. *J. Geophys. Res. Atmos.* **2001**, *106*, 17359–17373. [CrossRef]

97. Li, X.; Bao, H.; Gan, Y.; Zhou, A.; Liu, Y. Multiple oxygen and sulfur isotope compositions of secondary atmospheric sulfate in a mega-city in central China. *Atmos. Environ.* **2013**, *81*, 591–599. [CrossRef]

98. Carn, S.A.; Clarisse, L.; Prata, A.J. Multi-decadal satellite measurements of global volcanic degassing. *J. Volcanol. Geotherm. Res.* **2016**, *311*, 99–134. [CrossRef]

99. Yang, K.; Krotkov, N.A.; Krueger, A.J.; Carn, S.A.; Bhartia, P.K.; Levelt, P.F. Improving retrieval of volcanic sulfur dioxide from backscattered UV satellite observations. *Geophys. Res. Lett.* **2009**, *36*, L03102. [CrossRef]

100. Galeazzo, T.; Bekki, S.; Martin, E. *Modeling the Pathways of Generation of O-MIF in Tropospheric Sulfates*; GOLDSCHMIDT: Paris, France, 2017.

101. Kroll, J.H.; Cross, E.S.; Hunter, J.F.; Pai, S.; Wallace, L.M.M.; Croteau, P.L.; Jayne, J.T.; Worsnop, D.R.; Heald, C.L.; Murphy, J.G.; et al. Atmospheric Evolution of Sulfur Emissions from Kīlauea: Real-Time Measurements of Oxidation, Dilution, and Neutralization within a Volcanic Plume. *Environ. Sci. Technol.* **2015**, *49*, 4129–4137. [CrossRef] [PubMed]

102. Dupart, Y.; King, S.M.; Nekat, B.; Nowak, A.; Wiedensohler, A.; Herrmann, H.; David, G.; Thomas, B.; Miffre, A.; Rairoux, P.; et al. Mineral dust photochemistry induces nucleation events in the presence of SO_2. *Proc. Natl. Acad. Sci. USA* **2012**, *109*, 20842–20847. [CrossRef] [PubMed]

103. Vallyathan, V.; Shi, X.; Dalal, N.S.; Irr, W.; Castranova, V. Generation of Free Radicals from Freshly Fractured Silica Dust: Potential Role in Acute Silica-induced Lung Injury. *Am. Rev. Respir. Dis.* **1988**, *138*, 1213–1219. [CrossRef] [PubMed]

104. Konecny, R. Reactivity of Hydroxyl Radicals on Hydroxylated Quartz Surface. 1. Cluster Model Calculations. *J. Phys. Chem. B* **2001**, *105*, 6221–6226. [CrossRef]

105. Narayanasamy, J.; Kubicki, J.D. Mechanism of Hydroxyl Radical Generation from a Silica Surface: Molecular Orbital Calculations. *J. Phys. Chem. B* **2005**, *109*, 21796–21807. [CrossRef] [PubMed]

106. Von Glasow, R.; Sander, R.; Bott, A.; Crutzen, P.J. Modeling halogen chemistry in the marine boundary layer 2. Interactions with sulfur and the cloud-covered MBL. *J. Geophys. Res. Atmos.* **2002**, *107*, 4323. [CrossRef]

107. Roberts, T.J.; Jourdain, L.; Griffiths, P.T.; Pirre, M. Re-evaluating the reactive uptake of HOBr in the troposphere with implications for the marine boundary layer and volcanic plumes. *Atmos. Chem. Phys.* **2014**, *14*, 11185–11199. [CrossRef]

108. McCormick, M.P.; Thomason, L.W.; Trepte, C.R. Atmospheric effects of the Mt Pinatubo eruption. *Nature* **1995**, *373*, 399–404. [CrossRef]

109. Castellano, E.; Becagli, S.; Jouzel, J.; Migliori, A.; Severi, M.; Steffensen, J.P.; Traversi, R.; Udisti, R. Volcanic eruption frequency over the last 45 ky as recorded in Epica-Dome C ice core (East Antarctica) and its relationship with climatic changes. *Glob. Planet. Chang.* **2004**, *42*, 195–205. [CrossRef]

110. Cole-Dai, J.; Mosley-Thompson, E.; Wight, S.P.; Thompson, L.G. A 4100-year record of explosive volcanism from an East Antarctica ice core. *J. Geophys. Res. Atmos.* **2000**, *105*, 24431–24441. [CrossRef]

111. Delmas, R.J.; Kirchner, S.; Palais, J.M.; Petit, J.-R. 1000 years of explosive volcanism recorded at the South Pole. *Tellus B* **1992**, *44*, 335–350. [CrossRef]

112. Langway, C.C.; Osada, K.; Clausen, H.B.; Hammer, C.U.; Shoji, H. A 10-century comparison of prominent bipolar volcanic events in ice cores. *J. Geophys. Res. Atmos.* **1995**, *100*, 16241–16247. [CrossRef]

113. Legrand, M.; Delmas, R.J. A 220-year continuous record of volcanic H_2SO_4 in the Antarctic ice sheet. *Nature* **1987**, *327*, 671–676. [CrossRef]

114. Hattori, S.; Schmidt, J.A.; Johnson, M.S.; Danielache, S.O.; Yamada, A.; Ueno, Y.; Yoshida, N. SO_2 photoexcitation mechanism links mass-independent sulfur isotopic fractionation in cryospheric sulfate to climate impacting volcanism. *Proc. Natl. Acad. Sci. USA* **2013**, *110*, 17656–17661. [CrossRef] [PubMed]

115. Danielache, S.O.; Hattori, S.; Johnson, M.S.; Ueno, Y.; Nanbu, S.; Yoshida, N. Photoabsorption cross-section measurements of32S, 33S, 34S, and 36S sulfur dioxide for the B1B1-X1A1 absorption band. *J. Geophys. Res. Atmos.* **2012**, *117*, D24301. [CrossRef]

116. Danielache, S.O.; Eskebjerg, C.; Johnson, M.S.; Ueno, Y.; Yoshida, N. High-precision spectroscopy of 32S, 33S, and 34S sulfur dioxide: Ultraviolet absorption cross sections and isotope effects. *J. Geophys. Res. Atmos.* **2008**, *113*, D17314. [CrossRef]

117. Whitehill, A.R.; Jiang, B.; Guo, H.; Ono, S. SO_2 photolysis as a source for sulfur mass-independent isotope signatures in stratosphehric aerosols. *Atmos. Chem. Phys.* **2015**, *15*, 1843–1864. [CrossRef]

118. Brasseur, G.P.; Solomon, S. *Aeronomy of the Middle Atmosphere, Chemistry and Physics of the Stratosphere and Mesosphere*; Springer: Dordrecht, The Netherlands, 2005.

119. Bao, H.; Yu, S.; Tong, D.Q. Massive volcanic SO_2 oxidation and sulphate aerosol deposition in Cenozoic North America. *Nature* **2010**, *465*, 909–912. [CrossRef] [PubMed]

120. Fraser, W.; Lomax, B.; Beerling, D.; James, D.; Pyle, J.; Self, S.; Sephton, M.; Wellman, C. Episodic perturbations of end-Permian atmosphere recorded in plant spore chemistry. In Proceedings of the EGU General Assembly 2016, Vienna, Austria, 17–22 April 2016; Volume 18, p. EPSC2016-17251.

121. Lomax, B.H.; Fraser, W.T. Palaeoproxies: Botanical monitors and recorders of atmospheric change. *Palaeontology* **2015**, *58*, 759–768. [CrossRef]

122. Shepherd, T.G. Issues in stratosphere-troposphere coupling. *J. Meteorol. Soc. Jpn.* **2002**, *80*, 769–792. [CrossRef]

123. Robock, A. Climatic impact of volcanic emissions. In *State of the Planet*; Sparks, R.S.J., Hawkesworth, C.J., Eds.; American Geophysical Union: Washington, DC, USA, 2004; Volume 19.

124. Bekki, S. Oxidation of Volcanic SO_2: A Sink for Stratospheric OH and H_2O. *Geophys. Res. Lett.* **1995**, *22*, 913–916. [CrossRef]

geoscienes

MDPI

Article

Aerosol Optical Properties of Pacaya Volcano Plume Measured with a Portable Sun-Photometer

Pasquale Sellitto [1,*], Letizia Spampinato [2], Giuseppe G. Salerno [2] and Alessandro La Spina [2]

[1]   Laboratoire de Météorologie Dynamique, Institut Pierre Simon Laplace, École Normale Supérieure, PSL Research University, École Polytechnique, Université Paris-Saclay, Sorbonne Université, UPMC Université Paris 6, CNRS, 75005 Paris, France

[2]   Istituto Nazionale di Geofisica e Vulcanologia, Osservatorio Etneo, I95123 Catania, Italy; letizia.spampinato@ingv.it (L.S.); giuseppe.salerno@ingv.it (G.G.S.); alessandro.laspina@ingv.it (A.L.S.)

*   Correspondence: psellitto@lmd.ens.fr; Tel.: +33-1-4432-2731

Received: 14 October 2017; Accepted: 22 December 2017; Published: 23 January 2018

Abstract: In this paper, Sun-photometer multichannel measurements of aerosol optical depths (AODs) in the visible and near-infrared spectral ranges, and Ångström parameters of the plume issued from the Pacaya volcano, Guatemala, are presented for the first time. These observations, made during a short-term campaign carried out on 29 and 30 January 2011, indicate a diluted (AODs lower than 0.1) volcanic plume composed of small particles (Ångström exponent ∼1.0 on 29 January and ∼1.4 on 30 January). Results are consistent with an ash-free plume. Finally, the impact of the choice of different wavelength pairs for the calculation of the Ångström parameters from the spectral AOD observations is tested and critically discussed.

Keywords: volcanic aerosols; portable photometry; aerosol optical properties

1. Introduction

Volcanic emissions have important impacts on the atmospheric composition (e.g., [1,2]), cloud distribution (e.g., [3]) and regional (e.g., [4]) to global climate (e.g., [5]). The direct climate forcing of volcanic plumes critically depends on the optical and micro-physical properties of the volcanic aerosols, that in turn depend on the evolution processes of the effluents in the atmosphere [6]. Despite sulphur dioxide (SO_2) being only the third most abundant gas species in the volcanic gas mixture (after water vapour and carbon dioxide), it may strongly perturb the atmospheric composition due to its conversion to sub-micron-sized secondary sulphate aerosols (SSA) (e.g., [7]). Over their atmospheric lifetime, SSA particles can undergo condensation, growth and chemical and micro-physical processes when interacting with volcanic ash. All these processes, that are generally scarcely characterised, contribute to determining the radiative properties of volcanic aerosols and hence the direct radiative forcing of volcanic emissions. The micro-physical properties of volcanic aerosols may also play a role in a number of other atmospheric processes, including their interaction with cloud fields (aerosol indirect climatic effect). The net indirect effect of volcanic aerosols is debated [8].

In many cases, micro-physical properties of aerosols are not directly accessible by observations or modelling. Optical proxies of these aerosol properties, for example the aerosol optical depth (AOD) or Ångström exponent (α), are, on the contrary, commonly observed both by satellite and ground-based photometers and spectrophotometers. Therefore, observing the optical properties of volcanic aerosols is crucial to assess their direct and indirect forcing on the atmospheric radiative balance at a number of spatial and temporal scales.

The optical properties of volcanic aerosols can be measured in the near-field, i.e., in proximity and/or in the surrounding area or emitting vents, using portable Sun-photometers such as the Microtops-II. Thanks to their small size and weight, these hand-held instruments are very well suited to field measurement of volcanic plumes near the source, in difficult access areas. In the past, Microtops-II "Sun Photometer" systems (hereafter referred to as MIISP), have been used to characterise the optical properties of plumes, i.e., to observe the spectral AOD in the visible and near infrared (NIR) spectral ranges, and to derive the Ångström coefficients α and β, from volcanoes such as Mount Etna [9,10], Kilauea [11], Masaya [12], Lascar and Villarica [13,14] and Eyjafjallajökull [15]. Recently, ultraviolet (UV) AOD and UV-to-NIR Ångström coefficient observations have been derived at Mount Etna by means of a Microtops-II "Ozone Monitor" system (hereafter referred to as MIIOM) [16]. These optical properties can be used to gather optical information on the burden and typology (AOD, β) and mean size (α) of the volcanic aerosols and can be used as inputs for dispersion and evolution models that could bridge the near-source characterisation of the plume to the downwind impacts at larger scales (e.g., at the regional scale).

In this paper, near-source observations of the optical properties of the plume of Pacaya volcano (Guatemala) are presented for the first time. Remote MIISP measurements were carried out on 29 and 30 January 2011, during a non-eruptive passive degassing phase. The paper is organised as follows: the MIISP and the methods used in this work to retrieve the optical properties of the plume are introduced in Section 2; the volcanology and visual observations during the campaign are described in Section 3; results are shown and discussed in Section 4; finally, conclusions are given in Section 5.

2. Instruments and Methods

2.1. The Microtops-II "Sun Photometer"

The multichannel hand-held MIISP Sun-photometer used in this work measures direct Sun radiance (2.5° field of view) in five channels centred in the visible (Ch.1: 440.0 ± 1.5 nm, Ch.2: 675.0 ± 1.5 nm, Ch.3: 870.0 ± 0.3 nm), in a water vapour NIR absorption band (Ch.4: 936.0 ± 1.5 nm) and in the NIR spectral window (Ch.5: 1020.0 ± 1.5 nm), with nominal full band width at half maximum (FWHM) of 10.0 ± 1.5 nm [17,18]. The five channels are used to derive AOD spectra. The NIR Ch.4 is also used to derive water vapour vertical content. The instrument used in the present study was pre-calibrated applying a Langley method at Mauna Loa Observatory, Hawaii. The Sun-pointing alignment is performed manually, with the aid of a Sun target window which projects the Sun position with respect of the input optics.

2.2. Observations of Volcanic Plume Optical Properties with Portable Photometry

At each MIISP acquisition, the photometer measures direct Sun radiance at the ground, at the five spectral channels. Using the internal calibration constant and correcting for the Rayleigh absorption, the photometer automatically calculates the AODs at the five nominal wavelengths. Volcanic AOD data are collected in solar occultation mode by viewing the Sun through the plume. The total AOD of the observation will be made up of the aerosol optical depth of the plume AOD_p and the background aerosol optical depth AOD_b. Using quasi-simultaneous observations (within less than 1 h from in-plume observations, see Table 1) of the background atmosphere, e.g., by pointing the instrument towards the Sun in the absence of volcanic plume, the volcanic AOD is then isolated by applying background atmosphere correction for each individual in-plume observation:

$$AOD_p(\lambda) = AOD(\lambda) - AOD_b(\lambda) \tag{1}$$

Practically, we have performed one preliminary background observation session, each day before the in-plume session, and calculated AOD_b. Background atmosphere and in-plume conditions were identified by visual inspection and this identification is subsequently confirmed by the smaller AOD values and variability of the background. We assume that the atmosphere remains relatively homogeneous between background and plume observations and that the clear atmosphere aerosol optical depth in the volume occupied by the plume is negligible with respect to AOD_b.

The uncertainty of individual AOD retrievals with a MIISP, in the atmospheric window channels, has previously been estimated at 0.02, e.g., [19]. Uncertainties in the retrieved AODs mainly arise from manual Sun-pointing and internal calibration errors. Higher values are expected in spectral regions affected by the absorption of atmospheric gases, such as Ch.4 (sensitive to water vapour absorption) [16]. The standard deviation of the in-plume AOD σ_{AOD_p} is:

$$\sigma_{AOD_p}(\lambda) = \sqrt{\sigma_{AOD}^2 + \frac{\sigma_{AOD^i_b}^2}{n}} \tag{2}$$

with n the number of individual background measurements AOD^i_b made to compute the average background. As n is in the order of tens to hundreds, the uncertainty content of AOD_p is approximately σ_{AOD}, and then 0.02, as well.

The plume-isolated AOD spectral variability can be modelled using the empirical Ångström law, using the α and β parameters [20]:

$$AOD_p(\lambda) = \beta_p \lambda^{-\alpha_p} \tag{3}$$

The α parameter is the negative spectral slope of the optical depth in log-log scale and is an optical proxy for the mean size of the sampled aerosol particles. Small or negative α values are typical of bigger particles, and bigger values, from about 1.0 to approximately 2.5, are typical of smaller particles. The β parameter is the modelled AOD value at 1.0 μm and is related to the amount and chemical composition of the aerosol particles. Using Equation (3), the Ångström parameters for each in-plume MIISP acquisition are derived, in this work, using selected wavelength pairs, in the following way:

$$\alpha_p = -\frac{\ln\left[\frac{AOD_p(\lambda_1)}{AOD_p(\lambda_2)}\right]}{\ln\left[\frac{\lambda_1}{\lambda_2}\right]} \tag{4}$$

$$\beta_p = AOD_p(\lambda_1) \cdot \lambda_1^{\alpha_p} \tag{5}$$

The uncertainties of the derived α_p and β_p can be expressed as follows:

$$\sigma_{\alpha_p} = \left(1/\ln\left[\frac{\lambda_1}{\lambda_2}\right]\right) \sqrt{\left(\frac{\sigma_{AOD_p(\lambda_1)}}{AOD_p(\lambda_1)}\right)^2 + \left(\frac{\sigma_{AOD_p(\lambda_2)}}{AOD_p(\lambda_2)}\right)^2} \tag{6}$$

$$\sigma_{\beta_p} = \lambda_1^\alpha \sqrt{\sigma_{AOD_p(\lambda_1)}^2 + \left(AOD_p(\lambda_1) \cdot \ln \lambda_1\right)^2 \sigma_{\alpha_p}^2} \tag{7}$$

Considering the moderate values of the observed plume-isolated AODs during our campaign (typically 0.1 at 440 nm and 0.05 at longer wavelengths) and the mentioned uncertainties of about 0.02 for the AOD, using wavelength couples of 440/870 nm and 440/1020 nm, the uncertainties σ_{α_p} and σ_{β_p} are estimated at about 0.50 to 0.65 (α) and 0.02 to 0.04 (β).

3. Campaign Conditions

Remote photometric observations of the bulk plume's aerosols from Pacaya volcano (geographical position in Figure 1a) were made on 29 and 30 January 2011. Data were collected from two different sites on the WNW flank of the volcano at a mean altitude of 1700 m a.s.l. and 3.5 Km far from the vent (Figure 1b; 14°23′52.80″ N–90°37′51.69″ W, 14°23′30.26″ N–90°38′4.39″ W, respectively). These sites were chosen to locate the Sun behind the plume during the measurements and plume-sun occultation was ensured by manual adjustment of the tripod gears. Atmospheric background (AOD_b) was measured before each measurements in-plume session (AOD) and plume-isolated optical properties (AOD_p) were derived by applying the method described in Section 2. Plume opacity appeared to vary with variable intensity of degassing pulses. Measurements were taken during times when meteorological clouds were absent (Figure 1c). Wind speed and direction for the days of our field campaign were obtained at 700 and 750 mbar (2500 and 3000 m, respectively) from the NOAA (National Oceanic and Atmospheric

Administration, http://www.arl.noaa.gov) real-time environmental applications and display system (READY), running the Global Data Assimilation System (GDAS) reanalysis model. Data show that between the time of sampling (14:00 and 17:30 UTC Table 1), on 29 January 2011, mean wind speed and direction was 6.5 knots and 132° between 700 and 750 mbar. Instead, on 30 January, mean wind speed was 17 knots with a mean wind direction of 250° SW. Therefore, the plume transport direction, reported in Figure 1b, was retrieved according to the meteorological data together with visual observation in field during the sampling. The time intervals and the subdivision in background and in-plume observations for the two days are listed in Table 1. During the collection, the MacKenney cone of Pacaya Volcano [21] (Figure 1), was quietly degassing, producing pulses of plume steam-gas, which dispersed rapidly southwards in the atmosphere downwind. Unlike what was reported by INSIVUMEH (Instituto Nacional de SIsmología, VUlcanología, MEteorología e Hidrología) [22], no anomalous seismic activity was recorded by the geophysical monitoring network on 29 and 30 January. In Figure 1c, a sketch of the plume section intercepted by the MIISP in-plume observations is shown for the measurements taken on 30 January 2011. Considering a 2.5° field of view and a distance of about 3500 m between the sampling site and Pacaya summit, the intercepted circular area has a radius of about 150 m. This area is comparable with the plume's horizontal extension at summit altitude (see Figure 1c), thus assuring that the measured AODs are representative of the whole plume and not only of a subsection.

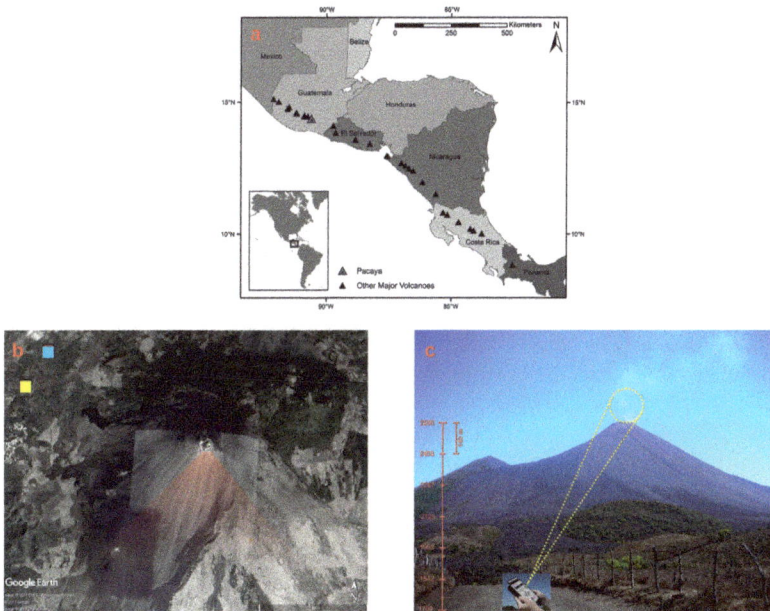

Figure 1. (a) Map of the Central American volcanic front [23] showing the location of the volcanoes along the front (black triangles), Pacaya is located in southeastern Guatemala (gray triangle). (b) Satellite images of Pacaya showing the direction of plume transport issued by McKenney Cone (reddish area; approximate locations), and the sampling locations from which aerosol optical depth (AOD) measurements were taken on 29 and 30 January 2011 (blue and yellow squares, respectively; Google Earth image CNES 2017). (c) View of Pacaya volcano from the western flank of the volcano showing the diluted, roughly vertical plume emitted on 30 January 2011 morning from the volcano summit crater (yellow square in (b)). The area of the plume captured by the instrument is also displayed, considering the distance of 3500 m between the sampling site and the summit of Pacaya. The calculated radius of the circular observed area is 150 m above the volcano. The dimension of the plume captured in the photo was scaled considering the altitude of the sampling site and that of Pacaya (1700 and 2550 m, respectively).

Table 1. Dates and time intervals for the background and in-plume observation sessions. All times are UTC.

Day	Type	Time Interval
29 January	Background	15:25–16:00
29 January	In-plume	16:45–17:30
30 January	Background	14:00–14:10
30 January	In-plume	15:10–16:40

4. Results and Discussion

In Figure 2, the AOD observations of Pacaya volcano plume, on 29 and 30 January, at the operating wavelengths of MIISP, are shown. Average values for the background and in-plume sessions are also shown as dotted and solid lines. Observations at 936 nm are excluded due to the water vapour interference in this band. Observations at 675 nm are excluded from the plot to enhance the clarity of the figure: these data points and mean values would have appeared very close to data points for 870 and 1020 nm. As shown in Figure 2, the AOD observations for the in-plume session are systematically larger than the background, at all wavelengths, although both observations are taken at a very small spatio-temporal distance from each other. This suggests that an additional aerosol source is present in the in-plume observations, related to the volcanic source. In addition, the in-plume observations are more variable than the background. While the variability of these latter observations is systematically confined between 0.05 and 0.15, in-plume AOD observations reach values up to 0.6–0.7, depending on the wavelength, indicating an inhomogeneous volcanic aerosol layer.

The spectral variability of average AOD values, for the two days of the campaign, the background, total (in-plume) and plume-isolated observations is shown in Figure 3. For both days, background observations have an almost flat spectral trend, with small variations between shorter and longer wavelengths. This is typical of an atmosphere with an aerosol layer dominated by bigger particles, such as mineral dust or marine aerosols. It should be considered that, although at relatively high altitude, the Pacaya region is only a few tens of kilometres from the Pacific Ocean and thus its background atmosphere could be largely affected by marine aerosols. On the other hand, due to the proximity of this area with Guatemala city (30 km), the impact from anthropogenic pollution, e.g., traffic exhaust emissions, cannot be excluded. The background AODs are larger on 30 January (between 0.10 and 0.12 depending on the wavelength) than 29 January (between 0.05 and 0.07 depending on the wavelength). The total AOD observations are characterised by bigger values than the background, at all wavelengths, thus indicating the presence of an additional aerosol layer (the volcanic plume). The average total AOD reaches values as high as about 0.20 (30 January) and 0.13 (29 January). There is a marked wavelength dependence of the average total AODs, thus indicating that the mentioned additional aerosol layer has smaller particles than the background aerosol layer. The plume-isolated AODs, calculated using Equation (1), are also shown in Figure 2. The marked wavelength dependence is even more apparent than for the total AOD observations, at least at shorter wavelengths. The average plume AOD reaches values as high as about 0.10 (30 January) and 0.07 (29 January) at 440 nm and quickly decreases with wavelength down to values of about 0.05 (30 January) and 0.04 (29 January).

The Ångström parameters α_p and β_p have been subsequently derived using Equations (4) and (5), using different wavelength pairs. Using sufficiently distant wavelengths is crucial to obtain small uncertainties on α_p [16]. Operational MIISP wavelengths in the spectral window region allow multiple choices for the mentioned wavelength pairs, i.e., 440/1020 nm or 440/870 nm. While both combinations are associated to limited uncertainties on α_p, selecting one pair with respect to another is not straightforward. Thus, we have analysed more in-depth the consistency of estimations of the Ångström parameters using these pairs. It has to be mentioned that differences between estimations with different wavelength pairs can be partially attributed to the expected spectral dependency of α_p (and, to a lesser extent, of β_p) (e.g., [24]). The individual α_p and β_p estimations for 29 and 30 January, using the wavelength pairs 440/1020 nm and 440/870 nm, are shown in Figure 4. The average values of α_p and β_p for both 29 and 30 January are reported in Table 2. The individual α_p estimations vary between about −0.5 and about 2.0, thus indicating the significant inhomogeneity of the plume, with the prevalence of alternatively very big and very small particles. An increase/decrease of over 100% can be observed in extremely short time intervals (e.g., of the order of a few minutes). As an example of this short-term variability, five cases of the occurrence of simultaneous extremely low values of α_p (lower than 0.5) and high values of β (bigger than 0.15) are indicated in Figure 4a. These cases are associated to extreme values of the AOD: 0.20, 0.40, 0.25, 0.50 and 0.20, at 320 nm, and 0.20, 0.45, 0.20, 0.55 and 0.20, at 1020 nm, for the five cases. The simultaneous low values of α and high values of AOD indicate the transitory perturbation of relevant burdens of bigger particles, like for small ash puffs. In any case, mean values of 1.4 ± 0.7 (440/1020 nm) and 1.5 ± 0.9 (440/870 nm), for 29 January, and 1.0 ± 0.5 (440/1020 nm) and 0.8 ± 0.4 (440/870 nm), for 30 January, indicate prevalently small to very small particles, with a significant short-term variability. Similar α mean values have been associated to ash-free plumes in the past, at Mount Etna [9,10,16] and Lascar and Villarica volcanoes [13,14]. The individual β_p estimations vary between near zero to over 0.1. These estimations are inversely correlated with simultaneous α_p estimations. Observations of bigger β and smaller α can be associated with short-term overpasses of ash-bearing plume sections [10,13]. Our results, 0.05 ± 0.07 (440/1020 nm) and 0.05 ± 0.07 (440/870 nm), for 29 January, and 0.03 ± 0.04 (440/1020 nm) and 0.04 ± 0.05 (440/870 nm), for 30 January, denote prevalently ash-free plumes. Nevertheless, it must be mentioned that smaller values of β (0.001 to 0.007) have been observed at Lascar and Villarica volcanoes, while our estimations are more in line with ash-free plumes at, e.g., Mount Etna [9,10].

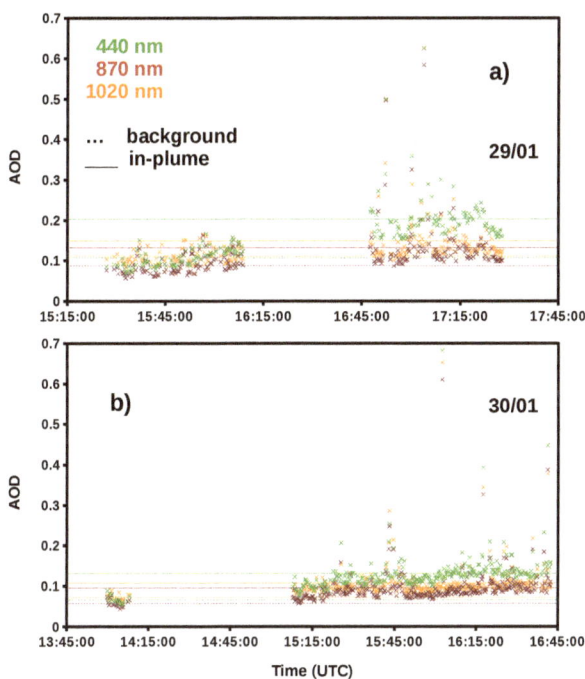

Figure 2. MIISP AOD observations at 440, 870 and 1020 nm on 29 (**a**) and 30 January 2011 (**b**) at Pacaya volcano. Average background (dotted lines) and in-plume AODs (solid lines) are also shown. Background and in-plume measurements are taken during the time intervals of Table 1 (see text for details). All times are UTC.

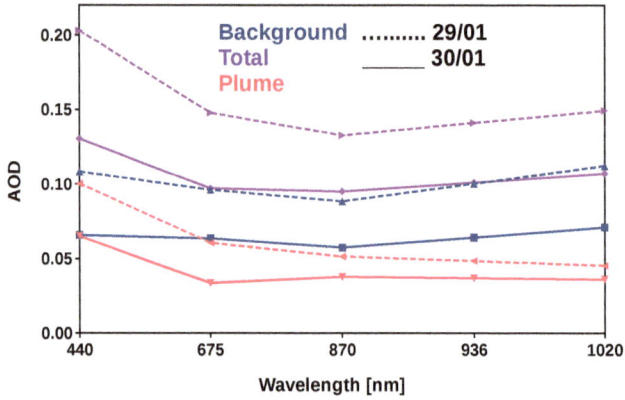

Figure 3. Spectral average MIISP AODs for background (blue symbols and lines) and total atmosphere (violet symbols and lines), and isolated volcanic plume (red symbols and lines), for 29 January (dotted lines) and 30 January (solid lines).

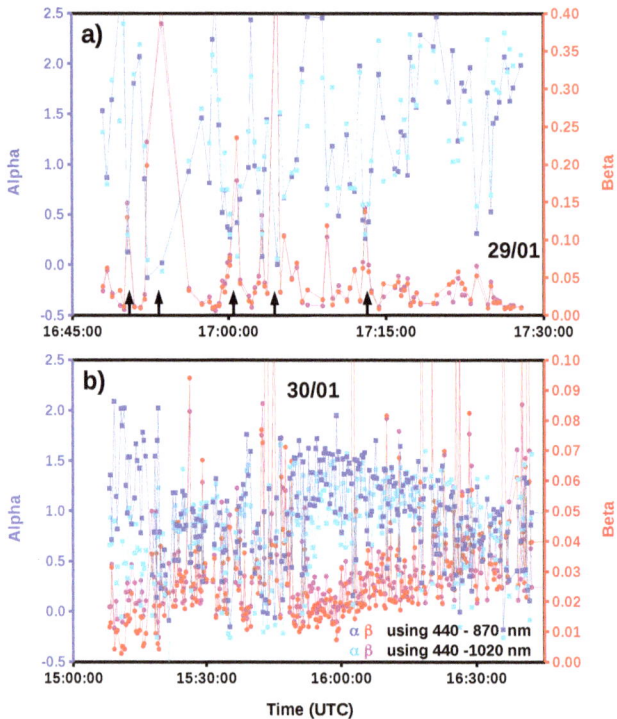

Figure 4. Volcanic plume α_p and β_p observations for 29 January (**a**) and 30 January 2011 (**b**). Determinations with different wavelength pairs are shown, 440/1020 nm (α_p: blue, β_p: red) and 440/870 nm (α_p: sky blue, β_p: pink).

Table 2. Average Ångström coefficients α_p and β_p for the two days of MIISP acquisitions. Average values obtained using different wavelength pairs (440/1020 and 440/870 nm) are reported.

Day	Alpha (440–1020 nm)	Alpha (440–870 nm)	Beta (440–1020 nm)	Beta (440–870 nm)
29 January	1.4 ± 0.7	1.5 ± 0.9	0.05 ± 0.07	0.05 ± 0.07
30 January	1.0 ± 0.5	0.8 ± 0.4	0.03 ± 0.04	0.04 ± 0.05

In the range of our observations, the choice of the wavelength pairs seems not to be a crucial factor in the determination of the mean values of α_p and β_p. The mean values of both parameters, calculated using AODs at 440–1020 nm and 440–870 nm, are well enveloped into each other's statistical uncertainty (measured as 1 standard deviation of the mean, Table 2). In order to obtain more insight into the retrieved data, scatter plots of the individual α_p and β_p measurements, obtained with different wavelength pairs, are shown in Figure 5. In addition, Table 3 shows the Pearson coefficient, root mean square error (RMSE) and mean bias for the four scatter plots. These results reveal that, though a general agreement exists between the mean values of Table 2, the individual observations of α_p can be significantly over/underestimated when using different wavelength pairs. This is the case of of 30 January, with an RMSE of nearly 50%. The β_p determinations, using different wavelength pairs, are more consistent.

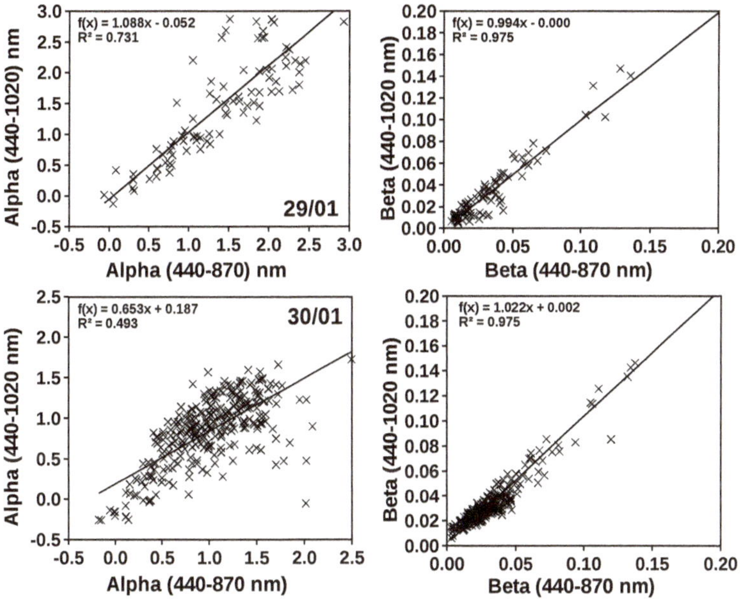

Figure 5. Scatter plots of α_p and β_p measurements with different wavelength pairs, for 29 January (top panels) and 30 January (bottom panels).

Table 3. Statistical parameters for the comparison of α and β with different wavelength pairs (440/1020 nm with respect to 440/870 nm).

	R^2	RMSE	Bias
α—29 January	0.73	28.2%	+14.7%
α—30 January	0.49	47.8%	−17.9%
β—29 January	0.97	25.7%	+6.9%
β—30 January	0.97	22.2%	−2.7%

5. Conclusions

In this paper, the optical characterisation of the volcanic aerosol of Pacaya volcano has been presented. Observations were taken during a field campaign, carried out on 29 and 30 January 2011, using a hand-held MIISP sun-photometer. The volcanic plume was characterised in terms of its spectral AOD in the visible and NIR spectral ranges, and of the subsequently derived Ångström parameters α and β. Overall, moderate plume-isolated AOD values were found. The plume-isolated AOD at 440 nm did not exceed 0.1 during the observation sessions of this campaign. The average α_p (β_p) values of the two measurement sessions are relatively big (small), consistent with an ash-free plume, theoretically composed of a mixture of small secondary aerosols. A potential influence of both marine and anthropogenic aerosol on the aerosol signature of Pacaya region could not be excluded. The use of different wavelength pairs (440–1020 and 440–870 nm) was tested in Ångström parameters retrieval. While the choice of wavelength pairs has a negligible impact on the daily averaged α_p and β_p, individual determinations of these two parameters can be strongly affected by this choice.

Acknowledgments: Letizia Spampinato, Alessandro La Spina, Giuseppe G. Salerno, and Pasquale Sellitto acknowledge the INSIVUMEH director Eddy Sánchez and Gustavo Chigna for scientific collaboration and logistic support, A.J.L. Harris, R.E. Wolf, K. Brill for logistic suggestions, and V. Tsanev for his scientific support. The Natural Environmental Research Council (NERC) is acknowledged for providing the Microtops II Sun photometer. This study was supported by grants from the INGV Osservatorio Etneo, Fondi Studi e Ricerche.

Author Contributions: Letizia Spampinato, Alessandro La Spina and Giuseppe G. Salerno conceived, designed and performed the experiments; Pasquale Sellitto analyzed the data; all authors wrote the paper.

Conflicts of Interest: The authors declare no conflict of interest.

References

1. Von Glasow, R.; Bobrowski, N.; Kern, C. The effects of volcanic eruptions on atmospheric chemistry. *Chem. Geol.* **2009**, *263*, 131–142.
2. Sellitto, P.; Zanetel, C.; di Sarra, A.; Salerno, G.; Tapparo, A.; Meloni, D.; Pace, G.; Caltabiano, T.; Briole, P.; Legras, B. The impact of Mount Etna sulfur emissions on the atmospheric composition and aerosol properties in the central Mediterranean: A statistical analysis over the period 2000–2013 based on observations and Lagrangian modelling. *Atmos. Environ.* **2017**, *148*, 77–88.
3. Gassó, S. Satellite observations of the impact of weak volcanic activity on marine clouds. *J. Geophys. Res. Atmos.* **2008**, *113*, D14S19.
4. Sellitto, P.; di Sarra, A.; Corradini, S.; Boichu, M.; Herbin, H.; Dubuisson, P.; Sèze, G.; Meloni, D.; Monteleone, F.; Merucci, L.; et al. Synergistic use of Lagrangian dispersion and radiative transfer modelling with satellite and surface remote sensing measurements for the investigation of volcanic plumes: the Mount Etna eruption of 25–27 October 2013. *Atmos. Chem. Phys.* **2016**, *16*, 6841–6861.
5. Ridley, D.A.; Solomon, S.; Barnes, J.E.; Burlakov, V.D.; Deshler, T.; Dolgii, S.I.; Herber, A.B.; Nagai, T.; Neely, R.R.; Nevzorov, A.V.; et al. Total volcanic stratospheric aerosol optical depths and implications for global climate change. *Geophys. Res. Lett.* **2014**, *41*, 7763–7769.
6. Sellitto, P.; Briole, P. On the radiative forcing of volcanic plumes: modelling the impact of Mount Etna in the Mediterranean. *Ann. Geophys.* **2015**, *58*, doi:10.4401/ag-6879.
7. McCormick, M.P.; Thomason, L.W.; Trepte, C.R. Atmospheric effects of the Mt Pinatubo eruption. *Nature* **1995**, *373*, 399–404.
8. Malavelle, F.F.; Haywood, J.M.; Jones, A.; Gettelman, A.; Clarisse, L.; Bauduin, S.; Allan, R.P.; Karset, I.H.H.; Kristjánsson, J.E.; Oreopoulos, L.; et al. Strong constraints on aerosol-cloud interactions from volcanic eruptions. *Nature* **2017**, *546*, 485–491.
9. Watson, I.M.; Oppenheimer, C. Particle size distributions of Mount Etna's aerosol plume constrained by Sun photometry. *J. Geophys. Res. Atmos.* **2000**, *105*, 9823–9829.
10. Watson, I.M.; Oppenheimer, C. Photometric observations of Mt. Etna's different aerosol plumes. *Atmos. Environ.* **2001**, *35*, 3561–3572.

11. Porter, J.N.; Horton, K.A.; Mouginis-Mark, P.J.; Lienert, B.; Sharma, S.K.; Lau, E.; Sutton, A.J.; Elias, T.; Oppenheimer, C. Sun photometer and lidar measurements of the plume from the Hawaii Kilauea Volcano Pu'u O'o vent: Aerosol flux and SO_2 lifetime. *Geophys. Res. Lett.* **2002**, *29*, 30-1–30-4.

12. Martin, R.S.; Mather, T.A.; Pyle, D.M.; Power, M.; Tsanev, V.I.; Oppenheimer, C.; Allen, A.G.; Horwell, C.J.; Ward, E.P.W. Size distributions of fine silicate and other particles in Masaya's volcanic plume. *J. Geophys. Res. Atmos.* **2009**, *114*, D09217.

13. Mather, T.A.; Tsanev, V.I.; Pyle, D.M.; McGonigle, A.J.S.; Oppenheimer, C.; Allen, A.G. Characterization and evolution of tropospheric plumes from Lascar and Villarrica volcanoes, Chile. *J. Geophys. Res. Atmos.* **2004**, *109*, D21303.

14. Sawyer, G.; Salerno, G.; Blond, J.L.; Martin, R.; Spampinato, L.; Roberts, T.; Mather, T.; Witt, M.; Tsanev, V.; Oppenheimer, C. Gas and aerosol emissions from Villarrica volcano, Chile. *J. Volcanol. Geotherm. Res.* **2011**, *203*, 62–75.

15. Ilyinskaya, E.; Tsanev, V.; Martin, R.; Oppenheimer, C.; Blond, J.L.; Sawyer, G.; Gudmundsson, M. Near-source observations of aerosol size distributions in the eruptive plumes from Eyjafjallajokull volcano, March-April 2010. *Atmos. Environ.* **2011**, *45*, 3210–3216.

16. Sellitto, P.; Salerno, G.; La Spina, A.; Caltabiano, T.; Terray, L.; Gauthier, P.J.; Briole, P. A novel methodology to determine volcanic aerosols optical properties in the UV and NIR and Ångström parameters using Sun photometry. *J. Geophys. Res. Atmos.* **2017**, *122*, 9803–9815.

17. Morys, M.; Mims, F.M.; Anderson, S.E. *MICROTOPS II, Ozone Monitor and Sunphotometer: User's Guide*; Technical Report; Solar Light Company: Glenside, PA, USA, 2001.

18. Morys, M.; Mims, F.M.; Hagerup, S.; Anderson, S.E.; Baker, A.; Kia, J.; Walkup, T. Design, calibration, and performance of MICROTOPS II handheld ozone monitor and Sun photometer. *J. Geophys. Res. Atmos.* **2001**, *106*, 14573–14582.

19. Porter, J.N.; Miller, M.; Pietras, C.; Motell, C. Ship-Based Sun Photometer Measurements Using Microtops Sun Photometers. *J. Atmos. Ocean. Technol.* **2001**, *18*, 765–774.

20. Angström, A. The parameters of atmospheric turbidity. *Tellus* **1964**, *16*, 64–75.

21. Kitamura, S.; Matias, O. *Tephra Stratigraphic Approach to the Eruptive History of Pacaya Volcano, Guatemala*; Seventh Series: Geography 45; Technical Report; Tohoku University: Sendai, Japan, 1995; pp. 1–41.

22. Global Volcanism Program. Report on Pacaya (Guatemala). In *Smithsonian Institution and US Geological Survey Weekly Volcanic Activity Report, 5 January–11 January 2011*; Sennert, S.K., Ed.; Smithsonian Institution and US Geological Survey: Washington, WA, USA, 2011.

23. Carr, M.J.; Rose, W.I., Jr.; Stoiber, R.E. Volcanism in Central America. In *Orogenic Andesites and Related Rocks*; Thorpe, R.S., Ed.; Wiley and Sons: New York, NY, USA, 1982; pp. 149–166.

24. O'Neill, N.T.; Eck, T.F.; Smirnov, A.; Holben, B.N.; Thulasiraman, S. Spectral discrimination of coarse and fine mode optical depth. *J. Geophys. Res. Atmos.* **2003**, *108*, 4559.

geosciences

MDPI

Article

Ground-Based Measurements of the 2014–2015 Holuhraun Volcanic Cloud (Iceland)

Melissa A. Pfeffer [1,*], Baldur Bergsson [1], Sara Barsotti [1], Gerður Stefánsdóttir [1], Bo Galle [2], Santiago Arellano [2], Vladimir Conde [2], Amy Donovan [3], Evgenia Ilyinskaya [4], Mike Burton [5], Alessandro Aiuppa [6], Rachel C. W. Whitty [4], Isla C. Simmons [7], Þórður Arason [1], Elín B. Jónasdóttir [1], Nicole S. Keller [8], Richard F. Yeo [9], Hermann Arngrímsson [1], Þorsteinn Jóhannsson [8], Mary K. Butwin [1,10], Robert A. Askew [10], Stéphanie Dumont [10,11], Sibylle von Löwis [1], Þorgils Ingvarsson [1], Alessandro La Spina [12], Helen Thomas [13], Fred Prata [14], Fausto Grassa [15], Gaetano Giudice [15], Andri Stefánsson [10], Frank Marzano [16], Mario Montopoli [17] and Luigi Mereu [16]

[1] Icelandic Meteorological Office, Bústaðavegur 7-9, 108 Reykjavík, Iceland; bb@vedur.is (B.B.); sara@vedur.is (S.B.); gerdur@vedur.is (G.S.); arason@vedur.is (Þ.A.); elin@vedur.is (E.B.J.); hermann@vedur.is (H.A.); mkb5@hi.is (M.K.B.); sibylle@vedur.is (S.v.L.); thorgils@vedur.is (Þ.I.)
[2] Department of Space, Earth and Environment, Chalmers University of Technology, SE-412 96 Göteborg, Sweden; bo.galle@chalmers.se (B.G.); santiago.arellano@chalmers.se (S.A.); vladimir.conde@chalmers.se (V.C.)
[3] Department of Geography, King's College London, WC2R 2LS London, UK; amy.donovan@kcl.ac.uk
[4] School of Earth and Environment, University of Leeds, LS2 9JT Leeds, UK; e.ilyinskaya@leeds.ac.uk (E.I.); eercww@leeds.ac.uk (R.C.W.W.)
[5] School of Earth and Environmental Sciences, University of Manchester, M13 9P Manchester, UK; mike.burton@manchester.ac.uk
[6] Dipartimento DiSTeM, University of Palermo, Via Archirafi 36, 90123 Palermo, Italy; alessandro.aiuppa@unipa.it
[7] School of GeoSciences, University of Edinburgh, EH8 9XP Edinburgh, UK; isla.simmons@ed.ac.uk
[8] Environment Agency of Iceland, Suðurlandsbraut 24, 108 Reykjavík, Iceland; nicole.keller@umhverfisstofnun.is (N.S.K.); thorsteinnj@umhverfisstofnun.is (Þ.J.)
[9] AUV Consultants, 52Vesturás, Reykjavík 101, Iceland; richard@auvconsultants.com
[10] Institute of Earth Sciences, University of Iceland, Sæmundargata 2, 101 Reykjavík, Iceland; raa22@hi.is (R.A.A.); ; sdumont@segal.ubi.pt (S.D.); as@hi.is (A.S.)
[11] Instituto Dom Luiz, University of Beira Interior, Rua Marques d´Avila e Boloma, 6201-001 Covilhã, Portugal
[12] Istituto Nazionale di Geofisica e Vulcanologia- Sezione di Catania, 95125 Catania, Italy; alessandro.laspina@ingv.it
[13] School of Earth Sciences, University of Bristol, BS8 1TH Bristol, UK; helen.thomas@bristol.ac.uk
[14] AIRES Pty Ltd, PO Box 156, Mt Eliza, Victoria 3930, Australia; fred_prata@hotmail.com
[15] Istituto Nazionale di Geofisica e Vulcanologia- Sezione di Palermo, Via Ugo la Malfa 153, 90146 Palermo, Italy; fausto.grassa@ingv.it (F.G.); gaetano.giudice@ingv.it (G.G.)
[16] Sapienza, University of Rome, 00184 Rome, Italy; marzano@diet.uniroma1.it (F.M.); mereu@diet.uniroma1.it (L.M.)
[17] National Research Council of Italy, Institute of Atmospheric Sciences and Climate, Via Fosso del Cavaliere 100, 00133 Rome, Italy; m.montopoli@isac.cnr.it
* Correspondence: melissa@vedur.is; Tel.: +354-893-5157

Received: 13 November 2017; Accepted: 10 January 2018; Published: 18 January 2018

Abstract: The 2014–2015 Bárðarbunga fissure eruption at Holuhraun in central Iceland was distinguished by the high emission of gases, in total 9.6 Mt SO_2, with almost no tephra. This work collates all ground-based measurements of this extraordinary eruption cloud made under particularly challenging conditions: remote location, optically dense cloud with high SO_2 column amounts, low UV intensity, frequent clouds and precipitation, an extensive and hot lava field, developing ramparts, and high-latitude winter conditions. Semi-continuous measurements of SO_2 flux with three scanning DOAS instruments were augmented by car traverses along the ring-road and along the lava. The ratios of other gases/SO_2 were measured by OP-FTIR, MultiGAS, and filter packs. Ratios of SO_2/HCl = 30–110 and SO_2/HF = 30–130 show a halogen-poor eruption cloud. Scientists on-site reported extremely minor tephra production during the eruption. OPC and filter packs showed low particle concentrations similar to non-eruption cloud conditions. Three weather radars detected a droplet-rich eruption cloud. Top of eruption cloud heights of 0.3–5.5 km agl were measured with ground- and aircraft-based visual observations, web camera and NicAIR II infrared images, triangulation of scanning DOAS instruments, and the location of SO_2 peaks measured by DOAS traverses. Cloud height and emission rate measurements were critical for initializing gas dispersal simulations for hazard forecasting.

Keywords: Holuhraun; Bárðarbunga; gas; SO_2; cloud height; eruption monitoring; fissure eruption

1. Introduction

The 2014–2015 fissure eruption of Bárðarbunga (also known as Veiðivötn) lasted six months, from 31 August 2014–27 February 2015. This was the largest Icelandic eruption in over 200 years: 1.6 ± 0.3 km³ of lava and prodigious amounts of gases were released [1]. The Bárðarbunga volcanic system includes a central volcano capped by the Vatnajökull glacier in the highlands of central Iceland, and also includes a 190 km long fissure swarm extending to the northeast and southwest from the central volcano. Bárðarbunga erupts frequently, with an average of two eruptions per century over the last 11 centuries [2]. The greatest amount of lava known to have been produced during a Bárðarbunga eruption is >20 km³, so while the 2014–2015 eruption was extraordinary in recent times, it is well within the known behavior of this volcanic system.

The eruption was preceded by seven years of increased seismicity within the volcanic system, which escalated for two months, followed by two weeks of migration of seismic swarms and associated ground deformation manifesting as a rifting event, finally culminating in a small, few-hours long lava effusion on 29 August. Two days later, on 31 August, the six-month long Holuhraun eruption started. The geophysical changes were closely monitored in real-time as potential precursors to an eruption. A segmented dyke intrusion originated at the Bárðarbunga central volcano that propagated laterally over 45 km. This intrusion culminated in an effusive fissure eruption at the end of the dyke [3], a few km north of the Vatnajökull glacier, where a lava field of the same name, Holuhraun, had erupted previously in 1862–1864, also originating from Bárðarbunga [4].

This eruption was one of the most polluting volcanic eruptions in centuries. The remote location and winter-season timing of the eruption, however, reduced its potential impact on people and the environment in Iceland [1]. The prodigious emissions of gases and the sulfate aerosol formed as the eruption cloud aged impacted the air quality in populated areas of Iceland significantly throughout the course of the eruption [5]. The remote location, however, meant that the concentrations of gases were diluted before reaching population centers. The dry atmosphere and weak winter sunlight conditions during most of the eruption slowed down the formation of sulfate aerosol, which, despite these dampening effects, exceeded Icelandic health standards far above legal limits [1,5]. If the dyke had breached the surface beneath the glacier as opposed to north of it, ash and floods would likely have been produced [6].

The anticipatory period allowed for the continued development of gas and particle monitoring instrumentation and techniques suitable for Icelandic conditions, benefiting from the EU-FP7 FUTUREVOLC project. This project fostered instrumentation development, deployment strategies, data processing techniques, and strengthened relationships between Icelandic and foreign collaborators for a better response during volcanic eruptions, and therefore contributed to the success of the eruption cloud monitoring.

The eruption occurred in a remote, very difficult to access location, so in spite of the instrumental improvements made prior to the eruption, there were serious challenges for acquiring data and maintaining the instruments. The nearest farm and municipality are each about 100 km away. The eruption site is located within Dyngjusandur, the most extensive dust source area in Iceland [7], an active sandy desert where dust storms are very common. Tremendous efforts were made to install continuous monitoring instrumentation; however, because of these harsh field conditions, there are many temporal gaps in the data. Field campaigns for non-continuous instrumentation overcame many difficulties, mainly pertaining to weather and the high concentration of gas near the eruption vent. Traveling to the field and maintaining instrumentation was a major undertaking during winter conditions.

The aim of this paper is to bring together all of the ground-based measurements of the volcanic cloud. We report on the results and discuss what the combined data sets tell us about this extraordinary event and how to optimize the monitoring of volcanic clouds from future fissure eruptions in Iceland and elsewhere.

2. Materials and Methods

2.1. DOAS

The primary monitoring tool for this long-lasting, gas-rich volcanic cloud was Differential Optical Absorption Spectrometry (DOAS) [8]. Ultraviolet light from the sun, scattered from aerosols and molecules in the atmosphere, is collected by a telescope. Light is transferred from the telescope to the grating spectrometer by a quartz optical fiber. In-cloud spectra are analyzed against clear-sky and dark spectra and the differential slant column of various gases, primarily SO_2, is derived.

2.1.1. ScanDOAS

Through the support of the FUTUREVOLC project, a version of the NOVAC ScanDOAS instrument [9,10] was developed that is adapted to high latitudes with low UV radiation and severe meteorological conditions. Two major developments were made: the standard Ocean Optics SD2000 spectrometer was replaced by the more UV-sensitive Ocean Optics Maya2000 Pro spectrometer, and the scanning device was modified to avoid external moving parts to make it more robust in freezing conditions (Figure 1). The scanning device was modified by replacing the rotating hood with a quartz window with a closed scanner with a cylindrical quartz tube, and a cylindrical lens was included in the optical system. This changed the field of view (FOV) of the instrument to be rectangular instead of circular, covering the full 7.2° angle used as the scan interval. A fixed exposure time of 200 ms was used. Co-adding 15 spectra resulted in a total time of 2 min for one scan to be completed.

At the onset of the fissure eruption at Holuhraun, a ScanDOAS instrument, DOAS 25, was prepared at the Icelandic Meteorological Office (IMO) and installed at the eruption site on the second day of the eruption (Figure 2). In the first week of September, a second ScanDOAS instrument, DOAS 27, was installed, which was made available through cooperation with Prof. Konradin Weber at Fachhochschule Düsseldorf. Data transfer and real-time evaluation was fully implemented at IMO within the first couple of days of the eruption. After about two weeks, DOAS 27 was surrounded by active lava flows and eventually stopped transmitting data in the absence of sufficient power. During and after the remaining six months of the eruption, one or two ScanDOAS instruments, DOAS 25 and DOAS 26, were operational. DOAS 26 was moved around the eruption site in response to the advancement of the lava. Its final location is shown in Figure 2.

Figure 1. OPC, ScanDOAS, and NicAIR II at Þorvaldshraun, 10 km northeast of the main eruption vent. The OPC and NicAIR II were moved from this site to their final locations (Möðrudalur and Vaðalda, respectively) due to problems operating them at this site. The visibility is poor because of a dust storm, which are common here.

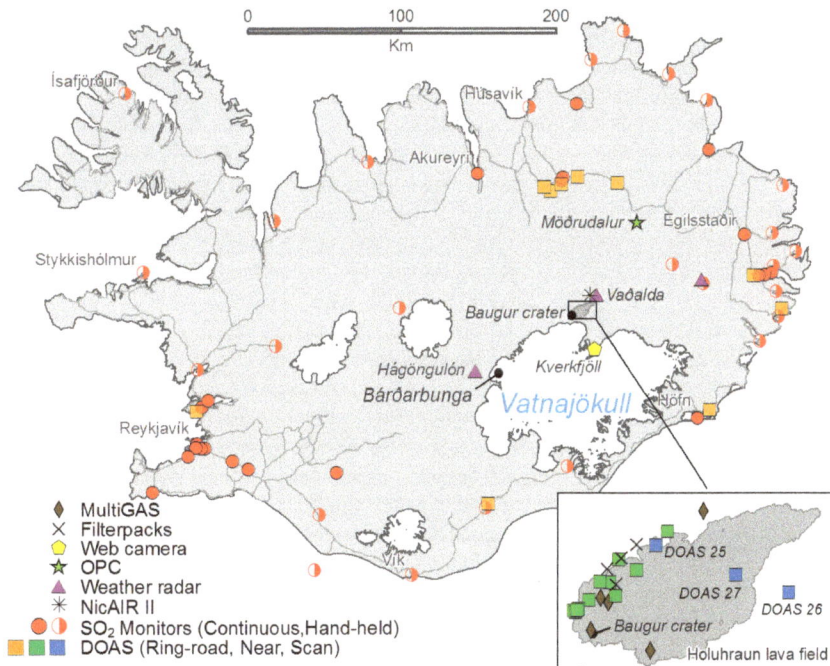

Figure 2. Map of Iceland showing the locations of the ground-based volcanic cloud monitoring instruments. An inset of the area around the eruption site is enlarged. The DOAS instruments are identified by number as described in the main text of this section. The ring-road DOAS traverses are marked at the location with the maximum SO_2 column amount. The near DOAS traverses are marked at the midpoint of the traverse.

To calculate emission rates from the ScanDOAS data, wind speed, wind direction, and cloud height must be known. Wind direction and wind speed at the eruption cloud height were obtained from the HARMONIE numerical weather prediction model utilized by IMO [11]. HARMONIE runs on a regional scale over Iceland with a horizontal resolution of 2.5 km and an hourly forecast provided every six hours. Atmospheric parameters calculated by the model at an altitude of 850 hPa (the model level closest to 1387 m elevation, explanation below) were used as representative of the conditions at the eruption cloud height for processing the ScanDOAS data.

When two ScanDOAS instruments simultaneously measured the eruption cloud, cloud height was derived by triangulation. After about two weeks, the second ScanDOAS instrument was trapped by an active lava flow and alternative methods were necessary to determine the cloud height. Direct observations from the field, observations from air craft, and web cam images were used during the eruption to estimate cloud heights. These showed a high temporal variance and disagreement between techniques. As a result, the average cloud height of 1387 m measured while the two ScanDOAS instruments were both operating, which is within the 1–3 km frequently reported by the various other cloud height observation methods listed above, was used for the processing of the DOAS data for the duration of the eruption.

Because of the extremely high emissions of SO_2 from the fissure eruption at Holuhraun, in combination with severe atmospheric scattering [12], it was not possible to apply standard evaluation procedures for processing the ScanDOAS data. The spectral evaluation window was changed to 319–325 nm, where the absorption by SO_2 is weaker, light intensity is stronger, and atmospheric scattering reduced, compared with the usual 310–325 nm interval. This did not remove all the effects of atmospheric scattering, so further data filtering was required to select data least affected by scattering.

Figure 3 analyzes the bias in SO_2 flux caused by wind direction and wind speed based on all data collected during the eruption by DOAS 25. Wind direction produced the strongest bias in the data (Figure 3a). The greater the angle of the wind from line-of-sight from the main vent to the DOAS, the greater the amount of intervening atmosphere is included between the eruption cloud and the instrument, and the biggest impact from the so-called dilution of the absorption signal from scattering was found. This bias was removed by restricting the acceptable wind direction to +/−15° from the line-of-sight from the main vent to the instrument and filtering out data collected at other wind directions (Figure 3b). Work is on-going to develop an algorithm for the spectral data to make a first order compensation for the scattering effect of clean atmosphere outside of the eruption cloud [9], which could, in the future, make this filtering of the data for non-optimal wind directions unnecessary. Sensitivity analysis of the error in flux related to the scanning geometry shows that for conical scanning, uncertainties in wind direction of up to 40% produce errors in flux <5%, not considering scattering effects [13]. Figure 3c shows the bias from wind speed. At low wind speeds, <7.5 m/s, calculated fluxes increase as the wind speed increases. Restricting the wind speed to ≥7.5 m/s removes this bias (Figure 3d). Days with fewer than three acceptable scans at an individual ScanDOAS were removed.

The ScanDOAS data was further used to detect SO_2 emitted by the cooling lava field, as distinct from the emissions from the main vent. SO_2 emissions from a lava field will typically form a broadly dispersed low-level haze [14]. The optical path through a uniform haze, and hence the SO_2 column density, will be greater at low elevation angles compared with high elevation angles with a shorter path through the haze. The distribution of column densities of SO_2 in all DOAS scans collected during and after the eruption were visually inspected to identify the characteristic symmetrical trough-shape anticipated for a measurement through a uniform haze. Only scans with sufficient symmetry to indicate the presence of a uniform haze were used to calculate the SO_2 flux from the lava field. These data were filtered for wind direction (±20° from line-of-sight from the lava field to the instrument), and the SO_2 flux was calculated assuming the width of the haze was equal to the width of the lava field as viewed from the ScanDOAS instrument. The uncertainty of these measurements is estimated to be 45%. The daily average value of SO_2 flux from the lava field was assessed for the months following the end of the eruption. The lava field emission rate data has been previously published in [14].

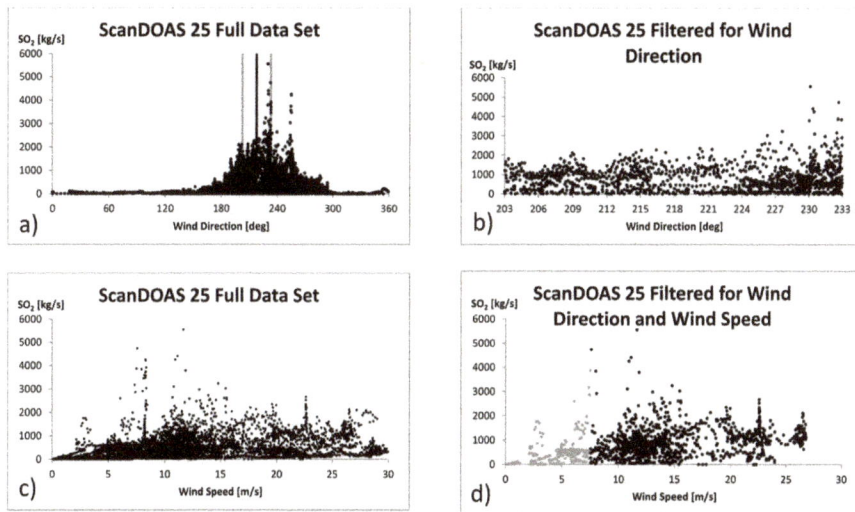

Figure 3. (**a**) Impact of wind direction on SO_2 flux. The vertical black line shows the direct line of sight from the main eruption vent to the instrument. The vertical gray lines show the direct line of sight ±15°; (**b**) SO_2 flux filtered for wind direction; (**c**) Impact of wind speed on SO_2 flux; (**d**) SO_2 flux filtered for wind direction and wind speed. The measurements in gray were made at weak wind speeds and have been filtered out.

2.1.2. Down-Wind MobileDOAS and In Situ SO_2 Traverses

Iceland is encircled by the so-called ring-road highway (Figure 2). Traverses of aged, diluted, down-wind eruption clouds were made by mounting a MobileDOAS system on a car that drove along the ring-road of Iceland with the intention of transecting the transportation path of the eruption cloud. The MobileDOAS instrument is described in [15]. The location and time of the traversing instrument was obtained from a GPS, from which the integration of the cloud cross section column densities and cloud transport direction could be determined. A Thermo Scientific in situ SO_2 analyzer was also transported by car during some of the traverses. The closest the ring-road comes to the eruption site is just over 100 km, where the approximate age of the volcanic cloud would be 2.8 h at a 10 m/s wind speed. This distance meant that the gases were diluted and therefore the DOAS spectra were not saturated as they sometimes were near-vent, and the ring-road is outside of the dust-producing region close to the eruption site, providing for easier, more immediate data processing. Aged clouds, however, may have lost SO_2 due to gas-to-particle conversion and deposition. Successful MobileDOAS traverses were made during the eruption on seven days when the winds were conducive to producing a coherent eruption cloud and it was not raining or snowing. After a traverse was completed, the spectra were analyzed using the MobileDOAS software developed at Chalmers University of Technology, following standard DOAS procedures [15,16] and the slant column of SO_2 was derived. The ring-road traverses, a subset of the full DOAS data collected during the eruption, have been published previously in [1].

Near-lava field traverses were made on five days by mounting a DOAS on a car like for the ring-road traverses. The Thermo Scientific in situ SO_2 analyzer is a rather large and delicate instrument that was never taken off of the main highways to the eruption site. The eruption cloud was rarely grounded near to the eruption site, as the large temperature gradient between the lava field and the ambient air encouraged the eruption cloud to remain aloft near-source. The near-field traverses were retrieved at 360–390 nm rather than the 319–325 nm used for the ScanDOAS retrievals. They were retrieved using DOASIS software and scripts available from [17]. A near-lava field scan was made on 21 January 2015 using the Avoscan system to drive the spectrometer [18].

The emission rate of SO_2 was successfully measured by DOAS on 33 days during the 181 days of the eruption. The ScanDOAS installations, ring-road MobileDOAS traverses, and near-source traverses measured the flux on 23, seven, and five days, respectively. On the two days when SO_2 flux was measured by more than one DOAS method, all flux measurements were averaged.

The uncertainty in the SO_2 flux measurement by the ScanDOAS method has been given to be 54% for "fair" conditions, meaning situations where spectroscopic errors, atmospheric scattering, uncertainties in wind, and measurement geometry are not ideal, but it is not raining or snowing and the cloud is not strongly meandering [10]. This estimation pertains to measurements with the standard instrument deployed in the NOVAC network. For the instruments used in this work, the spectroscopic error (including spectrometer noise, errors in reference cross sections, changes of instrument line-shape, and fitting errors) is considered to be similar to the standard one and less than 15%. While the signal to noise of a Maya2000 Pro is better than for a SD2000 spectrometer, the temperature sensitivity is larger. The measurement geometry error is kept low (estimated to be less than 10%) in our measurements since the data is carefully selected to minimize sampling errors. The cloud speed error is unknown, but good correspondence between forecasted and observed gas transport suggests that this uncertainty may be less than 15%. The cloud height error can be considered to be less than 20%. The most important source of uncertainty is the effect from atmospheric scattering. UV DOAS measurements of an optically thick cloud surrounded by a hazy environment are affected by dilution of the absorption signal caused by scattering of light before it reaches the cloud, leading to an underestimation of the flux, and by multiple-scattering within the cloud, leading to an overestimation of the flux [19]. In addition to these complexities, strong absorption by SO_2 causes a suppression of large optical paths, resulting in an underestimation of the flux. The bias from wind direction (Figure 3) suggests that the scattering effect of dilution of the absorption signal prior to reaching the cloud was significant and probably the dominant source of uncertainty for our measurements. This follows from the environment where the eruption took place being so dusty. The data filtering removes the ScanDOAS measurements most affected by atmospheric scattering. From analysis of the ring-road traverse measurements [9], an underestimation due to dilution of up to 40% was found in the distant cloud. For the near-vent measurements, the distance to the cloud was shorter, but the atmosphere was hazier and the cloud was more concentrated. We think these effects together result in a net underestimation of the flux that increases with increasing distance from the cloud and attribute a value of at least −40% to +10%. The total flux uncertainty in our scanning and near and far traverse DOAS measurements is estimated to be −50% to +30%.

2.2. Icelandic Environmental Agency Network

Prior to the eruption, air quality in Iceland was monitored in real-time by a network of 11 automatic stations operated by the Environment Agency of Iceland (EAI) [1,5]. The network measured the ground-level concentration of SO_2, H_2S, NO, NO_2, and particulate matter (PM_{10} and $PM_{2.5}$), located in areas exposed to pollution from anthropogenic sources including factories and aluminum smelters. The number of SO_2-monitoring stations was increased as the eruption progressed and 21 stations were operating at the end of the eruption. These were installed in communities around the country to monitor populated areas. The data were streamed in real-time to EAI and the data was made publicly-available on their web-site and at IMO.

In addition to this permanent network, hand-held personal sensors were distributed to local police. These were set up to activate an acoustic alarm if gas concentration thresholds were exceeded.

Measurements from the automatic stations and hand-held sensors, in conjunction with SO_2 ground-level concentration forecasts generated by IMO, were used by Icelandic Civil Protection and Emergency Management (NCIP DCP) to warn the public about unhealthy concentrations of gases and to advise them to stay indoors when high concentrations of SO_2 were detected or forecasted.

Monitoring data from the automatic stations were used for validating the SO_2 dispersal forecasts provided by IMO using the CALPUFF model [1].

2.3. Open-Path Fourier Transform IR

An open-path Fourier transform (OP-FTIR) spectrometer (MIDAC model M4401-S-E) was used successfully on seven days during the eruption. The OP-FTIR has been used intensively in Italy for monitoring gas emissions from Stromboli [20,21] and Etna [22], both between and during eruptions. Successful measurements have been made previously in Iceland during the 2010 eruption of Fimmvörðuháls [23] and the 2010 Eyjafjallajökull summit eruption [24]. During Holuhraun, gas compositions were measured with OP-FTIR by pointing the instrument directly towards the volcanic plume (vertically rising part of the volcanic cloud) or cloud with either lava or the sun as the IR source. The OP-FTIR spectra were analyzed using a forward model and non-linear fitting algorithm [21] after collecting the data. Each spectrum records the slant column amounts of gases contained in the atmosphere, in the volcanic cloud, or both. For example, H_2O vapor, typically the most abundant magmatic gas, is also abundant in the atmosphere, as is CO_2, while magmatic species such as SO_2, HCl, and HF only exist in trace amounts in the non-eruption atmosphere. The ratios of the column densities are equal to the molar ratio of the measured gases. A subset of the FTIR data has been previously published in [23], and model calculations based on a subset of the data have been previously published in [25].

2.4. MultiGAS

The Multi-component Gas Analysis system (MultiGAS) instrument was developed by the University of Palermo and INGV-Palermo and modified for use in Iceland [26–28]. It measures in situ (at 0.1 Hz rate) the concentrations of major volcanic gas species (H_2O, CO_2, SO_2, H_2S) in the atmosphere, by integrating (i) an infrared spectrometer for CO_2 (Gascard II, calibration range 0–10,000 ppmv (0–1%); accuracy ±2%, resolution, 3 ppmv); (ii) two specific electrochemical sensors for the measurement of SO_2 (CityTechnology, sensor type 3ST/F, calibration range, 0–50 ppmv, accuracy, ±5%, resolution, 0.1 ppmv) and H_2S (CityTechnology, sensor type 2E, calibration range, 0–50 ppmv, accuracy, ±5%, resolution, 0.1 ppmv); and (iii) temperature, pressure, and relative humidity (Galltec sensor, measuring range, 0–100% Rh, accuracy, ±2%) sensors for the calculation of H_2O concentrations. Gas ratios measured with the MultiGAS are calculated using the Ratiocalc software [29]. The collected data are de-trended to adjust the baselines to zero to correct for instrument drift (largely due to increasing temperature in the instrument as it operates), and then ratios between species are calculated. Acceptable ratios were measured for at least five minutes when the gas concentration was at least 0.8 ppm and have an R^2 value greater than 0.5.

During the Holuhraun eruption, one MultiGAS measurement was made inside the active crater on the day between the first minor lava effusion and the main eruption (Figure 2). Measurements were made on the edge of the advancing lava field on four days during the eruption. In January 2015, a continuous monitoring MultiGAS was installed at Þorvaldshraun, 10 km from the main eruption vent, and measurements were obtained of the aged, dilute cloud on eight days until the end of the eruption. A subset of this data has been previously published in [1] and model calculations based on a subset of the MultiGAS data have been previously published in [25].

2.5. Filter Pack

Filter pack samples to collect acidic gases, primarily SO_2, HF, and HCl, in the near-source eruption cloud were collected on four days during the eruption (Figure 2). Three days were previously published in [30] and one day was previously published in [5]. Filter pack samples have also been reported in [31], independent from the official monitoring of the eruption. All samples were drawn through the filter pack apparatus by a pump and acidic gases were collected on base-impregnated Whatman filters. All samples were leached in deionized water in a laboratory and the solutions were analyzed for anion concentration using ion chromatography. All samples were blank corrected by treating filters in the same way as for samples followed by chemical analysis. The measurements included here are those where the F measured on the base-impregnated filter was ≥ 5 µg/m^3; two previously-published samples are excluded due to the too low concentration of F collected. The filter pack samples collected by [5] also included a PM filter (Millipore, 47 mm, AAWP, pore size 0.8 µm) in series before the gas filters. The mass of the PM was calculated by extracting elements from the PM using sequential leaching. The sum of the mass of all analyzed elements is calculated to represent the mass of the PM.

2.6. OPC

An Optical Particle Counter/Sizer (OPC/OPS) (OPS 3330, TSI Inc.) was installed at Möðrudalur, 72 km from the main eruption vent, from October 2014 until after the end of the eruption (Figure 2). Particle counts in 16 size bins from 0.3 to 10 µm provided a sum every five minutes, and can be used to provide different temporal means or maxima. The OPC was deployed to monitor the concentration and size distribution of any ash or other particulate matter produced by the eruption, with the maximum detectable particles being 10 m, so coarser ash would not be measured by this instrument.

A research campaign, independent from the official monitoring of the eruption, launched a balloon 9 km from the main crater to fly through the Holuhraun eruption cloud carrying a Light Optical Aerosol Counter (LOAC) [32].

2.7. Weather Radar

One permanent C-band radar at Fljótsdalsheiði (86 km from the eruption site), close to the city of Egilsstaðir, operated throughout the eruption (Figure 2). Two mobile X-band radars were moved close to their eruption targets within the first week of the eruption. One X-band radar was located in Vaðalda with a clear view of the eruptive fissure (20 km distance) and the other one was located in Hágöngulón, ready to detect an eruption cloud originating from the central Bárðarbunga volcano (36 km distance). There were persistent technical issues involved with operating the mobile radars in the highlands, including frequent radio communication disruptions, fuel consumption, sand storms harming the generators, and cooling and overheating problems. The C-band radar, part of the continuous weather monitoring of Iceland, operated throughout the entire eruption.

2.8. Web Cam

A web camera located in Kverkfjöll (Figure 2), about 25 km south of the main eruption vent, provided images every 10 min. The camera image area was scaled using seven mountains visible on the images with elevations from 741–1682 m asl. The eruption cloud top seen in the images was transformed to the height profile above sea level vs. distance along the cloud assuming no lens distortion and that the cloud was transported in the direction 80° east. The eruption cloud was detectable by this camera during the first 19 km from the eruption site. The maximum eruption cloud height often was detected a couple of kilometers down-wind from the eruption site. A subset of this data has been previously published in [33].

2.9. NicAIR II

A NicAIR II multi spectral infrared imaging camera was installed at Vaðalda, approximately 20 km from the main eruption vent, from 20 November 2014 until the end of the eruption (Figure 2). Gases and particles emit and absorb thermal radiation, which is detected by a microbolometer array fitted in the NicAIR infrared camera [34,35]. Radiation counts are then converted to radiance, and then to brightness temperatures (Kelvin) through a pre-determined instrument calibration scheme [34]. The NicAIR II camera operates in the 8–13 µm region of the infrared spectrum. A filter wheel contains one broadband (8–13 µm) and three narrow band filters centered at 8.62 µm, 10.0 µm, and 10.87 µm, selected to be sensitive to specific signatures in the cloud. Approximately one composite data image was recorded per minute due to the number of filters in use. The 8.62 and 10.87 µm channels were processed to retrieve SO_2 and the 10.0 µm channel was used to determine the cloud temperature [35]. Similar instruments have been used previously on eruption clouds at Karymsky and Stromboli volcanoes to monitor SO_2 and ash [36,37]. Good quality data was obtained on seven days in November and December 2014, as other days were hindered due to hardware problems and meteorological challenges (e.g., clouds obscuring the view, icing on the camera window).

Brightness temperature data were post-processed into SO_2-sensitive images, including static and dynamic parameters. Dynamic parameters were assessed at least every 10 min. The cloud temperature was obtained from the 10.0 µm wavelength image, extracted using Fits Liberator software in an opaque section of the cloud. The ground level of the image was identified and all horizontal rows of data below this level were set to SO_2 = null to remove noise and interference coming from the lava field with its high brightness temperatures. Background infrared radiation levels were determined by taking vertical profiles where the sky was clear, i.e., there was no coverage of the sky by the volcanic cloud or meteorological clouds. Whenever there was no suitable clear section, if low-lying background clouds were responsible, the horizontal rows including the low-lying clouds were set to SO_2 = null, allowing processing of the image, otherwise the image was excluded. Parameters which remained constant include the distance from the camera to the main eruption vent and the angle of the camera above the horizontal. These two static parameters were used to calculate the dimensions of the NicAIR field of view (FOV).

Both cloud height and SO_2 amount were calculated from the good quality images; however, the SO_2 retrieval was severely impacted by the high concentrations of H_2O in the eruption cloud. This caused the region closest to the eruption vent, where SO_2 concentrations are expected to be greatest, to be opaque. SO_2 retrievals were therefore significant underestimations and are not reported here, but can be found in [38]. Cloud heights were calculated following a revised methodology from [39] and [40]. The 10.0 µm channel data was processed with pixel heights as defined by the camera geometry and distance from the camera to eruption vent. The apparent cloud top height in each image was then identified by the thermal contrast at the cloud's leading edge (Figure 4). Only data where the thermal contrast of the cloud's leading edge was clearly visible were used to estimate the apparent cloud heights. The recorded apparent cloud top height was taken at the edge of the FOV, at which point the cloud was furthest from the vent and at the maximum height recorded by the NicAIR camera. Due to large variations in cloud heights, an average hourly apparent cloud height was calculated for each hour of suitable data.

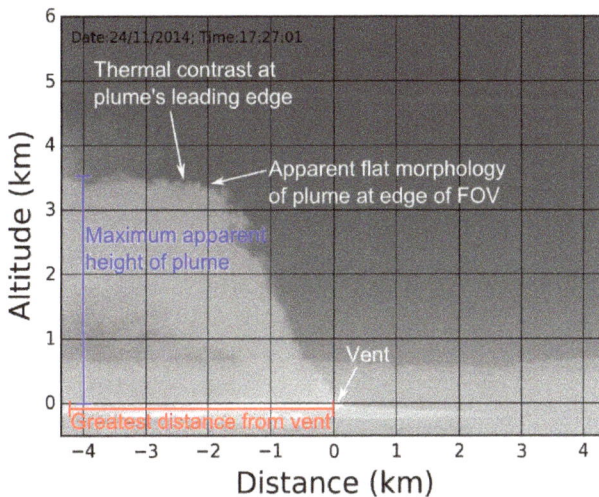

Figure 4. Apparent cloud top height measured from NicAIR II 10 µm images.

The apparent cloud height derived from the NicAIR image is not equivalent to the actual cloud top altitude. Due to the viewing geometry, the actual top of the cloud was hidden from the camera's view and the apparent cloud top often appeared flat in morphology as a result of this (Figure 5). In addition, wind transports the cloud away from the eruption vent and the altitude scale at the vent is not equivalent to the altitude scale at the cloud's location of maximum height. In order to calculate the actual altitude of the cloud top, the location of the cloud in relation to the camera was required so that a new altitude scale derived from the camera geometry and the distance between the camera and the cloud location could be calculated.

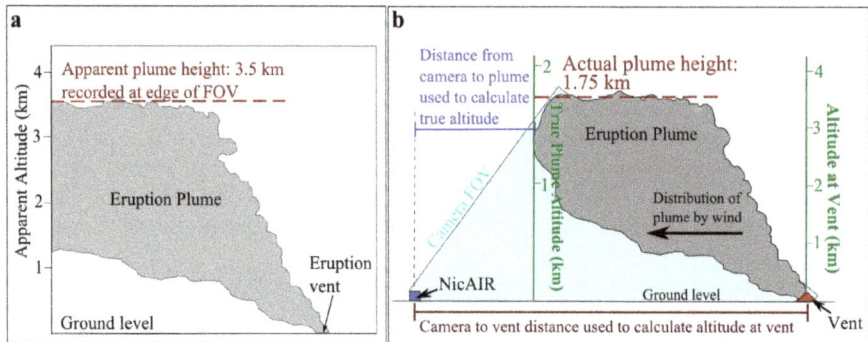

Figure 5. Actual cloud height derived from apparent cloud height and cloud location. (**a**) Schematic field of view from the perspective of the camera showing apparent plume height and (**b**) schematic side-view illustrating the effect of transport towards the camera on actual versus apparent plume height.

In order to determine the location of the cloud in relation to the camera, wind dispersal of the cloud was analyzed from meteorological data. Meteorological conditions at the approximate height of the eruption cloud above sea level at the eruption site were extracted from the HARMONIE model and radiosonde data. The HARMONIE meteorological model predicted weather conditions for the eruption site every hour, providing detailed modeled data. Radiosonde data from the Egilsstaðir weather station (located ~120 km east north east of the eruption site) provided measurements but at a low temporal resolution (measurements once every 24 h). Wind directions from both were used, depending on the visual analysis of the cloud distribution in the NicAIR data.

For each hour of averaged apparent cloud height, the wind direction was mapped out in relation to the bearing of the camera line of sight to the vent, and the angle between the two directions was calculated. The NicAIR camera FOV was considered as a cone, where the cloud at the edge of the FOV may be considered as being located somewhere along the edge of the FOV cone. By combining the wind direction and the bearing of the camera line of sight, the approximate location of the apparent cloud height was estimated. The distance between the camera and the apparent cloud height location was then determined and new pixel heights were calculated using this new distance. The new pixel heights were combined with the average hourly apparent cloud height to give the actual cloud height as recorded by the NicAIR camera (Figure 5b).

2.10. Observations from Ground and Aircraft

Observations were made throughout the eruption by scientists working at the eruption site and scientists making airborne observations from the TF-FMS and TF-SIF aircrafts, owned by Isavia and the Icelandic Coast Guard, respectively. Visual observations of eruption cloud height made by scientists in the field were intended to be the maximum eruption cloud height. These ground-based observations were recorded in their field notes and/or were called into IMO. The airborne observations were called into IMO and stored in its database.

3. Results

3.1. Eruption Cloud Composition

3.1.1. Gases

The average daily value of SO_2 flux measured by DOAS is in Table 1, and the time series of the measurements throughout the eruption, linearly interpolated between measurement dates, is shown in Figure 6. The total over the duration of the eruption is calculated to be 9.6 Mt SO_2 with an uncertainty of −50−+30%, or 6.7–14.3 Mt SO_2. The average emission rate during the eruption is 610 kg/s, with an uncertainty of 430–920 kg/s. The maximum daily average emission rate during the eruption was in excess of 2100 kg/s and the maximum high-quality scan included in the data set was in excess of 5500 kg/s. The overall trend of the eruption was a decrease in emissions of SO_2 as the eruption progressed (September 2014–February 2015), with the exception of November 2014, when higher emission rates alternated with lower ones. The total of the emissions over the duration of the eruption is less than, but within the uncertainty of, the 11.8 ± 4 Mt previously published in [1]. This previous publication only included the ring-road traverses (marked as Ring in Table 1).

The post-eruptive outgassing of the lava field remained above detection limits by the ScanDOAS instruments for three months following the end of the eruption. During these three months, the average post-eruption outgassing of SO_2 was 3 ± 1.9 kg/s. The post-eruptive degassing, interpolated for the three months, equals 24 kt SO_2 [14], which is less than 1% of the degassing during the eruption.

Table 1. Daily average DOAS measurements of SO_2 flux, distinguished by measurement technique and if the measurement occurred during or after the eruption.

Date (DDMMYY)	SO_2 Flux (kg/s)	Technique	Syn- or Post-Eruption
02/09/2014	520 ± 100	Scan	Syn
08/09/2014	1330 ± 440	Scan	Syn
10/09/2014	1050 ± 370	Scan	Syn
11/09/2014	1140 ± 230	Scan	Syn
12/09/2014	1290 ± 290	Scan	Syn
13/09/2014	1120 ± 220	Scan	Syn
14/09/2014	610 ± 180	Scan	Syn
18/09/2014	820 ± 310	Near	Syn
19/09/2014	250 ± 100	Near	Syn
20/09/2014	680 ± 260	Near	Syn
21/09/2014	2170 ± 1720	Near; Ring	Syn
22/09/2014	1130 ± 490	Scan	Syn
24/09/2014	710 ± 210	Scan	Syn
25/09/2014	960 ± 320	Scan	Syn
30/09/2014	1200 ± 230	Scan	Syn
01/10/2014	1180 ± 240	Scan	Syn
02/10/2014	840 ± 460	Scan	Syn
06/10/2014	890 ± 340	Ring	Syn
05/11/2014	1450 ± 550	Ring	Syn
18/11/2014	220 ± 30	Scan	Syn
21/11/2014	1070 ± 410	Ring	Syn
25/11/2014	250 ± 50	Scan	Syn
26/11/2014	990 ± 1810	Scan	Syn
02/12/2014	300 ± 110	Scan	Syn
21/01/2015	250 ± 50	Scan; Near	Syn
22/01/2015	410 ± 70	Scan	Syn
25/01/2015	520 ± 120	Scan	Syn
27/01/2015	320 ± 100	Scan	Syn
30/01/2015	40 ± 20	Ring	Syn
31/01/2015	410 ± 160	Ring	Syn

Table 1. *Cont.*

Date (DDMMYY)	SO$_2$ Flux (kg/s)	Technique	Syn- or Post-Eruption
04/02/2015	240 ± 90	Ring	Syn
06/02/2015	220 ± 30	Scan	Syn
16/02/2015	90 ± 30	Scan	Syn
07/03/2015	2 ± 0.5	Scan	Post
13/03/2015	1 ± 0.2	Scan	Post
21/03/2015	2 ± 0.5	Scan	Post
22/03/2015	2 ± 0.5	Scan	Post
26/03/2015	2 ± 0.5	Scan	Post
27/03/2015	2 ± 0.5	Scan	Post
28/03/2015	3 ± 0.7	Scan	Post
15/04/2015	4 ± 1	Scan	Post
22/05/2015	4 ± 1	Scan	Post
23/05/2015	4 ± 1	Scan	Post
24/05/2015	5 ± 1	Scan	Post

Scan = ScanDOAS; Near = near-lava traverses and near-lava scan; Ring = ring-road traverses.

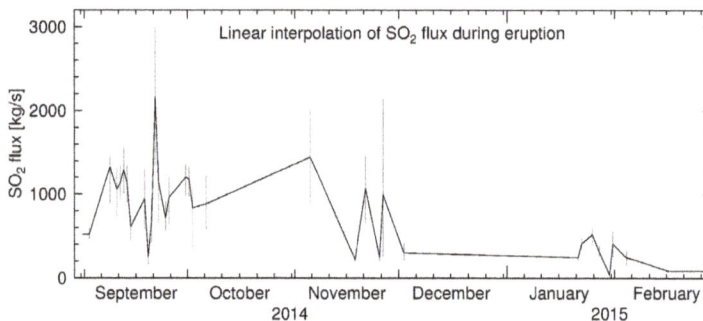

Figure 6. Time series of the daily average SO$_2$ flux measured by DOAS connected with a solid black line showing the linear interpolation of the data between measurement dates. The gray vertical bars show the uncertainty for each day.

The SO$_2$ peaks measured down-wind by the car-mounted in situ instrument and the atmospheric column amount measured by the MobileDOAS were sometimes concurrent and sometimes shifted from one another (Figure 7). These measurements are shown as raw data, where the baseline of the measurements has not been shifted to zero. Sometimes the traverses needed to be extremely long, up to two hours driving time, to obtain a measurement of background clean air (undetectable SO$_2$) before and after both instruments detected SO$_2$. Due to the short daylight in winter, only three hours of daylight on the shortest day of the year in the north of Iceland, some attempted traverses were not completed.

The changes in the gas ratios measured by OP-FTIR, MultiGAS, and filter pack over the duration of the eruption are seen in Figure 8. Exponential trends are only seen for CO$_2$/SO$_2$ (R^2 = 0.43), H$_2$O/CO$_2$ (R^2 = 0.47), and SO$_2$/H$_2$S (R^2 = 0.78). For CO$_2$ we utilize this trend to calculate the emission of CO$_2$ relative to SO$_2$. Using the exponential change for the SO$_2$/H$_2$S ratio is not appropriate, because while there is an excellent fit, this fit is based on only four measurements which were all made at the beginning of the eruption. The extrapolation to the rest of the eruption is not grounded in measurements. In the absence of a trend, but with the high variance observed in the gas ratios, for H$_2$O, HCl, and HF, as well as for H$_2$S, we calculate the gas ratios using the 25–75% percentiles of the data (Table 2). We then calculate the emissions over the duration of the eruption of the non-SO$_2$ gases based on this range of ratios and the emission rate of SO$_2$ extrapolated from the DOAS measurements (Table 2).

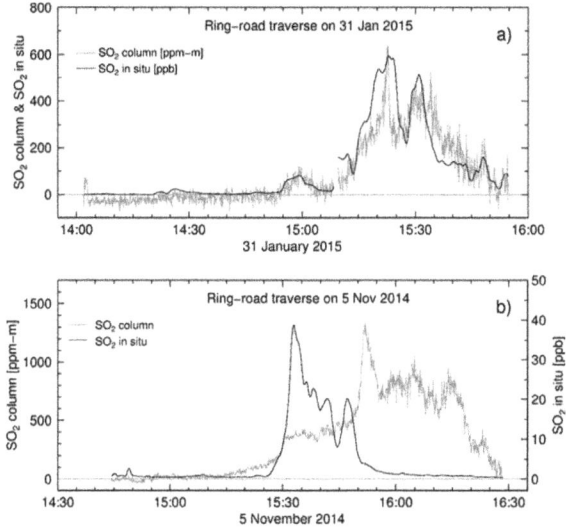

Figure 7. Examples of ring-road traverses with concurrent MobileDOAS SO$_2$ column and in situ SO$_2$ measurements. (**a**) 31/01/15 with peaks corresponding very closely and (**b**) 05/11/14 with incongruent peaks.

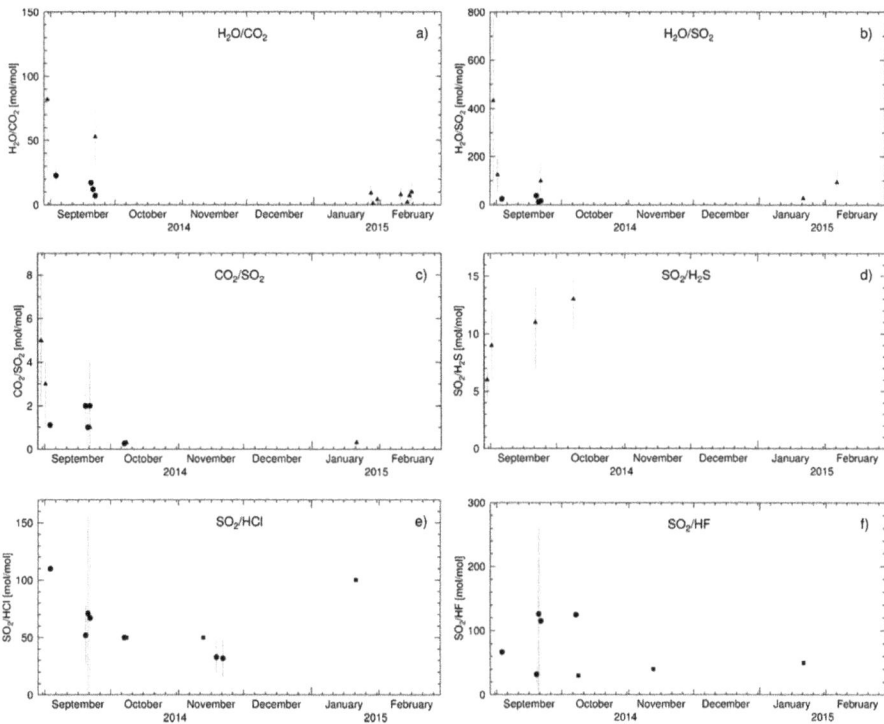

Figure 8. The time series of daily averaged gas ratios (**a**) H$_2$O/CO$_2$; (**b**) H$_2$O/SO$_2$; (**c**) CO$_2$/SO$_2$; (**d**) SO$_2$/H$_2$S; (**e**) SO$_2$/HCl; (**f**) SO$_2$/HF. Measurements made by OP-FTIR, MultiGAS, and Filter Pack are shown as circles, triangles, and squares, respectively. Vertical gray bars indicate uncertainty.

Table 2. Representative gas ratios over the duration of the eruption and the calculated emission of each gas relative to SO_2 based on the DOAS flux rate of 9.6 Mt SO_2 and in parentheses with the uncertainty range of 6.7–14.3 Mt SO_2.

Gas Ratios (Mol/Mol)	25%–75% Percentiles	Emissions (Mt)
H_2O/CO_2	5–16	
H_2O/SO_2	18–98	H_2O: 49–263 (34–394)
CO_2/SO_2	0.3–2	CO_2: 5.1 (3.6–7.6)
SO_2/H_2S	9–13	H_2S: 0.4–0.6 (0.3–0.9)
SO_2/HCl	46–79	HCl: 0.07–0.1 (0.05–0.2)
SO_2/HF	34–122	HF: 0.02–0.09 (0.02–0.1)

3.1.2. Particles

Minor tephra production and fall out, including Pele´s hair, was reported by scientists in the field during the first week of the eruption.

The monthly mean OPC particle count measurements during the Holuhraun eruption, October 2014–February 2015, ranged from 6.12×10^4–1.17×10^5 cm^{-3} (14–69 µg/m^3), with the greatest measured in December 2014 and the least in October 2014. The maximum instantaneous concentration was recorded in December 2014 with 4.52×10^6 cm^{-3} (162 µg/m^3). The highest concentrations were relatively short lived and daily averages remained quite consistent. In the five months following the eruption, March–July 2015, the monthly means were very similar to the values measured during the eruption (4.81×10^4–7.86×10^4 cm^{-3}). Comparing the measurements made during and after the eruption, there was no increase in the strength of individual maxima or the frequency of maxima.

All filter pack samples except one were collected on the ground. The filter pack samples showed very low masses of particles, except for the airborne sample collected in the eruption cloud. With the exception of this one sample, the mass of particles remained relatively constant (57–191 µg/m^3), regardless of the mass of SO_2 (5–4300 µg/m^3) collected on the gas filters. The particle concentrations collected on the ground-based filter packs were slightly enhanced relative to the particle concentrations measured by the continuous OPC both during and after the eruption. For the airborne sample, the mass ratio of SO_2/particles was 2.

The radar network intermittently detected the eruption cloud. This manifested in the radar data as a cloud with increased values of reflectivity (displayed as maximum dBZ) up to 30 dBZ that persisted close to the eruption site (Figure 9). Meteorological clouds were also detected by the radar during the eruption period as clusters of higher reflectivity that were dynamically transported by winds. An increase in reflectivity can be due to droplets, such as are found in meteorological clouds or eruption-induced droplets [41], and particles, such as ice, volcanic ash [42,43], or suspended dust [44]. The radar-detected eruption cloud was often enhanced by precipitation, and sometimes formed a precipitating cloud in the absence of other weather clouds in the region. When the eruption cloud was enhanced by transitory precipitating clouds, these clouds would move above the eruption site and a cloud would develop over the eruption site, and this cloud would continue to be visible after the meteorological clouds had moved past. The eruption cloud was most consistently observed by the C-band radar at Fljótsdalsheiði. This often showed the cloud in the layer of the atmosphere closest to the ground, which at the distance from the radar to the eruption site, means within the lowest 1 km of the atmosphere. The cloud was most frequently observed during low wind speeds and when there was a change in wind direction, and often in the morning. A persistent cloud above the eruption site remained detectable by the radar after the end of the eruption (Figure 9d,e), when the lava field continued to give off heat and to outgas. The reflectivity of the radar-detectable eruption cloud was greatest at the start of the eruption, and became weaker later in the eruption and in the post-eruptive period.

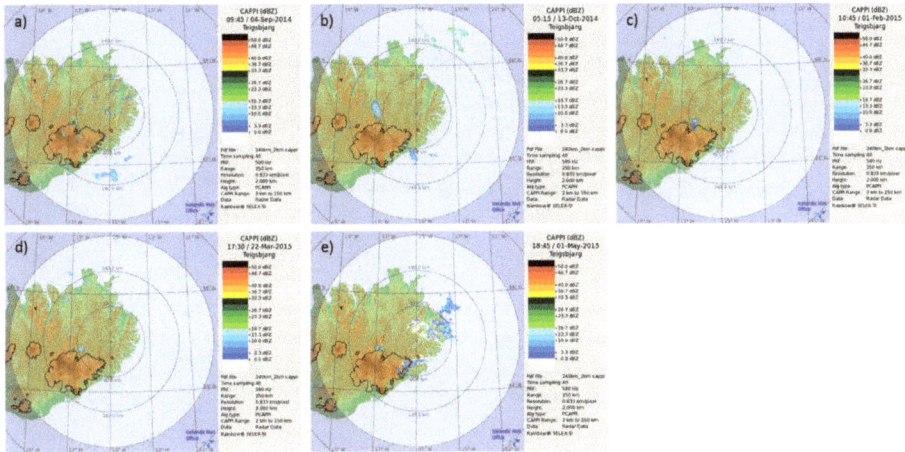

Figure 9. The radar-detectable cloud above the eruption site (indicated in red) detected by the C-band radar at Fljótsdalsheiði. (**a**) 04/09/14; (**b**) 13/10/14; (**c**) 01/02/15; (**d**) 22/03/15; (**e**) 01/05/15.

3.2. Cloud Height

The daily average, minimum, and maximum of the height of the top of the eruption cloud measured by ground- and aircraft-based observations, web cam, ScanDOAS, MobileDOAS, and NICAIR II IR camera are found in Table 3. The ScanDOAS and MobileDOAS approaches, however, do not provide the height of the top of the cloud, but rather the height of the center of mass of the cloud, so all days including these techniques will be under-detections of the top of cloud height. In general, the start of the eruption was the strongest with the highest cloud heights, and then for the duration of the eruption, until the end, the height varied mainly between 1–3 km agl. The variance of the daily averages is high. A diurnal variation was evident with higher maximum cloud heights in the afternoons, indicating that atmospheric stability was influencing the volcanic cloud [45]. On 22 January 2015, a balloon was launched carrying a miniature optical particle counter [32]. The balloon-borne particle counter found the top of the eruption cloud to be between 2.7 and 3.1 km, which is in excellent agreement with the average value of 2.8 km made by field observations on this day.

Table 3. Observations of height of eruption cloud made by ground- and aircraft-based observations, web camera, ScanDOAS, MobileDOAS, and NICAIR II IR camera.

Date (DDYYMM)	Daily Average Top of Cloud Height (km·AGL)	Daily Min Top of Cloud Height (km·AGL)	Daily Max Top of Cloud Height (km·AGL)	Technique
29/08/2014	1.0			Cam
01/09/2014	4.5			Flight
03/09/2014	5.3	4.5	6.0	Flight
04/09/2014	4.1	1.7	5.5	Field; DOAS
05/09/2014	3.8			Field
06/09/2014	3.1			Field
07/09/2014	3.5			Field
08/09/2014	3.0	1.9	4.3	DOAS
09/09/2014	2.1	1.3	3.7	DOAS
10/09/2014	1.6	1.2	2.5	DOAS
11/09/2014	2.0	1.1	3.1	Cam; DOAS
12/09/2014	1.8	1.3	2.7	Cam; DOAS

Table 3. *Cont.*

Date (DDYYMM)	Daily Average Top of Cloud Height (km·AGL)	Daily Min Top of Cloud Height (km·AGL)	Daily Max Top of Cloud Height (km·AGL)	Technique
13/09/2014	2.6	1.6	3.5	Cam; Flight; Field; DOAS
14/09/2014	2.3	2.0	2.5	Cam; Field; DOAS
15/09/2014	1.6			Cam
16/09/2014	2.3			Cam
17/09/2014	2.0			Cam
19/09/2014	3.0			Field
20/09/2014	2.3			Cam; Field
21/09/2014	1.6	1.0	2.0	Cam; Field; Mobile
22/09/2014	2.5	2.1	3.0	Cam; Flight; Field
23/09/2014	3.9	3.5	4.2	Field
24/09/2014	2.1	1.9	2.8	Field; DOAS
26/09/2014	2.0			Field
27/09/2014	2.5			Field
28/09/2014	3.0			Field
02/10/2014	1.7			Field
06/10/2014	1.2			Mobile
08/10/2014	1.0			Field
17/10/2014	2.6	2.4	3.0	Flight; Field
21/10/2014	1.3			Field
22/10/2014	1.2			Field
23/10/2014	1.2			Field
28/10/2014	2.0			Field
29/10/2014	2.6	1.8	3.5	Field
30/10/2014	2.7	2.3	2.9	Field
04/11/2014	2.9	2.8	3.0	Flight
05/11/2014	1.4	1.0	1.5	Field; Mobile
11/11/2014	2.2	1.3	3.0	Field
13/11/2014	3.5			Field
14/11/2014	2.8	2.5	3.0	Flight
18/11/2014	1.7	0.4	2.3	Flight; Field
19/11/2014	1.3	1.1	1.5	Field
20/11/2014	1.0	0.5	1.4	Field
21/11/2014	0.8	0.5	1.5	Field; Mobile
23/11/2014	2.1	1.9	2.2	NICAIRII
24/11/2014	0.5			NICAIRII
25/11/2014	2.6	0.8	4.0	Field; NICAIRII
26/11/2014	1.8	0.4	3.1	Field; NICAIRII
27/11/2014	1.9	0.9	3.1	Field
30/11/2014	2.5	2.4	2.6	NICAIRII
01/12/2014	1.0			NICAIRII
02/12/2014	2.5			Field
03/12/2014	1.3			Field
04/12/2014	2.1	1.5	2.7	Flight; Field
05/12/2014	2.0	1.6	2.3	Field
09/12/2014	0.5			NICAIRII
30/12/2014	2.6	2.5	2.7	Flight
10/01/2015	2.5			Flight
21/01/2015	3.5	3.0	4.2	Mobile
22/01/2015	2.8	2.8	2.8	Field
27/01/2015	1.3			Field
28/01/2015	1.4	1.3	1.5	Field
29/01/2015	2.8	2.3	3.2	Field
30/01/2015	0.7	0.5	1.0	Mobile
31/01/2015	1.7	1.5	2.0	Mobile
03/02/2015	1.5			Flight
04/02/2015	0.9	0.7	1.0	Mobile
19/02/2015	0.9	0.7	1.0	Field

Cam = Webcam; Flight = Aircraft observations; Field = Ground-based observations; DOAS = two ScanDOAS triangulation; Mobile = MobileDOAS peaks with HARMONIE winds; NICAIRII = NICAIRII IR Camera.

4. Discussion

The SO_2 flux from Holuhraun was enormous at the start of the eruption (an average during September 2014 of 1007 kg/s) and diminished over the course of the eruption as shown by our ground-based measurements. An exponential decay curve fit to the SO_2 flux measurements gives an $R^2 = 0.38$, as there are days with high emissions that do not follow a simple decay. An exponential decay has been identified as characterizing the rate of change of the Bárðarbunga caldera volume, which was directly associated with the mass eruption rate at Holuhraun during the six months of the eruption [25]. The rate of lava effusion, based on the thermal emissivity of the lava field, similarly exponentially decayed like the caldera volume change, until a rapid cessation during the last month of the eruption [46].

The gas flux can deviate from the eruption vigor (such as lava effusion rate) if there is a change in the physical processes responsible for releasing SO_2 into the atmosphere and/or if there is a change in the composition of the magma. Few of the atmospheric measurements of SO_2 ratio relative to other gas species have a statistically significant trend over the course of the eruption, although the CO_2/SO_2 ratio does exponentially decay in a coherent, albeit statistically weak, fashion. As there is no systematic change in melt inclusion compositions measured over the course of the eruption [47], physical processes related to SO_2 degassing are therefore considered a likely contributor to the discrepancy between the SO_2 flux rate and other qualities of the eruption such as the lava effusion rate and thermal emissivity.

SO_2 measured in the cloud was released from magma as it rose through the conduit and erupted at the vent; from non-erupted magma; and from the lava flow during and after emplacement [14]. The degassing of the erupted magma is expected to follow the same curve as the lava effusion rate; gas emitted from non-erupted magma and the lava flow would contribute to deviations from this trend.

Fracturing and cracking of the lava facilitates gas release, and can increase outgassing from the lava flow [14]. The lava outgassing can persist for months after the lava is emplaced, and outgassing from lava can be episodic, affected by variable rates of lava fractionation. In the first phase of the eruption, 31 August–12 October, 2014, the lava dynamically changed between 'a'ā and slabby and rubbly pāhoehoe lava types [48]. These continuous changes in lava texture, with the more fractured lava facilitating the release of gases, are potentially responsible for some of the high variance in the SO_2 flux data set. The second phase of the eruption, 12 October–30 November, 2014, was distinguished by the presence of a continually replenished lava lake, which is considered to be the supply for the lava flow [48]. There are six measurements in November and early December, alternating with high emissions of SO_2 (average 1169 kg/s on 5, 21, and 26 November) and low emissions (average 254 kg/s on 18, 25 November and 2 December), when there was no corresponding changes in the lava effusion rate or eruption intensity. The high values could be due to the contribution of outgassing from the lava lake overprinting the decay of emissions originating from the erupted magma. The SO_2 fluxes are therefore not reflecting solely the deep magmatic system: they also are affected by surface processes.

The total amount of SO_2 emitted by the eruption reported here is 9.6 Mt, which is less than, but within the uncertainty, of the 11.8 ± 4 Mt previously reported [1]. The very few measurements made in winter make the interpolated sum of emissions over the entire eruption very sensitive to two measurement days. The very low value of the last measurement day in December 2014 connected with the very low value of the first measurement day at the end of January makes the sum over the six months significantly lower than it would be if these had been higher values. The ScanDOAS measurements and near-lava traverses were often saturated and there were significant impacts, which also affected the long-distance traverses, from the scattering of light within and outside of the eruption cloud [9]. The daily-average fluxes, and therefore the sum over the eruption, should be interpreted as under-detections, with the near-source scans and traverses under-detecting the most. The ring-road traverses measured a more dilute cloud, however they were too infrequent to capture the true variance in emissions during the eruption. There were huge variations in SO_2 flux throughout each day and between different days. A traverse is a snapshot measurement and therefore not necessarily representative of the average for a longer time range. We therefore consider the full data set, with its far greater number of flux measurements, to provide a more accurate basis for calculating the sum of emissions over the duration of the six-month long eruption, rather than relying on a smaller subset of data including only the ring-road traverses. Near-lava traverse measurements were made when the meteorological conditions were optimal. Labor-intensive data processing was, however, necessary to account for the optical-thickness and scattering. The improved automation of DOAS in Iceland during the dark winter months, so as to minimize the time when UV measurements are impossible, requires the development of improved data processing techniques.

During the first months of the eruption, the emissions rates were greatest, the eruption cloud was highest in elevation, and satellite-borne instruments were most sensitive to the emissions. We therefore use measurements from this time to compare the ground-based DOAS measurements with satellite-derived SO_2 fluxes. Infrared Atmospheric Sounding Interferometer (IASI) and Ozone Monitoring Instrument (OMI) SO_2 mass burdens were integrated with simulations made with the Numerical Atmospheric-dispersion Modeling Environment (NAME) in [49]. This approach provides a total emission during September of 2.0 ± 0.6 Mt SO_2, while our ground-based measurements, linearly interpolated between days without measurements, find a total emission of 2.5 (1.8–3.8) Mt SO_2 during September. Thermal infrared (TIR) data from MSG-SEVIRI is used in [31] to calculate a time-averaged SO_2 mass flux. They report a total emission for 01 September–25 November of 8.9 ± 0.3 Mt SO_2, while our ground-based measurement approach finds 7.3 (5.1–11.0) Mt SO_2 over this same time period. Both the satellite and ground-based approaches have explanations for why the measurements should be viewed as minimum values, and all agree within the uncertainty of each approach. Only the ground-based DOAS instruments were able to make measurements of SO_2 flux throughout the eruption and in the post-eruptive outgassing period.

The ground-based measurements of the SO_2 flux, despite the significant temporal gaps, were important for the initialization of the gas dispersion model used for real-time forecasts and warnings during the eruption. As we improve our measurement and data processing techniques, this data will be used ever more reliably for this important mitigation tool.

The 9.6 Mt of SO_2 emitted during six months of eruption are extraordinary. Since 1978, UV satellites have been used to quantify SO_2 emissions from volcanic eruptions [50]. Only the explosive dacitic eruption of Pinatubo in 1991 released more SO_2 than this eruption on an annual basis. The Holuhraun eruption is the effusive eruption with the highest emissions of SO_2 in this annually-based record. Hawaii Island's Kilauea Volcano has erupted continuously since January 1983, with sporadic eruptive behavior preceding this. USGS measurements show a sum over the period 1979–1997 of 9.45–9.93 Mt SO_2 [51], approximately what Holuhraun emitted in six months. The Kilauea eruption has emitted more SO_2 than the Holuhraun eruption over a much longer time. Within Iceland, a total of 0.06 Mt of anthropogenic SO_2 was emitted in 2015 [52]. The emissions of CO_2 from Holuhraun are also substantial. A total of 5.1 Mt CO_2 is calculated to have been emitted during the eruption, while within Iceland in 2015, a total of 3 Mt of anthropogenic CO_2 was emitted [52].

The SO_2/HF molar ratio measured in the Holuhraun eruption cloud was 109–392. This is quite high: most SO_2/HF ratios measured at volcanoes around the world exhibit much lower ratios (i.e., are richer in HF). Many papers, including [53–62], report values significantly less than those measured during the Holuhraun eruption with the exception of Kilauea Pu'O with an SO_2/HF molar ratio of 108 and Poás volcano with 190–greater than 200 (important to note that Poás is in a very different setting than Bárðarbunga). Despite the prolific gas-rich nature of the eruption, precipitation samples collected around the country show that the majority of the Cl in the samples came from sea spray as opposed to the volcanic eruption cloud [30]. This was a gas-rich but halogen-poor eruption.

The H_2O/CO_2 and H_2O/SO_2 ratios tend to show higher H_2O content in the MultiGAS measurements compared with the OP-FTIR measurements. There are too few measurements to study this closely, but it is possible that there is additional meteoric water in the grounded eruption cloud measured by the MultiGAS compared with the younger plume measured by the OP-FTIR using the lava as the IR source.

The SO$_2$ column amounts measured by the MobileDOAS system during the ring-road traverses and the in situ SO$_2$ measurements were sometimes congruent and sometimes not (Figure 7). We identified two factors that had the largest impact on this congruence: (1) the surface concentration of gases, as measured with the in situ instrument, was not always reflecting the current state of the transporting eruption cloud. SO$_2$ was sometimes observed by the car-mounted in situ SO$_2$ instrument and by the hand-held personal sensors carried by local police to accumulate in valleys, particularly overnight, and particularly when winds were weak. It sometimes took hours of stronger winds to flush the older gases out of a region, meaning that the air quality was not always reflective of the location of the young eruption cloud; (2) The two kinds of measurements were most congruent when the elevated eruption cloud had smaller SO$_2$ column amounts and the surface SO$_2$ concentrations were strong. Weaker emissions were likely lofted to lower elevations than stronger emissions (the cloud height data is insufficiently resolved to definitively answer this). The surface measurements and the remotely detected column concentrations measured by DOAS were most likely measuring the same, coherent low-level eruption cloud. The DOAS measurements made under these conditions would have been impacted the least from atmospheric dilution.

The balloon-borne LOAC reported in [32] did not start collecting data until the instrument had reached about 1.7 km elevation. They found that in clean air, beneath and above the eruption cloud, the background number of particles was 5–10 cm^{-3}, and that this was enhanced by a factor of 10–100 for a particle count of 100–500 cm^{-3} within the eruption cloud. The values are not directly comparable with the ground-based OPC measurements as the flow rate (2 L/min vs 1 L/min) and sampling interval (10 s versus 5 min) are different for the LOAC and OPC, respectively. When these differences are accounted for, the converted monthly averages measured by the OPC (4080–7800 cm^{-3}) are 10–80 times greater than the number of particles measured in the eruption cloud by the LOAC. The two instruments were not co-located; the OPC was installed at Möðrudalur, 72 km from the main eruption vent; where the background conditions and the concentration of particles within the eruption cloud could be expected to be different. An important difference between the two is that the launched instrument did not count the particles at the surface of the earth. The background dusty conditions on the surface in this area produce such high particle counts that the additional particles within the eruption cloud have a very small impact on the total number of particles. This is why we find very little difference in the particle counts by the OPC during and after the eruption.

The passing over of meteorological clouds triggered the development of a radar-detectable eruption cloud. The conditions when the eruption cloud was most frequently observed by the radar are the same conditions that allow for the pooling of gases as measured by the in situ SO$_2$ sensor, specifically during low wind speeds and when there was a change in wind direction, and often in the morning. The timing, environmental conditions, and behavior of the radar-detectable cloud all suggest that the cloud above the eruption site detected by the radar mostly consisted of droplets, with some particles, and it was coupled in behavior with the behavior of the gas cloud. The droplet-rich nature of the eruption cloud is supported by the balloon-launch described in [32]. They found that most of the particles measured in the eruption cloud were consistent with the typology of droplets and also found evidence for increased humidity and the slowing down of the balloon due to condensation on the balloon as it traveled through the eruption cloud.

The particles within the eruption cloud likely included the minor ash produced by the eruption and dust lofted by the strong thermal gradients induced by the lava field. Particles from the eruption likely served as seeds for cloud droplets in conditions conducive to cloud formation, such as when passing rain storms induced the formation of a persistent cloud over the eruption site. The water vapor and other volcanic gases injected in the atmosphere by the eruption became detectable droplets due to condensation above the eruption site. The radar reflectivity, which is sensitive to droplets and particles, but not gases, was indirectly monitoring the gases in this gas-rich, particle-poor eruption cloud. In the future, this might allow for the development of radar algorithms suitable for initializing gas dispersion models and for plume rise speed quantification.

From the perspective of optimizing observation/measurement frequency during future eruptions, we should continue to cultivate the use of as many different techniques as possible. In all data sets: SO_2 fluxes, the ratios of other gases to SO_2, and cloud heights, it is seen that the tables are populated due to the diversity of instrumentation and techniques. The environmental challenges were so great that few measurements were systematically obtained by any one single approach. In the future, we aim to augment our ground-based volcanic cloud eruption monitoring instrumentation with the use of a portable lidar system that could help with measuring the height of the eruption cloud and potentially describing the particle- and/or droplet rich nature of the eruption cloud.

Over the course of this eruption, we improved the use of several techniques. We optimized data filtering for the ScanDOAS measurements in order to remove the greatest impacts from atmospheric scattering. In the future, ScanDOAS data will be able to be processed and used much quicker than it was during this eruption, because sub-optimal wind directions will be automatically filtered from the real-time data analysis. The techniques used to calculate the fluxes from the near-vent traverses under high-emission, low-UV conditions can be automated, and we will work towards this in the future. We advanced the use of IR camera data for determining eruption cloud height, and we will attempt to automate this to retrieve cloud heights when other techniques are "blind" due to conditions such as darkness.

We will attempt to increase the frequency of gas ratio measurements. All three techniques used here, FTIR, MultiGAS, and filter pack sampling, should be attempted with as high frequency as possible. This will enable us to learn more, in the future, about the impacts of plentiful ground water, including rivers or glacier melt, affecting the eruption cloud, and differences in the emissions at the vent (important for constraining our physical models of the magmatic system) versus the emissions in the eruption cloud itself (important for constraining our dispersion models). These enhancements in future monitoring will enable us to improve our advice for people on potential mitigation actions to reduce societal harm during future eruptions.

Acknowledgments: Konradin Weber loaned a DOAS essential to the work. Bogi Brynjar Björnsson helped by making maps. Guðrún Nína Petersen helped by discussing results. Nial Peters helped collect DOAS and FTIR data. Clive Oppenheimer helped process FTIR data. Nahum Clements helped collect MultiGAS data. Þorstein Jónsson and Sveinbjörn Steinþórsson helped with installing and maintaining instruments. Simon Carn helped with putting the eruption into a global context. Tamar Elias helped with contrasting the eruption with Kilauea's eruption. This work was greatly supported by the FUTUREVOLC project funded by the European Union's Seventh Program for research, technological development, and demonstration under grant agreement No 308377. BG and SA acknowledge a grant from the Swedish Research Council FORMAS. EI acknowledges NERC urgency grant NE/M021130/1. Thank you to three anonymous reviewers for their helpful suggestions and comments.

Author Contributions: Melissa A. Pfeffer contributed to all parts of the paper. Baldur Bergsson installed and maintained instruments, and collected and processed MultiGAS and FTIR data. Sara Barsotti coordinated field work, installed and maintained instruments, ran simulations, interpreted data, and wrote sections. Gerður Stefánsdóttir processed and interpreted DOAS data and collected and processed FTIR data. Bo Galle and Vladimir Conde installed ScanDOAS with IMO. Bo Galle, Santiago Arellano, and Vladimir Conde performed and processed MobileDOAS measurements. Bo Galle and Santiago Arellano analyzed DOAS data and wrote sections. Amy Donovan collected and processed DOAS and FTIR data and interpreted data. Evgenia Ilyinskaya collected and processed FTIR and filter pack data, interpreted data, and revised the paper. Mike Burton collected and processed FTIR data and interpreted data. Alessandro Aiuppa processed MultiGAS data and interpreted data. Rachel C. W. Whitty processed NicAIR II data, interpreted data, and wrote sections. Isla C. Simmons processed DOAS data, interpreted data, and wrote sections. Þórður Arason processed webcam data, interpreted data, and generated figures. Elín B. Jónasdóttir coordinated overflight data and interpreted data. Nicole S. Keller collected and processed FTIR and filter pack data. Richard F. Yeo and Hermann Arngrímsson installed and maintained instruments and collected and processed radar data. Þorsteinn Jóhannsson coordinated the Icelandic Environmental Agency network and interpreted data. Mary K. Butwin processed and interpreted OPC data. Robert A. Askew collected and processed FTIR data. Stéphanie Dumont interpreted data and analyzed the time series. Stéphanie Dumont installed the OPC and interpreted data. Þorgils Ingvarsson installed and maintained instruments. Alessandro La Spina collected and processed FTIR data. Helen Thomas and Fred Prata developed the NicAIR II and processed and interpreted IR camera data. Fausto Grassa analyzed direct samples. Gaetano Giudice developed the MultiGAS and interpreted data. Andri Stefánsson analyzed and interpreted filter pack samples. Frank Marzano, Mario Montopoli, and Luigi Mereu processed radar data and interpreted radar data.

Geosciences **2018**, *8*, 29

Conflicts of Interest: The authors declare no conflict of interest. The founding sponsors had no role in the design of the study; in the collection, analyses, or interpretation of data; in the writing of the manuscript, and in the decision to publish the results.

References

1. Gíslason, S.R.; Stefánsdóttir, G.; Pfeffer, M.A.; Barsotti, S.; Jóhannsson, Þ.; Galeczka, I.; Bali, E.; Sigmarsson, O.; Stefánsson, A.; Keller, N.S.; et al. Environmental pressure from the 2014–15 eruption of Bárðarbunga volcano, Iceland. *Geochem. Perspect. Lett.* **2015**, 84–93. [CrossRef]
2. Bárðarbunga. Alternative Name: Veiðivötn. Available online: http://www.icelandicvolcanoes.is/?volcano=BAR (accessed on 18 January 2018).
3. Sigmundsson, F.; Hooper, A.; Hreinsdóttir, S.; Vogfjörd, K.S.; Ófeigsson, B.G.; Heimisson, E.R.; Dumont, S.; Parks, M.; Spaans, K.; Gudmundsson, G.B.; et al. Segmented lateral dyke growth in a rifting event at Bárðarbunga volcanic system, Iceland. *Nature* **2014**, *517*, 191–195. [CrossRef] [PubMed]
4. Hartley, M.E.; Thordarson, T. The 1874–1876 volcano-tectonic episode at Askja, North Iceland: Lateral flow revisited: Askja 1874–1876 Volcano-Tectonic Episode. *Geochem. Geophys. Geosyst.* **2013**, *14*, 2286–2309. [CrossRef]
5. Ilyinskaya, E.; Schmidt, A.; Mather, T.A.; Pope, F.D.; Witham, C.; Baxter, P.; Jóhannsson, T.; Pfeffer, M.A.; Barsotti, S.; Singh, A.; et al. Understanding the environmental impacts of large fissure eruptions: Aerosol and gas emissions from the 2014–2015 Holuhraun eruption (Iceland). *Earth Planet. Sci. Lett.* **2017**, *472*, 309–322. [CrossRef]
6. The Scientific Advisory Board of the Icelandic Civil Protection. Volcanic Activity in the Bárðarbunga system. Available online: http://en.vedur.is/media/jar/Factsheet_Bardarbunga_20150127.pdf (accessed on 10 January 2018).
7. Arnalds, O.; Dagsson-Waldhauserova, P.; Olafsson, H. The Icelandic volcanic aeolian environment: Processes and impacts—A review. *Aeolian Res.* **2016**, *20*, 176–195. [CrossRef]
8. Platt, U.; Stutz, J. *Differential Optical Absorption Spectroscopy: Principles and Applications*, 1st ed.; Springer: Berlin, Germany, 2008; ISBN 978-3-540-21193-8.
9. Galle, B.; Pfeffer, M.A.; Arellano, S.; Bergsson, B.; Conde, V.; Barsotti, S.; Stefánsdóttir, G.; Ingvarsson, Þ.; Bergsson, B.; Weber, K. Measurements of the gas emission from Holuhraun volcanic fissure eruption on Iceland, using Scanning DOAS instruments 2016. In Proceedings of the European Geosciences Union General Assembly 2016, Vienna, Austria, 17–22 April 2016; p. 13892.
10. Galle, B.; Johansson, M.; Rivera, C.; Zhang, Y.; Kihlman, M.; Kern, C.; Lehmann, T.; Platt, U.; Arellano, S.; Hidalgo, S. Network for Observation of Volcanic and Atmospheric Change (NOVAC)—A global network for volcanic gas monitoring: Network layout and instrument description. *J. Geophys. Res.* **2010**, *115*. [CrossRef]
11. HARMONIE—Numerical Weather Prediction Model. Available online: http://en.vedur.is/weather/articles/nr/3232 (accessed on 10 January 2018).
12. Mori, T.; Mori, T.; Kazahaya, K.; Ohwada, M.; Hirabayashi, J.; Yoshikawa, S. Effect of UV scattering on SO_2 emission rate measurements. *Geophys. Res. Lett.* **2006**, *33*. [CrossRef]
13. Johansson, M. Application of Passive DOAS for Studies of Megacity Air Pollution and Volcanic Gas Emissions. Ph.D. Thesis, Chalmers University of Technology, Gothenburg, Sweden, 2009.
14. Simmons, I.C.; Pfeffer, M.A.; Calder, E.S.; Galle, B.; Arellano, S.; Coppola, D.; Barsotti, S. Extended SO_2 outgassing from the 2014–2015 Holuhraun lava flow field, Iceland. *Bull. Volcanol.* **2017**, *79*. [CrossRef]
15. Galle, B.; Oppenheimer, C.; Geyer, A.; McGonigle, A.J.S.; Edmonds, M.; Horrocks, L. A miniaturised ultraviolet spectrometer for remote sensing of SO_2 fluxes: A new tool for volcano surveillance. *J. Volcanol. Geotherm. Res.* **2003**, *119*, 241–254. [CrossRef]
16. Johansson, M.; Galle, B.; Zhang, Y.; Rivera, C.; Chen, D.; Wyser, K. The dual-beam mini-DOAS technique—Measurements of volcanic gas emission, plume height and plume speed with a single instrument. *Bull. Volcanol.* **2009**, *71*, 747–751. [CrossRef]
17. Tsanev, V.I. A Collection of JScripts for Retrieval of Gas Column Amounts Using DOAS Methodology. Available online: https://www.geog.cam.ac.uk/research/projects/doasretrieval/ (accessed on 10 January 2018).
18. *Avoscan*. Available online: https://code.google.com/archive/p/avoscan/ (accessed on 18 January 2018).

19. Kern, C.; Deutschmann, T.; Vogel, L.; Wöhrbach, M.; Wagner, T.; Platt, U. Radiative transfer corrections for accurate spectroscopic measurements of volcanic gas emissions. *Bull. Volcanol.* **2010**, *72*, 233–247. [CrossRef]

20. La Spina, A.; Burton, M.; Harig, R.; Mure, F.; Rusch, P.; Jordan, M.; Caltabiano, T. New insights into volcanic processes at Stromboli from Cerberus, a remote-controlled open-path FTIR scanner system. *J. Volcanol. Geotherm. Res.* **2013**, *249*, 66–76. [CrossRef]

21. Burton, M.; Allard, P.; Mure, F.; La Spina, A. Magmatic gas composition reveals the source depth of slug-driven strombolian explosive activity. *Science* **2007**, *317*, 227–230. [CrossRef] [PubMed]

22. La Spina, A.; Burton, M.; Salerno, G.G. Unravelling the processes controlling gas emissions from the central and northeast craters of Mt. Etna. *J. Volcanol. Geotherm. Res.* **2010**, *198*, 368–376. [CrossRef]

23. Burton, M.; Ilyinskaya, E.; Hartley, M.; La Spina, A.; Salerno, G.G.; Bali, E.; Barsotti, S.; Bergsson, B.; Pfeffer, M.A.; Kaasalainen, H.; et al. Mantle source controls gas emissions and impact of Icelandic basaltic eruptions. *Nat. Commun.* **2017**. submitted for publication.

24. Allard, P.; Burton, M.; Oskarsson, N.; Michel, A.; Polacci, M. Magmatic gas composition and fluxes during the 2010 Eyjafjallajökull explosive eruption: implications for degassing magma volumes and volatile sources 2011. In Proceedings of the European Geosciences Union General Assembly 2011, Vienna, Austria, 3–8 April 2011.

25. Gudmundsson, M.T.; Jónsdóttir, K.; Hooper, A.; Holohan, E.P.; Halldórsson, S.A.; Ófeigsson, B.G.; Cesca, S.; Vogfjörd, K.S.; Sigmundsson, F.; Högnadóttir, T.; et al. Gradual caldera collapse at Bárdarbunga volcano, Iceland, regulated by lateral magma outflow. *Science* **2016**, *353*, aaf8988. [CrossRef] [PubMed]

26. Aiuppa, A.; Bertagnini, A.; Métrich, N.; Moretti, R.; Di Muro, A.; Liuzzo, M.; Tamburello, G. A model of degassing for Stromboli volcano. *Earth Planet. Sci. Lett.* **2010**, *295*, 195–204. [CrossRef]

27. Aiuppa, A.; Federico, C.; Giudice, G.; Giuffrida, G.; Guida, R.; Gurrieri, S.; Liuzzo, M.; Moretti, R.; Papale, P. The 2007 eruption of Stromboli volcano: Insights from real-time measurement of the volcanic gas plume CO_2/SO_2 ratio. *J. Volcanol. Geotherm. Res.* **2009**, *182*, 221–230. [CrossRef]

28. Ilyinskaya, E.; Aiuppa, A.; Bergsson, B.; Di Napoli, R.; Fridriksson, T.; Óladóttir, A.A.; Óskarsson, F.; Grassa, F.; Pfeffer, M.; Lechner, K.; et al. Degassing regime of Hekla volcano 2012–2013. *Geochim. Cosmochim. Acta* **2015**, *159*, 80–99. [CrossRef]

29. Tamburello, G. Ratiocalc: Software for processing data from multicomponent volcanic gas analyzers. *Comput. Geosci.* **2015**, *82*, 63–67. [CrossRef]

30. Stefánsson, A.; Stefánsdóttir, G.; Keller, N.S.; Barsotti, S.; Sigurdsson, Á.; Thorláksdóttir, S.B.; Pfeffer, M.A.; Eiríksdóttir, E.S.; Jónasdóttir, E.B.; von Löwis, S.; et al. Major impact of volcanic gases on the chemical composition of precipitation in Iceland during the 2014–2015 Holuhraun eruption: impact of volcanic gas on precipitation. *J. Geophys. Res. Atmos.* **2017**, *122*, 1971–1982. [CrossRef]

31. Gauthier, P.-J.; Sigmarsson, O.; Gouhier, M.; Haddadi, B.; Moune, S. Elevated gas flux and trace metal degassing from the 2014-2015 fissure eruption at the Bárðarbunga volcanic system, Iceland: Degassing at Holuhraun. *J. Geophys. Res. Solid Earth* **2016**, *121*, 1610–1630. [CrossRef]

32. Vignelles, D.; Roberts, T.J.; Carboni, E.; Ilyinskaya, E.; Pfeffer, M.A.; Dagsson Waldhauserova, P.; Schmidt, A.; Berthet, G.; Jegou, F.; Renard, J.-B.; et al. Balloon-borne measurement of the aerosol size distribution from an Icelandic flood basalt eruption. *Earth Planet. Sci. Lett.* **2016**, *453*, 252–259. [CrossRef]

33. De Michele, M.; Raucoules, D.; Arason, Þ. Volcanic Plume Elevation Model and its velocity derived from Landsat 8. *Remote Sens. Environ.* **2016**, *176*, 219–224. [CrossRef]

34. Prata, A.J.; Bernardo, C. Retrieval of volcanic ash particle size, mass and optical depth from a ground-based thermal infrared camera. *J. Volcanol. Geotherm. Res.* **2009**, *186*, 91–107. [CrossRef]

35. Prata, A.J.; Bernardo, C. Retrieval of sulfur dioxide from a ground-based thermal infrared imaging camera. *Atmos. Meas. Tech.* **2014**, *7*, 2807–2828. [CrossRef]

36. Lopez, T.; Thomas, H.E.; Prata, A.J.; Amigo, A.; Fee, D.; Moriano, D. Volcanic plume characteristics determined using an infrared imaging camera. *J. Volcanol. Geotherm. Res.* **2015**, *300*, 148–166. [CrossRef]

37. Lopez, T.; Fee, D.; Prata, F.; Dehn, J. Characterization and interpretation of volcanic activity at Karymsky Volcano, Kamchatka, Russia, using observations of infrasound, volcanic emissions, and thermal imagery: Characterization of volcanic activity. *Geochem. Geophys. Geosyst.* **2013**, *14*, 5106–5127. [CrossRef]

38. Whitty, R.C.W. The Use of an Infrared Camera to Analyse a Volcanic Eruption Extraordinarily Rich in SO_2 and H_2O. Master's Thesis, The University of Edinburgh, Edinburgh, 2016.

39. Scollo, S.; Prestifilippo, M.; Pecora, E.; Corradini, S.; Merucci, L.; Spata, G.; Coltelli, M. Eruption column height estimation of the 2011–2013 Etna lava fountains. *Ann. Geophys.* **2014**. [CrossRef]

40. Arason, Þ.; Petersen, G.N.; Bjornsson, H. Observations of the altitude of the volcanic plume during the eruption of Eyjafjallajökull, April–May 2010. *Earth Syst. Sci. Data* **2011**, *3*, 9–17. [CrossRef]
41. Björnsson, H.; Petersen, G.N.; Arason, Þ.; Jónasdóttir, E.B.; Ólafsson, H.; Pfeffer, M.A.; Barsotti, S.; Palmason, B.; von Löwis, S.; Dürig, T. Interaction Between the Eruption at Holuhraun and the Ambient Atmosphere 2015. In Proceedings of the European Geosciences Union General Assembly 2015, Vienna, Austria, 12–17 April 2015.
42. Marzano, F.S.; Lamantea, M.; Montopoli, M.; Di Fabio, S.; Picciotti, E. The Eyjafjöll explosive volcanic eruption from a microwave weather radar perspective. *Atmos. Chem. Phys.* **2011**, *11*, 9503–9518. [CrossRef]
43. Marzano, F.S.; Barbieri, S.; Vulpiani, G.; Rose, W.I. Volcanic ash cloud retrieval by ground-based microwave weather radar. *IEEE Trans. Geosci. Remote Sens.* **2006**, *44*, 3235–3246. [CrossRef]
44. Mereu, L.; Marzano, F.S.; Barsotti, S.; Montopoli, M.; Yeo, R.; Arngrimsson, H.; Björnsson, H.; Bonadonna, C. Ground-based microwave weather radar observations and retrievals during the 2014 Holuhraun eruption (Bárðarbunga, Iceland). In Proceedings of the European Geosciences Union General Assembly 2015, Vienna, Austria, 12–17 April 2015.
45. Arason, Þ.; Björnsson, H.; Petersen, G.N.; Jónasdóttir, E.B.; Oddsson, B. Plume height during the 2014–2015 Holuhraun volcanic eruption. In Proceedings of the European Geosciences Union General Assembly 2015, Vienna, Austria, 12–17 April 2015.
46. Coppola, D.; Ripepe, M.; Laiolo, M.; Cigolini, C. Modelling satellite-derived magma discharge to explain caldera collapse. *Geology* **2017**, *45*, 523–526. [CrossRef]
47. Bali, E.; Hartley, M.; Halldórsson, S.A.; Gudfinnsson, G.H.; Jakobsson, S. Melt inclusion constraints on volatile systematics and degassing history of the 2014–2015 Holuhraun eruption, Iceland. *Contrib. Mineral. Petrol.* **2017**. submitted for publication. [CrossRef]
48. Pedersen, G.B.M.; Höskuldsson, Á.; Dürig, T.; Thordarson, T.; Jónsdóttir, I.; Riishuus, M.S.; Óskarsson, B.V.; Dumont, S.; Magnusson, E.; Gudmundsson, M.T.; et al. Lava field evolution and emplacement dynamics of the 2014–2015 basaltic fissure eruption at Holuhraun, Iceland. *J. Volcanol. Geotherm. Res.* **2017**, *340*, 155–169. [CrossRef]
49. Schmidt, A.; Leadbetter, S.; Theys, N.; Carboni, E.; Witham, C.S.; Stevenson, J.A.; Birch, C.E.; Thordarson, T.; Turnock, S.; Barsotti, S.; et al. Satellite detection, long-range transport, and air quality impacts of volcanic sulfur dioxide from the 2014–2015 flood lava eruption at Bárðarbunga (Iceland): SO_2 from 2014 to 2015 eruption at Bárðarbunga. *J. Geophys. Res. Atmos.* **2015**, *120*, 9739–9757. [CrossRef]
50. Carn, S.A. Global Sulfur Dioxide Monitoring Home Page. Available online: https://so2.gsfc.nasa.gov/ (accessed on 14 September 2017).
51. Sutton, A.J.; Elias, T.; Gerlach, T.M.; Stokes, J.B. Implications for eruptive processes as indicated by sulfur dioxide emissions from Kilauea Volcano, Hawai'i, 1979–1997. *J. Volcanol. Geotherm. Res.* **2001**, *108*, 283–302. [CrossRef]
52. Hellsing, V.Ú.L.; Ragnarsdóttir, A.S.; Jónsson, K.; Keller, N.S.; Jóhannsson, Þ.; Guðmundsson, J.; Snorrason, A.; Þórsson, J. *National Inventory Report: Emissions of Greenhouse Gases in Iceland from 1990 to 2015*; The Environment Agency of Iceland: Reykjavík, Iceland, 2017.
53. Aiuppa, A.; Baker, D.R.; Webster, J.D. Halogens in volcanic systems. *Chem. Geol.* **2009**, *263*, 1–18. [CrossRef]
54. Aiuppa, A.; Bellomo, S.; D'Alessandro, W.; Federico, C.; Ferm, M.; Valenza, M. Volcanic plume monitoring at Mount Etna by diffusive (passive) sampling. *J. Geophys. Res. Atmos.* **2004**, *109*. [CrossRef]
55. Butz, A.; Dinger, A.S.; Bobrowski, N.; Kostinek, J.; Fieber, L.; Fischerkeller, C.; Giuffrida, G.B.; Hase, F.; Klappenbach, F.; Kuhn, J.; et al. Remote sensing of volcanic CO_2, HF, HCl, SO_2, and BrO in the downwind plume of Mt. Etna. *Atmos. Meas. Tech.* **2017**, *10*, 1–14. [CrossRef]
56. Granieri, D.; Salerno, G.; Liuzzo, M.; La Spina, A.; Giuffrida, G.; Caltabiano, T.; Giudice, G.; Gutierrez, E.; Montalvo, F.; Burton, M.R.; et al. Emission of gas and atmospheric dispersion of SO_2 during the December 2013 eruption at San Miguel volcano (El Salvador, Central America): SO_2 emission at San Miguel volcano. *Geophys. Res. Lett.* **2015**, *42*, 5847–5854. [CrossRef]
57. Martin, R.S.; Sawyer, G.M.; Spampinato, L.; Salerno, G.G.; Ramirez, C.; Ilyinskaya, E.; Witt, M.L.I.; Mather, T.A.; Watson, I.M.; Phillips, J.C.; et al. A total volatile inventory for Masaya Volcano, Nicaragua. *J. Geophys. Res.* **2010**, *115*. [CrossRef]
58. Oppenheimer, C.; Scaillet, B.; Martin, R.S. Sulfur degassing from volcanoes: Source conditions, surveillance, plume chemistry and earth system impacts. *Rev. Mineral. Geochem.* **2011**, *73*, 363–421. [CrossRef]

59. Pennisi, M.; Le Cloarec, M.-F. Variations of Cl, F, and S in Mount Etna's plume, Italy, between 1992 and 1995. *J. Geophys. Res. Solid Earth* **1998**, *103*, 5061–5066. [CrossRef]

60. Sawyer, G.M.; Carn, S.A.; Tsanev, V.I.; Oppenheimer, C.; Burton, M. Investigation into magma degassing at Nyiragongo volcano, Democratic Republic of the Congo: Magma degassing at Nyiragongo volcano. *Geochem. Geophys. Geosystems* **2008**, *9*. [CrossRef]

61. Spampinato, L.; Salerno, G.G.; Martin, R.S.; Sawyer, G.M.; Oppenheimer, C.; Ilyinskaya, E.; Ramirez, C. Thermal and geochemical signature of Poas volcano, Costa Rica. *Rev. Geol. Am. Cent.* **2010**, *43*, 171–189.

62. Symonds, R.B.; Rose, W.I.; Bluth, G.J.S.; Gerlach, T.M. Volcanic gas studies: Methods, results and applications. *Rev. Mineral. Geochem.* **1994**, *30*, 1–66.

geosciences

MDPI

Article

A New Degassing Model to Infer Magma Dynamics from Radioactive Disequilibria in Volcanic Plumes

Luca Terray [1,2,*], Pierre-J. Gauthier [1], Giuseppe Salerno [3], Tommaso Caltabiano [3], Alessandro La Spina [3], Pasquale Sellitto [4] and Pierre Briole [5]

[1] Laboratoire Magmas et Volcans, CNRS, IRD, OPGC, Université Clermont Auvergne, F-63000 Clermont-Ferrand, France; p.j.gauthier@opgc.fr
[2] Département de Géosciences, École Normale Supérieure, PSL Research University, 75005 Paris, France
[3] Istituto Nazionale di Geofisica e Vulcanologia, Sezione di Catania, Osservatorio Etneo, 95125 Catania, Italy; giuseppe.salerno@ingv.it (G.S.); tommaso.caltabiano@ingv.it (T.C.); alessandro.laspina@ingv.it (A.L.S.)
[4] Laboratoire de Météorologie Dynamique, CNRS-UMR8539, École Normale Supérieure, PSL Research University, 75005 Paris, France; psellitto@lmd.ens.fr
[5] Laboratoire de Géologie, CNRS-UMR8538, École Normale Supérieure, PSL Research University, 75005 Paris, France; briole@ens.fr
* Correspondence: luca.terray@uca.fr or terray@phare.normalesup.org

Received: 15 November 2017; Accepted: 14 January 2018; Published: 18 January 2018

Abstract: Mount Etna volcano (Sicily, Italy) is the place where short-lived radioactive disequilibrium measurements in volcanic gases were initiated more than 40 years ago. Almost two decades after the last measurements in Mount Etna plume, we carried out in 2015 a new survey of ^{210}Pb-^{210}Bi-^{210}Po radioactive disequilibria in gaseous emanations from the volcano. These new results [$(^{210}$Po$/^{210}$Pb$) =$ 42 and $(^{210}$Bi$/^{210}$Pb$) = 7.5$] are in fair agreement with those previously reported. Previously published degassing models fail to explain satisfactorily measured activity ratios. We present here a new degassing model, which accounts for ^{222}Rn enrichment in volcanic gases and its subsequent decay into ^{210}Pb within gas bubbles en route to the surface. Theoretical short-lived radioactive disequilibria in volcanic gases predicted by this new model differ from those produced by the former models and better match the values we measured in the plume during the 2015 campaign. A Monte Carlo-like simulation based on variable parameters characterising the degassing process (magma residence time in the degassing reservoir, gas transfer time, Rn-Pb-Bi-Po volatilities, magma volatile content) suggests that short-lived disequilibria in volcanic gases may be of use to infer both magma dynamics and degassing kinetics beneath Mount Etna, and in general at basaltic volcanoes. However, this simulation emphasizes the need for accurately determined input parameters in order to produce unambiguous results, allowing sharp characterisation of degassing processes.

Keywords: radioactive disequilibria ^{210}Pb-^{210}Bi-^{210}Po; volcanic gases; degassing processes; geochemical modelling; Mount Etna

1. Introduction

Many active basaltic open-conduit volcanoes emit a persistent gas plume, even during quiescence stages without eruptive activity at the surface. This behaviour provides evidence for ongoing magma degassing beneath volcanic centers. Degassing budgets at such active volcanoes may be inferred from both long time-series of SO_2 flux measurements [1] and analysis of the sulfur content preserved in melt inclusions [2]. At subduction-zone-related volcanoes, the amount of degassing magma usually exceeds by one or several orders of magnitude the volume of lava actually erupting during the same period [3], which is notably the case of Mount Etna [4]. Moreover, the budget of magma entering the Mount Etna plumbing system, inferred from gravity changes and deformation data, matches the volume of degassing magma and exceeds by far the erupted lava volume [5]. It thus emphasizes the fact that degassing and actually erupted volumes of magma can be significantly unbalanced, a feature that is not observed at non-subduction-related volcanoes [3,6]. This feature also appears to depend on the eruptive style over a given temporal window [7]. At a volcano like Mount Etna (subduction-zone-related stratovolcano fed with volatile-rich-alkali basalts, e.g., [8]), it could be explained by a dynamic regime of magma redistribution beneath the volcano during which degassed magma is continuously removed by convection from the degassing reservoir and replaced by fresh undegassed magma [9–11].

This observation raises questions about the geometry and dynamic properties of degassing magma reservoirs, which remains poorly known and might have a strong control on eruptive activity [12,13]. For instance, high gas fluxes could be explained by the degassing of either large volumes of deep stagnant magma (>1 km^3 stored at a few kilometres depth) or else small batches (10^5–10^6 m^3) of quickly overturned magma brought at shallow levels (hundreds of meter). It is thus of primary importance to set constraints on the characteristic timescales of both magma degassing and gas phase transfer from the degassing reservoir to the surface active craters.

The study of radioactive disequilibria in volcanic gases has proved relevant in constraining magmatic degassing dynamics [14,15]. This method focuses on the three last radionuclides of the ^{238}U decay chain: ^{210}Pb, ^{210}Bi and ^{210}Po (see Figure 1). Because they all have short half-lives (22 years, 5 days and 138 days, respectively), these isotopes are suitable to study recent fractionations (younger than two years based on ^{210}Po half-life, the longest-lived ^{210}Pb decay product) associated with pre-eruptive and syn-eruptive magmatic processes. Furthermore, lead, bismuth and polonium are strongly fractionated upon degassing (polonium being more volatile than bismuth, which is in turn more volatile than lead), which gives birth to large radioactive disequilibria between ^{210}Pb-^{210}Bi-^{210}Po in the gas phase [14,16,17]. These properties have enabled radioactive disequilibrium measurements in volcanic plumes to be linked to degassing activity through the use of two models. Lambert et al. [14] first developed a static degassing model for which radionuclide exsolution takes place in a degassing cell containing a proportion μ of deep undegassed magma in radioactive equilibrium. They also considered the transfer time of radionuclides between the time of exsolution from the magma and the time of emission at the surface, as ^{210}Bi (five days half-life) is short-lived enough to significantly decay during gas transfer towards the surface. This approach has been extensively used to characterise gaseous emissions at Mount Etna [14,18,19]. More recently, Gauthier et al. [15] proposed a dynamic degassing model for which radioactive disequilibria in the gas phase also depend on the magma residence time in the degassing reservoir because of continuous regeneration of highly volatile ^{210}Po by decay of its less volatile parent (^{210}Bi) within the reservoir. This latter model has been successfully applied to persistently degassing, open-conduit, basaltic volcanoes like Stromboli [15] or Ambrym [20]. At these two volcanoes, both the magma residence time in the degassing reservoir and the transfer time of the gas phase towards the surface were estimated and were shown to vary according to eruptive activity.

Almost two decades after the last radioactivity survey in Mount Etna gaseous emissions [19], we went back to Mount Etna in 2015 in order to investigate radioactive disequilibria in the plume. The new measurements presented in this paper have benefited from recent methodological and analytical improvements [21], and they are in fair agreement with previously published data ([19], and references therein). Based on the concept of the previous dynamic degassing model [15], a new theoretical framework for the degassing of radionuclides is presented here. We show that the very short-lived radon isotope ^{222}Rn can be significantly enriched in the gas phase when magma residence time in the degassing reservoir increases. Although it has been neglected so far, mostly because of its short half-life of 3.8 days, we show that ^{222}Rn plays a major role in controlling the magnitude of ^{210}Pb-^{210}Bi-^{210}Po disequilibria by producing, through its radioactive decay, a new generation of ^{210}Pb atoms within gas bubbles. By using a dataset of previously published values describing both trace element volatilities and volatile content in magmas from Etna, we present a Monte Carlo-like simulation that explains radioactive disequilibria measured in 2015 in Mount Etna gases well. Implications for the retrieval of quantitative information on degassing dynamics from radioactive disequilibria in a volcanic plume are presented at the end of the paper.

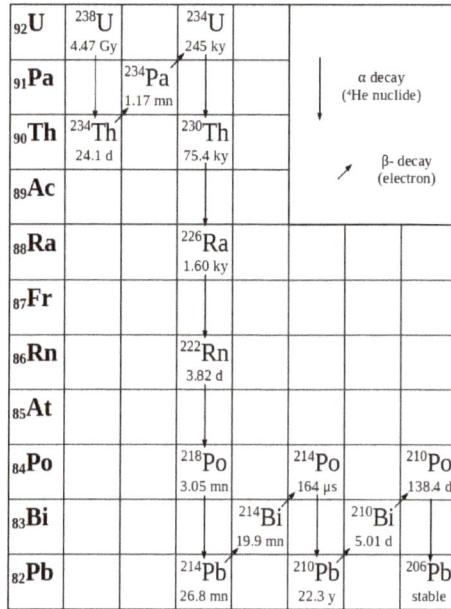

$_{92}$U	^{238}U 4.47 Gy		^{234}U 245 ky				
$_{91}$Pa		^{234}Pa 1.17 mn				α decay (^{4}He nuclide)	
$_{90}$Th	^{234}Th 24.1 d		^{230}Th 75.4 ky				
$_{89}$Ac						β- decay (electron)	
$_{88}$Ra			^{226}Ra 1.60 ky				
$_{87}$Fr							
$_{86}$Rn			^{222}Rn 3.82 d				
$_{85}$At							
$_{84}$Po			^{218}Po 3.05 mn		^{214}Po 164 µs		^{210}Po 138.4 d
$_{83}$Bi				^{214}Bi 19.9 mn		^{210}Bi 5.01 d	
$_{82}$Pb			^{214}Pb 26.8 mn		^{210}Pb 22.3 y		^{206}Pb stable

Figure 1. ^{238}U decay chain. Half-lives are indicated beneath the symbol of the element. Minor embranchments are not reproduced.

2. Short-Lived Radioactivity Measurements in Mount Etna Plume

2.1. Field Description and Sampling Techniques

Mount Etna, one of the most active volcanoes worldwide, is an easily accessible volcano with a persistent degassing activity, making it a strong gas emitter. It has been the focus of many studies aiming at surveying short-lived disequilibria in volcanic gases [14,16,18,19,22,23]. During these two decades of regular survey at the end of the 20th century, the Mount Etna summit area was quite different than it is today. At that time, the volcano had only four summit craters (northeast Crater (NEC), Voragine (VOR), Bocca Nuova (BN) and the late-born southeast Crater (SEC); Figure 2). Since 2001, Mount Etna has erupted frequently, almost on a yearly basis, producing large lava flows and powerful paroxysmal events associated with lava fountaining episodes that significantly remodeled the summit area and ultimately gave birth to the New South-East Crater (NSEC, see Figure 2) [24]. In May 2015, we carried out a new survey of ^{210}Pb-^{210}Bi-^{210}Po radioactive disequilibria in the volcanic plume. The field campaign took place during a brief, 5-day-long eruption at NSEC. After several days of tremor increase, eruptive activity started at NSEC on 12 May 2015 with loud explosions producing reddish to dark grey ash-rich plumes, followed by sustained strombolian activity at NSEC summit while a fissure opened on the eastern flank of the cone, emitting a lava flow that travelled towards and inside Valle del Bove. On 15 May, the intensity of the eruption gradually decreased until it reached an end on 16 May. The eruptive activity at NSEC during this short eruptive episode was too intense to grant safe access to the summit. We collected the diluted plume of Mount Etna downwind of the crater on the southern slope of the volcano, at remote sites near Torre del Filosofo (Figure 2). Samples collected at these locations were mostly from the main ash-rich gas plume emitted by the NSEC summit, although a minor contribution of the small gas plume released from the eruptive fissure cannot be ruled out.

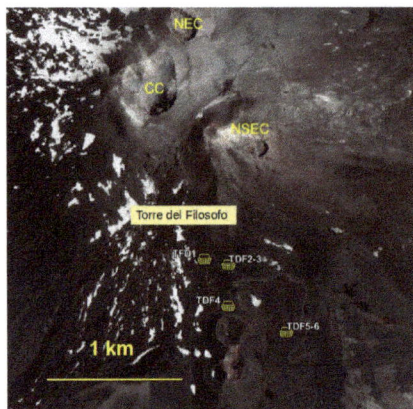

Figure 2. Map of Mount Etna summital zone in 2012 (aerial photography from Italian Geoportal). CC stands for Central Craters (Voragine and Bocca Nuova), NEC for northeast Crater and NSEC for new southeast Crater. Sampling sites (TDF1, TDF2-3, TDF4 and TDF5-6, where TDF stands for Torre del Filosofo) are pointed on the map.

The three investigated radionuclides are engaged in chemical compounds (mostly halides and sulfates) that are gaseous at magma temperature but are quenched into solid particles once in the cold atmosphere. They are consequently born in the aerosol fraction of the gas plume and can be easily sampled by filtration of the diluted plume through a membrane. For this study, we used high flowrate pumps (110 L/min) fed by a 12V-35Ah battery and connected to a home-designed polyvinyl chloride (PVC) filter-holder (exposition diameter of 50 mm) containing a single cellulose acetate filter (Poelmann–Schneider blue type, 0.2 µm mean porosity). Sampling times ranged between 15 min and 1 h (corresponding to a volume of filtrated air in the range 1.7–6.6 m^3, Table 1) to ensure that enough diluted plume passed through the filters so that radionuclide activities on filter samples can be analysed.

Table 1. ^{210}Pb-^{210}Bi-^{210}Po radioactivity in Mount Etna plume. Sample names correspond to sampling sites as described in Figure 2. Activities are reported in mBq/m^3 with 1-σ uncertainties. "bdl" stands for below detection limit. Sampling starting time is expressed in local time.

Sample	Date and time	Volume (m^3)	^{210}Pb	^{210}Bi	^{210}Po
Atmospheric blank	11/05/2015 12:00	3.9	bdl	bdl	bdl
TDF1	12/05/2015 12:07	3.9	bdl	bdl	bdl
TDF2	13/05/2015 11:04	6.6	0.8 ± 0.2	3.1 ± 1	14.6 ± 0.7
TDF3	13/05/2015 12:15	6.6	bdl	bdl	bdl
TDF4	14/05/2015 10:51	1.7	3.5 ± 1.0	29 ± 5	212 ± 3
TDF5A	14/05/2015 11:38	3.9	4.0 ± 0.4	31 ± 3	162 ± 2
TDF5B	14/05/2015 11:38	3.0	8.2 ± 0.4	62 ± 2	349 ± 2
TDF6	14/05/2015 12:31	6.6	5.9 ± 0.2	41 ± 2	217 ± 1

2.2. Analytical Techniques

Because ^{210}Bi (5.01 days half-life) thoroughly decays away in about one month, filter samples were taken back to Laboratoire Magmas et Volcans in Clermont-Ferrand within a few days after collection. Untreated filters were analysed with a low-background noise alpha-beta counting unit (IN20 Canberra) following the procedure described in Gauthier et al. [21]. Repeated measurements were carried on over a month after collection in the field, each analytical cycle lasting 48 h (one cycle comprises eight 6-hour-long counting blocks that are ultimately averaged, the mean activity being considered as the "instantaneous" activity at the time $t_i + 24$ h, where t_i is the starting date and time of a given cycle). Both alpha and beta activities were simultaneously determined on filter samples. Alpha counts provide a direct measure of ^{210}Po activity while beta counts correspond to the detection of ^{210}Bi beta decay particles. Beta emissions of ^{210}Pb cannot be directly measured because of their energies that are too low, so that ^{210}Pb is measured via ^{210}Bi one month after sampling when both isotopes have reached radioactive equilibrium [21]. Radionuclide activities on filter samples were determined by subtracting the detector background (electronic noise and some unblocked cosmic rays) and by taking into account the detector efficiency and an attenuation factor for alpha particles [25]. Initial activities at the time of sampling were then retrieved by fitting radioactive decay trend during the one-month period of analysis by using classical radioactivity laws.

2.3. Analytical Results

Analytical results are reported in Table 1 along with 1-σ uncertainties based on the counting statistics. Uncertainties deriving from the variability of the volume of sampled plume are not propagated since they are quite difficult to estimate while working in the field. However, the three radionuclides are sampled and measured on the same filter and, hence, this uncertainty does not impact radionuclide ratios (i.e., radioactive disequilibria). In addition to volcanic aerosol samples, a sample of the atmospheric background at Mount Etna was collected in clear sky conditions, outside of the plume influence, close to the Sapienza touristic area. As expected, activities in the atmospheric background are below the detection limit for the three radionuclides. Considering the volume of atmosphere sampled for the blank, and the minimum detectable activity (depending on the duration of a counting cycle and on the instrument background noise) in both alpha and beta modes with the IN20 counter, both ^{210}Pb and ^{210}Bi detection limit can be quantified at 0.8 mBq/m^3 and that of ^{210}Po at 0.3 mBq/m^3. Although activities are below detection limits for samples TDF1 and TDF3, and similar to the bêta detection limit for sample TDF2, all other samples appear significantly enriched in ^{210}Pb, ^{210}Bi and ^{210}Po compared to the atmospheric blank. As previously shown ([19], and references therein), this suggests that the plume of Mount Etna is considerably enriched in radionuclides over a standard atmosphere and that high-quality samples can be obtained at safe distances from the summit area. Figure 3 shows ^{210}Bi (Figure 3a) and ^{210}Po (Figure 3b) activities plotted against ^{210}Pb activities for all aerosol samples. Activities follow linear trends passing through the origin, which are interpreted as dilution trends of the volcanic gas (considerably enriched in ^{210}Pb, ^{210}Bi and ^{210}Po) into the atmosphere for which radionuclide activities are negligible. The two linear correlations in Figure 3 are well defined ($R^2 = 0.99$ for ^{210}Bi vs. ^{210}Pb; $R^2 = 0.91$ for ^{210}Po vs. ^{210}Pb), which suggests that the dilution of volcanic gases in the atmosphere does not significantly affect their pristine isotopic signature within at least 1.5 km distance from the summit area. Furthermore, it was reported that radioactive disequilibria in gases released at summit craters significantly differ from those in gas emanations from eruptive vents along eruptive fissures [19]. Therefore, the well-defined linear correlations observed in Figure 3 suggest that the contribution of gases released at the eruptive fissure to the main plume is negligible or alternatively steady through time, which would be highly fortuitous. Samples collected in the diluted plume are thus taken to be representative of the chemistry of volcanic gases at the source. Radioactive disequilibria in the volcanic plume of the NSEC are retrieved by linear regression of the whole dataset at $(^{210}\text{Bi}/^{210}\text{Pb}) = 7.5 \pm 0.4$ and $(^{210}\text{Po}/^{210}\text{Pb}) = 42 \pm 6$. These values are in fair agreement with those previously reported for Mount Etna's summit craters, in the range 10–30 for $(^{210}\text{Bi}/^{210}\text{Pb})$ and between 20 and up to 90 for $(^{210}\text{Po}/^{210}\text{Pb})$ ([19], and references therein).

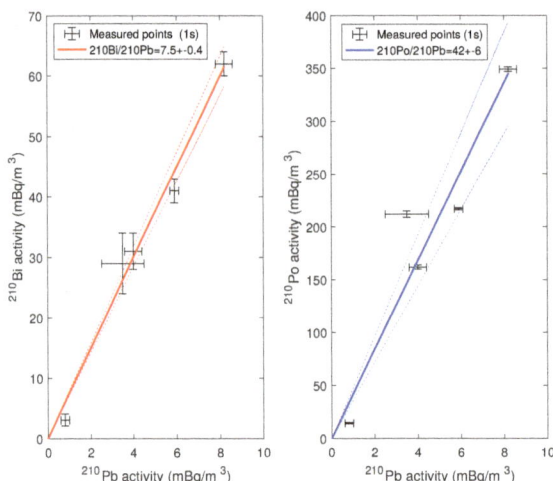

Figure 3. Volumic activities in the plume. Bars represent one sigma errors and are derived from the uncertainty of radioactivity analysis. Full lines represent the linear regression between the points and the origin, and dotted lines represent two sigma standard error on the slope of the regression trend.

3. Modelling of Radionuclide Degassing and ^{210}Pb-^{210}Bi-^{210}Po Radioactive Disequilibria in Volcanic Plumes

Lambert et al. [14] and then Gauthier et al. [15] linked ^{210}Pb-^{210}Bi-^{210}Po radioactive disequilibria in volcanic gases to degassing processes. Although radioactive disequilibria in volcanic gases from Mount Etna have been successfully explained by the model of Lambert et al. [14], Gauthier et al. [15] showed that Lambert's model can be used only if the magmatic vapour is released from a rapidly overturned batch of deep magma in radioactive equilibrium prior to degassing. This is due to the fact that the model proposed by Lambert et al. [14] neglects, in the degassing magma, the radioactive ingrowth of ^{210}Bi and ^{210}Po (both moderately to highly volatile at magma temperature) from their parent ^{210}Pb, which is weakly volatile and mostly remains in the melt. If the magma residence time in the degassing reservoir is long enough for ^{210}Pb to decay, then both ^{210}Bi and ^{210}Po atoms are regenerated in the melt. The longer is the residence time, the more efficient is the regeneration. Because ^{210}Pb decay products have greater affinity for the gas phase than for the melt, they preferentially partition into the vapour phase according to their own volatility. When the magma residence time in the degassing reservoir increases, magmatic gases consequently become more and more enriched in the most volatile ^{210}Po and, to a lesser extent, in the moderately volatile ^{210}Bi over ^{210}Pb [15]. By neglecting radioactive ingrowth within the degassing magma, Lambert et al. [14] considered that the ratio of two radionuclides in the magmatic vapour at the time of exsolution could reach a maximum value equal to the ratio of their emanation coefficients ϵ, the widely used parameter describing trace element volatility [14,26]. Gauthier et al. [15] showed that the ϵ ratio of two radionuclides corresponds instead to the minimum value for activity ratios in the magmatic vapour at the time of volatile exsolution. This minimum value is reached when magma residence time is negligible compared to ^{210}Po half-life i.e., less than about 10 days. For increasing values of the magma residence time, activity ratios in the gas phase increase up to a theoretical value, which is defined by the ratio of the liquid–gas partition coefficients D for the considered radionuclides [15].

There is a broad agreement that the emanation coefficient of ^{210}Po for basaltic systems (including Mount Etna) is close to 100%, suggesting that most of polonium atoms are transferred to the magmatic vapour upon degassing [22,27–30]. The emanation coefficient of lead in basaltic systems appears to be higher in calk-alkaline systems than in other geodynamical settings [21]. At Mount Etna, like at other arc-related basaltic volcanoes, it has a value of approximately $1.0 \pm 0.5\%$ [14,17,31,32]. Therefore, the minimum value for (^{210}Po/^{210}Pb) activity ratios in Mount Etna gases should be around 100, which has never been measured, excepted at the beginning of the 1992 eruption [19]. In particular, the ratio of 42 ± 6 we find in the May 2015 plume is far below this minimum theoretical value and cannot be explained by using the dynamic degassing model with realistic parameters for metal volatility [15]. In order to reproduce our observations, the degassing model would indeed require either polonium emanation coefficients lower than 50%, which is in strong disagreement with analyses of freshly erupted lavas at Mount Etna [22], or else lead emanation coefficients up to 3% or even more, which has never been reported.

3.1. ^{222}Rn Enrichments in Volcanic Gases: Towards a New Degassing Model for Short-Lived Radionuclides

Recent studies have suggested that magmas having accumulated an ^{222}Rn-rich gas phase may present significant ^{210}Pb excesses over ^{226}Ra [29,33–36]. Volcanic gases are also characterised, in some cases, by ^{210}Pb/Pb ratios significantly higher than those measured in lavas [23]. Although the origin of these ^{210}Pb enrichments in volcanic gases has not been fully understood, we tentatively assume that ^{210}Pb excesses could result from radioactive decay of ^{222}Rn atoms in the gas phase.

It is worth noting that, in the previously published degassing models [14,15], the radioactive decay of ^{222}Rn was not taken into account. We present therefore a new degassing model that accounts for it. The conceptual framework of this new model matches that of Gauthier et al. [15]. Accordingly, we consider that volatile exsolution takes place in an open degassing reservoir, which has reached a dynamical and chemical steady-state. Dynamical steady-state implies that the degassing reservoir has a constant mass M (or volume V) through time. It means that any input flux ϕ_0 of deep undegassed magma (in radioactive equilibrium for ^{222}Rn and its daughters) has to be balanced by a flux of gas ϕ_G and a flux of lava ϕ_L leaving the reservoir. Let α be the fraction of volatiles initially dissolved in the deep magma and ultimately released, the fluxes ϕ_G and ϕ_L can be written:

$$\phi_G = \alpha \phi_0,$$
$$\phi_L = (1 - \alpha) \phi_0. \tag{1}$$

In such a steady-state reservoir, the replenishment rate is given by ϕ_0/M. Its reciprocal, M/ϕ_0, defines the magma residence time τ in the degassing reservoir. In such reservoir, the number of atoms $N_{k,L}$ of each radioactive isotope I_k (either ^{222}Rn, ^{210}Pb, ^{210}Bi, ^{210}Po, k depending on the position along the decay chain) in the degassing magma varies according to:

$$\frac{\partial N_{k,L}}{\partial t} = \lambda_{k-1} N_{k-1,L} - \lambda_k N_{k,L} + \phi_0 C_{k,0} - \phi_L C_{k,L} - \phi_G C_{k,G}, \tag{2}$$

where C_k stands for the mass concentration of I_k either in the undegassed magma (index 0), the degassed lava (index L) or the gas phase (index G), and where λ_k stands for the radioactive decay constant of I_k. From left to right, the right terms of Equation (2) correspond to the production of I_k by I_{k-1} decay, the loss of I_k according to its own decay, the input of I_k from the undegassed magma entering the reservoir, the output of I_k by magma withdrawal and finally the output of I_k by degassing. This equation can be expressed in terms of activity per unit of mass by multiplying each member by λ_k/M:

$$\frac{\partial (I_k)_L}{\partial t} = \lambda_k (I_{k-1})_L - \lambda_k (I_k)_L + \frac{\phi_0}{M} (I_k)_0 - (1 - \alpha) \frac{\phi_0}{M} (I_k)_L - \alpha \frac{\phi_0}{M} (I_k)_G, \tag{3}$$

where parentheses denote activity per unit of mass, and ϕ_L and ϕ_G have been replaced by their expression in Equation (1). With the assumption that the reservoir has also reached a chemical steady-state, which is justified for short-lived volatile isotopes [15], Equation (2) is equal to 0. By introducing a liquid gas partitioning coefficient D_{I_k} such as:

$$D_{I_k} = \frac{(I_k)_G}{(I_k)_L},$$

(4)

the activity of any radionuclide I_k in the gas phase may be obtained from Equation (3) such as:

$$(I_k)_G = \frac{\lambda_k (I_{k-1})_L + \frac{(I_k)_0}{\tau}}{\frac{\lambda_k}{D_{I_k}} + (\alpha + \frac{1-\alpha}{D_{I_k}})\frac{1}{\tau}},$$

(5)

Note that both liquid–gas partitioning coefficients D and emanation coefficients ϵ are linked through the following relationship [15]:

$$\epsilon_{I_k} = \left(1 + \frac{1-\alpha}{\alpha D_{I_k}}\right)^{-1},$$

(6)

3.2. Radioactive Disequilibria in Gases at the Time of Exsolution

3.2.1. Specific Case of ^{222}Rn Exsolution

As shown by Equation (5), the activity of a radionuclide I_k in the gas phase depends on the activity of its precursor in the decay chain, which means that a precursor for ^{210}Pb has to be taken into account so as to compute all other activities. All four radionuclides between ^{222}Rn and ^{210}Pb (^{218}Po, ^{214}Pb, ^{214}Bi and ^{214}Po) have half-lives of a few minutes at most (Figure 1), which appears very short compared to the expected residence time of the magma in the degassing reservoir. They can thus be neglected in the model and we consider here, mathematically speaking, that ^{210}Pb is directly regenerated by ^{222}Rn decay. In order to quantify ^{222}Rn degassing efficiency Gauthier and Condomines [37] introduced a parameter f, ranging between 0 (no radon degassed) and 1 (total degassing of radon), and used this previous parameter to calculate ^{222}Rn activity in the degassing melt at steady-state:

$$(^{222}\text{Rn})_L = (1-f)(^{226}\text{Ra})_L,$$

(7)

with

$$(^{226}\text{Ra})_L = \frac{(^{226}\text{Ra})_0}{1-\alpha},$$

(8)

since radium forms no volatile compounds in basaltic magmas [14]. Applying Equation (3) to the case of ^{222}Rn, replacing $(^{222}\text{Rn})_L$ and $(^{226}\text{Ra})_L$ by their expression in Equations (7) and (8), respectively, and ϕ_0/M by $1/\tau$, we obtain:

$$(^{222}\text{Rn})_G = (^{226}\text{Ra})_0 \frac{f}{\alpha}\left(1 + \frac{\tau \lambda_{Rn}}{1-\alpha}\right),$$

(9)

Note that, in order to clarify notations, $\lambda_{^{222}\text{Rn}}$ is written λ_{Rn}, and other radioactivity constants are noted similarly. This relation suggests that significant excesses of ^{222}Rn can be produced in the gas phase for long residence times. Providing that the escape time of gases is long enough (see Section 3.3), ^{222}Rn atoms could then decay and act as a significant additional source of ^{210}Pb in the gas phase.

3.2.2. ^{222}Rn-^{210}Pb-^{210}Bi-^{210}Po Fractionation upon Exsolution

^{210}Pb, ^{210}Bi, ^{210}Po activities in the gas phase after exsolution can now be derived iteratively using Equation (5). Because short-lived ^{226}Ra daughters are thought to be in radioactive equilibrium in deep magmas prior to degassing [37], the application of Equation (5) to ^{210}Pb, taking into account the expression of $(^{222}$Rn$)_L$ given by Equations (7) and (8) leads to:

$$(^{210}\text{Pb})_G = A_0 \frac{\lambda_{\text{Pb}} \frac{1-f}{1-\alpha} + \frac{1}{\tau}}{\frac{\lambda_{\text{Pb}}}{D_{\text{Pb}}} + (\alpha + \frac{1-\alpha}{D_{\text{Pb}}}) \frac{1}{\tau}}, \tag{10}$$

where A_0 refers to the equilibrium activity of all nuclides.

The application of Equation (5) to ^{210}Bi, replacing $(^{210}$Pb$)_L$ by $(^{210}$Pb$)_G/D_{\text{Pb}}$ and using the expression of $(^{210}$Pb$)_G$ given in Equation (10), leads to an expression of $(^{210}$Bi$)_G$. This expression is then used to determine $(^{210}$Po$)_G$, still using Equation (5). The obtained expressions are reproduced in Appendix A and, like Equations (9) and (10), they have the following form:

$$A_0 \mathcal{F}(\tau, \alpha, D_{\text{Pb}}, D_{\text{Bi}}, D_{\text{Po}}, f), \tag{11}$$

where $\mathcal{F}$ is a function of τ, α, f and D_{Pb}, D_{Bi} and D_{Po}. Therefore, radioactive disequilibria $(^{222}$Rn$/^{210}$Pb$)$, $(^{210}$Bi$/^{210}$Pb$)$ and $(^{210}$Po$/^{210}$Pb$)$ do not depend on the magma initial activity but only on these six parameters.

For instance, dividing Equation (9) by Equation (10), $(^{222}$Rn$/^{210}$Pb$)_G$ may be expressed:

$$\left(\frac{^{222}\text{Rn}}{^{210}\text{Pb}}\right)_G = \frac{f}{\alpha}\left(1 + \frac{\tau \lambda_{\text{Rn}}}{1-\alpha}\right)\left(\frac{\lambda_{\text{Pb}} \frac{1-f}{1-\alpha} + \frac{1}{\tau}}{\frac{\lambda_{\text{Pb}}}{D_{\text{Pb}}} + (\alpha + \frac{1-\alpha}{D_{\text{Pb}}}) \frac{1}{\tau}}\right)^{-1}, \tag{12}$$

$(^{210}$Bi$/^{210}$Pb$)_G$ and $(^{210}$Po$/^{210}$Pb$)_G$ are given by Equations (6) and (9) in Gauthier et al. [15].

3.3. Gas Phase Transfer towards the Surface: Radioactive Decay within Gas Bubbles

Activities of each radionuclide in the gas phase after a transfer time θ are noted $(I_k)_G^\theta$. Assuming that the gas phase behaves as a closed system as far as radionuclides are concerned, the activities after a transfer time θ can be obtained by solving the following system of coupled equations:

$$\begin{cases} \partial(^{222}\text{Rn})_G^\theta/\partial\theta &= -\lambda_{\text{Rn}}(^{222}\text{Rn})_G^\theta, \\ \partial(^{210}\text{Pb})_G^\theta/\partial\theta &= \lambda_{\text{Pb}}(^{222}\text{Rn})_G^\theta - \lambda_{\text{Pb}}(^{210}\text{Pb})_G^\theta, \\ \partial(^{210}\text{Bi})_G^\theta/\partial\theta &= \lambda_{\text{Bi}}(^{210}\text{Pb})_G^\theta - \lambda_{\text{Bi}}(^{210}\text{Bi})_G^\theta, \\ \partial(^{210}\text{Po})_G^\theta/\partial\theta &= \lambda_{\text{Po}}(^{210}\text{Bi})_G^\theta - \lambda_{\text{Po}}(^{210}\text{Po})_G^\theta, \end{cases} \tag{13}$$

with initial conditions $(I_k)_G^{\theta=0} = (I_k)_G$, where $(I_k)_G$ are the expressions established above.

This system has been solved by Bateman [38] in its generalized formulation. The solution has also been formalized in an algebraic way e.g., [39]. This approach is more suitable for a numerical implementation and is thus chosen:

$$\begin{pmatrix} (^{222}\text{Rn})_G \\ (^{210}\text{Pb})_G \\ (^{210}\text{Bi})_G \\ (^{210}\text{Po})_G \end{pmatrix}^\theta = M \times \begin{pmatrix} e^{-\lambda_{\text{Rn}}\theta} & 0 & 0 & 0 \\ 0 & e^{-\lambda_{\text{Pb}}\theta} & 0 & 0 \\ 0 & 0 & e^{-\lambda_{\text{Bi}}\theta} & 0 \\ 0 & 0 & 0 & e^{-\lambda_{\text{Po}}\theta} \end{pmatrix} \times M^{-1} \times \begin{pmatrix} (^{222}\text{Rn})_G \\ (^{210}\text{Pb})_G \\ (^{210}\text{Bi})_G \\ (^{210}\text{Po})_G \end{pmatrix}^{\theta=0}, \tag{14}$$

where M and M^{-1} are matrices for which coefficients are only a function of radioactive constants λ_k and are reproduced in Appendix B (Equations (A3) and (A4)). Because initial conditions have the form described by Equation (11), the solution described by Equation (14) also has the same form.

Therefore, radioactive disequilibria in the gas phase after a given transfer time do not depend on the initial activity of I_k in the magma. Their expression are not reproduced here, but they can be explicited replacing M, M^{-1}, $(^{222}\text{Rn})_G$, $(^{210}\text{Pb})_G$, $(^{210}\text{Bi})_G$ and $(^{210}\text{Po})_G$ by their expressions provided before or in the appendix.

3.4. Results and Discussion

Radioactive disequilibria between ^{210}Pb, ^{210}Bi and ^{210}Po in the gas phase according to our model are presented in Figure 4. Values of α, f and volatility for Pb, Bi and Po (D or equivalently ϵ) are fixed using reasonable estimates from the literature ($\alpha = 5$ wt.%, $f = 1$, $\epsilon_{Pb} = 1.5\%$, $\epsilon_{Bi} = 36\%$, $\epsilon_{Po} = 100\%$). The choice of these values, and their impact on the model predictions, are discussed later (see Section 4). Theoretical values of disequilibria are plotted against the residence time in the degassing reservoir for several values of the transfer time. These new results are systematically compared to radioactive disequilibria produced by the model of Gauthier et al. [15].

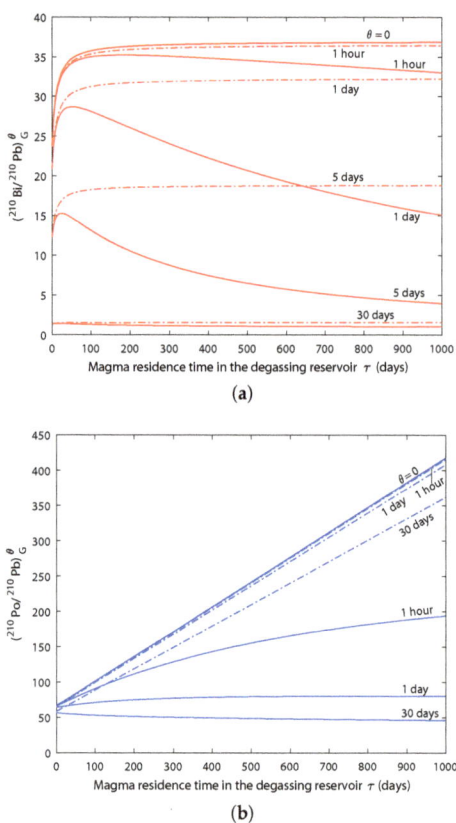

Figure 4. Radioactive disequilibria (**a**) $(^{210}\text{Bi}/^{210}\text{Pb})$ and (**b**) $(^{210}\text{Po}/^{210}\text{Pb})$ in the gas phase versus magma residence time in the degassing reservoir according to the new model (plain lines) and according the model of Gauthier et al. [15] (dashed lines). Curves for several values of transfer time θ (0, 1 h, 1, 5 and 30 days) are drawn. Fixed parameters of the model are $\alpha = 5$wt.%, $f = 1$, $\epsilon_{Pb} = 1.5\%$, $\epsilon_{Bi} = 36\%$ et $\epsilon_{Po} = 100\%$.

Radioactive disequilibrium $(^{222}\text{Rn}/^{210}\text{Pb})$ in the gas phase (not shown in Figure 4) upon exsolution ($\theta = 0$) dramatically increases with the residence time and can reach values as high as 1000 for residence times up to 100 days. Therefore, huge ^{222}Rn enrichments can be generated because of the difference of gas-melt partitioning coefficients between ^{222}Rn and its non-volatile precursor ^{226}Ra. This prediction confirms the potential for important ^{210}Pb ingrowth in the gas phase during its transfer.

Radioactive disequilibria (^{210}Bi/^{210}Pb) and (^{210}Po/^{210}Pb) (Figure 4) in the gas phase at the time of exsolution ($\theta = 0$) are identical between the two models, which was expected since ^{222}Rn has no time to decay within gas bubbles. The observed trends have been explained by Gauthier et al. [15]: (^{210}Po/^{210}Pb)$_G$ significantly increases with the residence time τ, as a result of ^{210}Po regeneration in the liquid phase of the degassing reservoir; (^{210}Bi/^{210}Pb)$_G$ also increases with τ but at a slower rate. This is because the regeneration of ^{210}Bi by ^{210}Pb decay in the liquid phase is not as important, which is due to the smaller difference of volatility between Pb and Bi than between Pb and Po.

In contrast, when the escape time of gases is not negligible ($\theta > 0$), our model predicts significantly lower (by a factor 2 or more) values than those derived from Gauthier et al. [15]. (^{210}Bi/^{210}Pb) activity ratio in the gas phase, as computed with the former model ([15], dashed lines in Figure 4a), decreases with increasing values of θ from 0 to 30 days owing to the short half-life of ^{210}Bi. In about one month (6 times ^{210}Bi half-life of 5.01 days), ^{210}Bi is back to equilibrium with its parent ^{210}Pb, leading to an activity ratio of 1. Our model suggests, however, that low values of (^{210}Bi/^{210}Pb)$_G$ can be produced for gas transfer time θ as short as a few days (solid lines, Figure 4a). This feature is explained by both the radioactive decay of ^{210}Bi according to its own half-life and the radioactive decay of ^{222}Rn within gas bubbles, which produces new ^{210}Pb atoms. Accordingly, (^{210}Bi)$_G^{\theta}$ decreases for increasing values of θ while (^{210}Pb)$_G^{\theta}$ increases, leading to a faster decrease in the (^{210}Bi/^{210}Pb)$_G$ activity ratio. However, for longer transfer times (e.g., $\theta = 30$ days), the two models produce again similar values close to the equilibrium ratio of 1, as expected since both ^{222}Rn and ^{210}Bi have similar half-lives (3.82 days and 5.01 days, respectively). In other words, the most important difference between the two models happens for magma residence times τ higher than a few hundred days (significant regeneration of ^{222}Rn from ^{226}Ra in the degassing melt and subsequent radon enrichments in the gas phase) and gas transfer times θ shorter than a few days (significant ^{222}Rn-driven production of novel atoms of ^{210}Pb within gas bubbles with limited return of ^{210}Bi to radioactive equilibrium).

The same conclusion can be drawn regarding (^{210}Po/^{210}Pb)$_G$ activity ratios (Figure 4b). The effect of ^{222}Rn decay is even more pronounced since the decrease in (^{210}Po)$_G^{\theta}$ according to the half-life of ^{210}Po (138.4 days) is rather limited for short transfer times θ. The radioactive decay of ^{222}Rn thus strongly controls the magnitude of ^{210}Pb-^{210}Po radioactive disequilibria. For instance, for values of τ higher than 150 days, it can be seen that our degassing model produces (^{210}Po/^{210}Pb)$_G$ activity ratios at $\theta = 1$ h lower than those derived from Gauthier et al. [15] for $\theta = 30$ days.

These results highlight the need for taking ^{222}Rn into account in dynamic degassing models. They also suggest that activity ratios in the gas phase as low as those measured in the plume of Mount Etna could be explained and modelled within this novel theoretical framework.

4. Model Application

It can be seen from Equations (11) and (14) that the model relies on two variables (i.e., the magma residence time τ and the gas transfer time θ) as well as on five parameters:

- the volatile weight fraction involved in the degassing process α,
- the fraction f of degassed radon [37],
- the emanation coefficients ϵ of lead, bismuth and polonium (ultimately converted to gas-melt partitioning coefficients using Equation (6)).

Provided that the five parameters α, f, ϵ_{Pb}, ϵ_{Bi}, and ϵ_{Po} can be estimated independently and the two (^{210}Po/^{210}Pb)$_G$ and (^{210}Bi/^{210}Pb)$_G$ activity ratios in the gas phase can be measured precisely, both τ and θ values can then be accurately determined from the model. However, these conditions are not necessarily met due to large analytical uncertainties on activity ratios and a broad range of values for emanation coefficients. Therefore, the mathematical system appears underdetermined.

To overcome this difficulty, we choose to use a range of likely values for the five input parameters and the two variables, based on previous results for Mount Etna and other basaltic systems. Then, we perform a Monte Carlo type simulation (see Section 4.2 for more details on the methodology) in order to determine which τ and θ values (or ranges of values) can explain activity ratios measured in the volcanic plume.

4.1. Estimation of Input Paramaters

4.1.1. Volatile Weight Fraction α

The volatile weight fraction involved in the degassing process, α, controls the relative proportion of gas and liquid phases coexisting in the degassing reservoir. It is a key parameter because gaseous trace element compounds are too scarce to nucleate gas bubbles in the melt, and they need a major volatile species (e.g., H_2O) to be flushed out of the magma [40]. The total volatile content of magmas is often estimated from volatile concentrations in melt inclusions trapped in crystals, assuming that these melt inclusions represent the deep undegassed magma. In magmatic systems, the main volatile species are, by decreasing order of importance, H_2O, CO_2, S-species (mostly SO_2 in basaltic systems), HCl and HF. Since these species have different solubilities in basalts, the depth-related pressure of inclusion entrapment has to be considered in order to derive a reliable total volatile content dissolved in the magma prior to degassing. At Mount Etna, olivine-hosted melt inclusions have been studied for long by different authors e.g., [41,42]. Their studies yield to close estimates of the volatile weight fraction α in the range 4–5 wt.%. Such high volatile content appears to be characteristic of alkali-rich basaltic magmas like those of Etna [42]. In other geodynamical settings and especially at non-arc-related volcanoes, the total amount of dissolved volatile usually is much lower [2]. Nevertheless, it must be pointed out that the model still applies to these volcanoes, provided that α is carefully quantified.

4.1.2. Fraction of Degassed Radon f

The fraction of radon released upon magma degassing, f, is a major parameter of our model since it directly controls the magnitude of ^{222}Rn enrichments in the gas phase and, subsequently, the radioactive ingrowth of ^{210}Pb within gas bubbles. Very few data exist in the literature about radon degassing from basaltic magmas. Nevertheless, analyses of freshly erupted basalts and andesites [27,43] provide evidence for almost thorough radon degassing from erupting magmas. Further experimental studies [40,44] confirm that radon is entirely flushed out of mafic magmas upon degassing, provided that a major gas species can act as a carrier. Although f is most likely close to 1 at Etna, we use here a conservative estimate with f varying between 90% and 100%.

4.1.3. Volatilities of Lead, Bismuth and Polonium (Emanation Coefficients ϵ and Gas-Melt Partitioning Coefficients D)

The volatility of lead and bismuth is not so easy to assess since both elements are much less volatile than radon. Lead and bismuth, as many other heavy metals, are not volatile in their pure metallic form but are rather engaged in chemical compounds (halides or sulfates) that can be degassed at magma temperature. However, near-equilibrium values of (^{226}Ra/^{210}Pb) in erupted basaltic lavas ([35], and references therein) including Etnean basalts [45] suggest a minimal loss of ^{210}Pb upon degassing and hence an emanation coefficient of lead of a few percents at most (1.5% according to [14]). Mather [26] computed a "volatility coefficient" (equivalent to a gas-melt partitioning coefficient) for lead using gas and lava data from Mount Etna, which is as low as 0.13. Using Equation (6) with $\alpha = 5$ wt.%, it corresponds to an emanation coefficient of 0.7%. Therefore, we take ϵ_{Pb} between 0.7 and 1.5% (corresponding to D_{Pb} values in the range 0.13–0.37).

When the emanation coefficient of lead is known, Pennisi et al. [17] suggested to obtain ϵ_{Bi} by scaling bismuth to lead with the ratio of their common stable isotopes according to:

$$\epsilon_{Bi} = \left(1 + \frac{1 - \epsilon_{Pb}}{\epsilon_{Pb}} \frac{[Pb]_G}{[Pb]_L} \frac{[Bi]_L}{[Bi]_G}\right), \tag{15}$$

where [X] represents concentration in lava (subscript L) and volcanic plume (subscript G). Using trace element analyses in both Etnean gas and erupted lava from Aiuppa et al. [32], we estimate $\epsilon_{Bi} = 36\%$. This value is in fair agreement with other estimates found in the literature in the case of Mount Etna: 45% in Lambert et al. [14]; 20% in Pennisi et al. [17]. A large range of variation of ϵ_{Bi} (20–45%) is thus chosen according to these few estimates. It corresponds to partitioning coefficients D_{Bi} in the range 4.75–19.6.

Almost nothing is known on the geochemical behaviour of polonium, apart from its affinity for the vapour phase. Because polonium has no stable isotope, its emanation coefficient cannot be calculated in the same way as bismuth. The value of ϵ_{Po} must therefore be determined from analyses of freshly erupted basaltic lavas e.g., [27–30]. All these studies show that erupting lavas are almost entirely Po-depleted and concur to a value of ϵ_{Po} close to 100%. At Etna, a similar method [22] suggested more incomplete polonium degassing, leading to an emanation coefficient as low as 80%. We thus use a range of ϵ_{Po} between 80% and 100%, which is $D_{Po} > 100$.

4.2. Inversion of the Model

4.2.1. Methodology

In order to assess whether the presented model can explain the measured values in the plume of Mount Etna, the following system of two equations has to be solved for τ and θ:

$$\begin{cases} \left(\frac{^{210}Bi}{^{210}Pb}\right)_G^{\tau,\theta} = 7.5 \pm 0.4, \\ \left(\frac{^{210}Po}{^{210}Pb}\right)_G^{\tau,\theta} = 42 \pm 6. \end{cases} \tag{16}$$

We prefer not to solve this system of theoretical equations for τ and θ since this would imply to know precisely each of the five parameters of the model (α, f, and emanation coefficients) while they are not so well constrained. Instead, we prefer a Monte Carlo type simulation according to the following procedure:

- each parameter (see Table 2) is chosen randomly in its range of variation according to an uniform law.
- the residence time τ and the transfer time θ are also chosen randomly (between 0 and 5000 days for τ, and between 0 and 15 days for θ). The upper limit for τ is in agreement with the order of magnitude of Mount Etna magma residence time in shallow reservoirs: a few tens of years in Condomines et al. [45], one year in Armienti et al. [46]. The upper limit for θ is coherent with maximum estimates of the gas phase transfer time at Mount Etna [14].
- radioactive disequilibria in the gas phase are computed according to the model equations.
- if the computed values match the measured ones, then the set of parameters and the dynamic variables (τ and θ) are stored in a database.
- these operations are repeated until a statistically relevant database (here 10,000 elements) is built. If enough combinations of parameters are simulated, the parameter space is sampled without any important gap.

The database is finally analysed using histograms or scatter plots (Figures 5–7).

Table 2. Ranges of value regarding each parameter for the simulation.

Parameter	Range	References
α (wt.%)	4–5	Métrich et al. [41], Spilliaert et al. [42]
f	0.9–1	Gill et al. [27], Gauthier et al. [40], Sato and Sato [43], Sato et al. [44]
ϵ_{Pb} (%)	0.7–1.5	Lambert et al. [14], Mather [26]
ϵ_{Bi} (%)	20–45	Lambert et al. [14], Pennisi et al. [17]
ϵ_{Po} (%)	80–100	Le Cloarec et al. [22], Gill et al. [27], Reagan et al. [28], Girard et al. [30]

4.2.2. Discussion of Results and Implications for Magma Dynamics at Mount Etna

The model can explain the radioactive disequilibria measured in Mount Etna plume for many combinations of parameters. Regarding input parameters (Figure 5), flat distributions observed for α and f means that, in the frame of our model, all values tested are equally compatible with the measured disequilibria. It thus appears that these two parameters have a rather limited influence on the mathematical modelling when they vary across the assigned range of values. The same conclusion holds true for ϵ_{Pb} although a significantly higher frequency is observed towards the highest values ($1.1\% < \epsilon_{Pb} < 1.5\%$). In contrast, non-flat distributions indicate that the capability of the model to reproduce measured disequilibria is very sensitive to both ϵ_{Bi} and ϵ_{Po} values, which are not constrained precisely enough. There is no satisfactory combination of parameters for $\epsilon_{Bi} < 34\%$ and $\epsilon_{Po} > 88\%$. It suggests that (i) ϵ_{Bi} might be as high as values reported by Lambert et al. [14] (45%); (ii) polonium degassing efficiency might be as low as reported by Le Cloarec et al. [22] (80%).

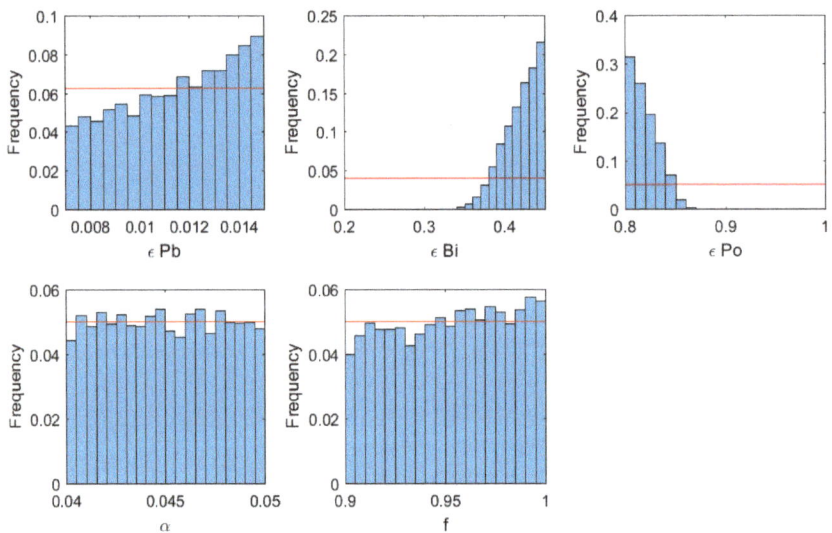

Figure 5. Histograms of the model parameters (α, f, ϵ_{Pb}, ϵ_{Bi}, ϵ_{Po}) for which the model can explain the measurements. The horizontal bar represents the uniform law used to generate random values. This line corresponds to the theoretical histogram that would be obtained for a parameter having no influence on the production of simulated results matching measured activity ratios. Its frequency is not relevant in itself and is merely equal to the reciprocal of the number of bars in the histogram.

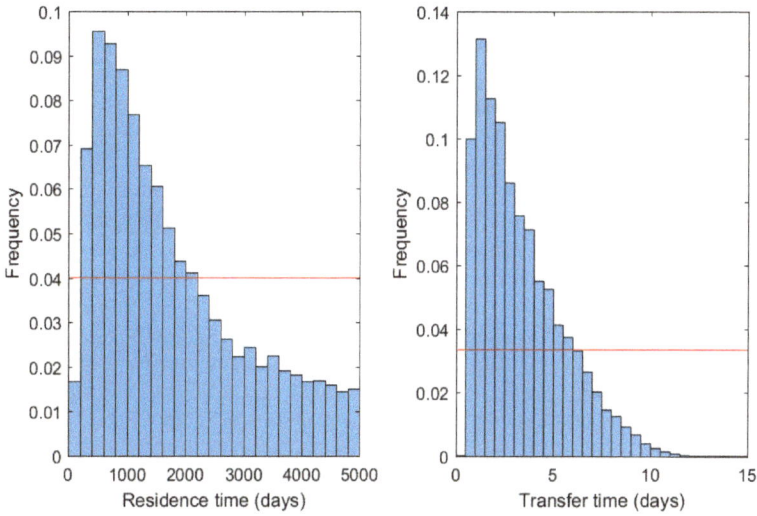

Figure 6. Histograms of the dynamic parameters (τ and θ) for which the model can explain the measurements. The horizontal bar represents the uniform law used to generate random values. See the caption of Figure 5 for further details.

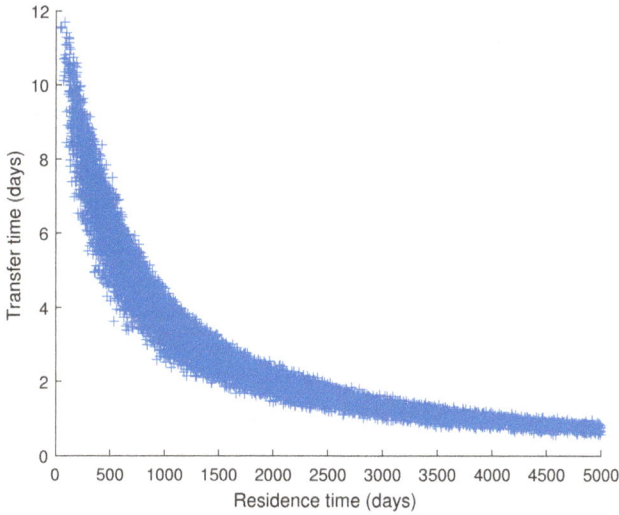

Figure 7. Scatter plot of the dynamic parameters (τ versus θ) for which the model can explain the measurements.

The histogram of the residence time values peaks between 500 and 1000 days but also extends to the maximum value of 5000 days (Figure 6). The histogram of the transfer time also peaks at low values (between 1 and 3 days) but shows instead that the computation produces no satisfactory results for $\theta > 12$ days. From a statistical perspective, it is thus most likely that measured radioactive disequilibria in the gas plume of Mount Etna are best explained by dynamical parameters in the range 500–1000 days and 1–3 days for τ and θ, respectively. Nevertheless, all satisfactory dynamic parameters align roughly on an hyperbole (Figure 7), which is logical since residence time and transfer time tend to act on disequilibria in opposed ways (disequilibria increasing with τ and decreasing with θ).

These two end-members have different geological implications. Because magma residence time is given by the relation $\tau = M/\phi_0$, higher residence times are associated with larger degassing reservoirs. At open conduit volcanoes, the transfer time of gas is likely dependent on the depth where gas exsolution takes places. Thus, longer transfer times likely correspond to deeper degassing reservoirs. The first above-mentioned end-member (high τ and low θ) is therefore associated with a large degassing reservoir stored at shallow depth. Such scenario appears unlikely since large magma bodies (>1 km^3) at very shallow depth (<1 km) would tend to erupt immediately. The second end-member (low τ and high θ) is instead associated with a smaller magma reservoir seated at a greater depth. Again, this appears unlikely as small volumes (<10^6 m^3) of magma stored at several kilometres depth would not be able to drive the present-day eruptive activity of Mount Etna.

An estimate of the magma flux entering the degassing reservoir ϕ_0 can be calculated from the mean SO$_2$ flux measured during the field campaign and the eruptive stage from 12 May to 16 May (5200 t/d). Scaled to the sulfur content of Etnean basalts (0.3 wt.% in [42]), considering complete SO$_2$ degassing and a magma density of 2700 kg/m^3, we find $\phi_0 = 8.7 \times 10^8$ kg/d $= 3.2 \times 10^5$ m^3/d. Owing to the definition of the magma residence time, the volume of the degassing reservoir is merely given by $V = \phi_0\tau$. Our estimate of ϕ_0 thus yields a volume of 0.15–0.30 km^3 for a degassing reservoir having a residence time of 500–1000 days (peak values in Figure 6). A residence time of 5000 days (highest value tested in our simulation; Figure 6) would correspond to a reservoir volume of 1.5 km^3.

These estimates can be compared to the erupted volumes of lava during recent major eruptive events at Mount Etna. From 1995 to 2005, a cumulative volume of ca. 230 $\times$ 10^6 m^3 (dense-rock equivalent magma) has been erupted during major lava flow events [47]. These events are thought to be responsible for the complete discharge of the shallow (and degassing) plumbing system of the volcano [47] whose maximum storage capacity is thus 0.23 km^3. This value compares well with previous estimates of the shallow plumbing system of Etna derived from degassing budgets during the previous eruptive cycle (0.3–0.6 km^3, [4]) or geochemical considerations (0.5 km^3, [45]). The 0.15–0.30 km^3 estimate derived from radioactive disequilibria in volcanic gases for τ = 500–1000 days thus appears more realistic than the 1.5 km^3 end-member obtained for longer residence times.

The transfer time of gases produced by the simulation falls in the range 1–12 days. Considering a likely value of τ between 500 and 1000 days, the range for θ is reduced to 2–7 days (Figure 7). These values of a few days are in good agreement with previous determinations made for Etna [14], but they are significantly higher than those determined for the shallow (<100 m) degassing system of Stromboli where gases escape the magmatic system in one hour or less [15]. Such gas transfer time cannot be easily quantified as a depth of degassing since the bubble ascent rate through the whole feeding system of Etna is mostly unknown. However, by analogy with Stromboli volcano which produces slightly more differentiated basaltic magmas, a transfer time of a few days could be tentatively related to an exsolution depth of a few kilometres at most. The bottom of the feeding system of Etna is often envisaged at a depth of about 5 km below sea level [47]. It thus appears that effective degassing-driven radionuclide fractionation takes place within the upper part of the magmatic system.

5. Conclusions

We carried out in 2015 a new survey of ^{210}Pb-^{210}Bi-^{210}Po radioactive disequilibria in the plume of Mount Etna. Measured activity ratios in the gas phase are in good agreement with values previously reported in the literature for the volcano. However, they can not be explained by existing theoretical models accounting for radionuclide degassing.

The presence of ^{222}Rn, ignored in previous models, in the exsolved magmatic gas phase is thought to play a major role on the ^{210}Pb-^{210}Bi-^{210}Po systematics by producing ^{210}Pb excesses in the gas phase during its transfer towards the surface. Here, this contribution has been modelled theoretically and it appears to produce radioactive disequilibria that can be twice as low as those predicted by the former degassing model of Gauthier et al. [15].

When applied to the case of Mount Etna, this novel degassing model can reproduce measured activity ratios. It strengthens the validity of our approach by underscoring the importance of ^{222}Rn enrichments in producing ^{210}Pb excesses in the gas phase. Precise quantification of magma dynamics (i.e., magma residence time τ and gas transfer time θ) beneath active volcanoes through the use of our degassing model necessitates a sharp characterisation of the different input parameters, especially radionuclide emanation coefficients ϵ. Nevertheless, using a range of published estimates for ϵ values, we found that measured activity ratios in the plume of Mount Etna are most likely explained by a magma residence time in the degassing reservoir of 500–1000 days and a transfer time no longer than seven days. These figures correspond to a volume of degassing magma of about 0.15 km^3 with an exsolution depth of no more than 5 km bsl. This volume of magma and its location in the shallowest part of the volcanic edifice suggests that most of the degassing process takes place within the shallow feeding system of Etna whose dynamics controls eruptive activity at the summit craters.

Further studies should now be devoted to direct measurements of ^{222}Rn activity in diluted volcanic plumes in order to provide evidence of radon enrichments. It will have further implications, notably in better deciphering ^{210}Pb-^{210}Bi-^{210}Po desequilibria in volcanic gases, including at basaltic volcanoes from other geodynamical settings.

Acknowledgments: This work was initiated in the frame of the MEDiterranean SUpersite Volcanoes (MED-SUV) project (funding from the European Union Seventh Framework Programme (FP7) under Grant agreement No. 308665). It was also part of EtnaPlumeLab-Radioactive Aerosols (EPL-RADIO) project funded by ENVRIplus project (EU Horizon 2020 research and innovation programme 498, grant agreement No. 654182). Département de Géosciences, Ecole Normale Supérieure, Laboratoire Magmas et Volcans (through the ANR DegazMag) and Observatoire de Physique du Globe de Clermont-Ferrand provided additional funding and support and are all acknowledged. This paper benefitted from comments from two anonymous reviewers who are greatly acknowledged. This is ClerVolc contribution No. 282.

Author Contributions: All authors equally contributed to data acquisition in the field either through direct sampling of volcanic aerosols, remote-sensing measurements, or logistic support. Luca Terray and Pierre-J. Gauthier conceived the study and designed the degassing model; sample analyses were carried out by Pierre-J. Gauthier and numerical simulations were ran by Luca Terray Furthermore, Luca Terray wrote the paper with the contribution of all authors.

Conflicts of Interest: The authors declare no conflict of interest.

Appendix A

Activities of ^{210}Bi and ^{210}Po in the gas phase are determined by following the procedure described in Section 3.2.2. Corresponding expressions are reproduced hereafter:

$$(^{210}\text{Bi})_G = \left[\frac{\lambda_{\text{Bi}}}{D_{\text{Pb}}} \times \frac{\lambda_{\text{Pb}}\frac{1-f}{1-\alpha} + \frac{1}{\tau}}{\frac{\lambda_{\text{Pb}}}{D_{\text{Pb}}} + \left(\alpha + \frac{1-\alpha}{D_{\text{Pb}}}\right)\frac{1}{\tau}} (A)_0 + \frac{1}{\tau}(A)_0 \right]$$
$$\times \left[\frac{\lambda_{\text{Bi}}}{D_{\text{Bi}}} + \left(\alpha + \frac{1-\alpha}{D_{\text{Bi}}}\right)\frac{1}{\tau} \right]^{-1}, \tag{A1}$$

$$(^{210}\text{Po})_G = \left[\frac{\lambda_{\text{Po}}}{D_{\text{Bi}}} \frac{\frac{\lambda_{\text{Bi}}}{D_{\text{Pb}}} \times \frac{\lambda_{\text{Pb}}\frac{1-f}{1-\alpha} + \frac{1}{\tau}}{\frac{\lambda_{\text{Pb}}}{D_{\text{Pb}}} + \left(\alpha + \frac{1-\alpha}{D_{\text{Pb}}}\right)\frac{1}{\tau}} (A)_0 + \frac{1}{\tau}(A)_0}{\frac{\lambda_{\text{Bi}}}{D_{\text{Bi}}} + \left(\alpha + \frac{1-\alpha}{D_{\text{Bi}}}\right)\frac{1}{\tau}} + \frac{1}{\tau}(A)_0 \right]$$
$$\times \left[\frac{\lambda_{\text{Po}}}{D_{\text{Po}}} + \left(\alpha + \frac{1-\alpha}{D_{\text{Po}}}\right)\frac{1}{\tau} \right]^{-1}. \tag{A2}$$

Appendix B

The system of equations describing the evolution in time of the activities in a decay chain is often referred to as Bateman equations. The solution of Bateman equations can be written using matrices [39]. This compact formalism is much more convenient to implement in a numeric code. Bateman equations solution is given by Equation (14). Two matrices (M and its inverse matrix M^{-1}) appear in the equation. These matrices are equal to:

$$M = \begin{pmatrix} 1 & 0 & 0 & 0 \\ \frac{\lambda_{Rn}}{\lambda_{Pb}-\lambda_{Rn}} & 1 & 0 & 0 \\ \frac{\lambda_{Pb}}{\lambda_{Bi}-\lambda_{Rn}}\frac{\lambda_{Rn}}{\lambda_{Pb}-\lambda_{Rn}} & \frac{\lambda_{Pb}}{\lambda_{Bi}-\lambda_{Pb}} & 1 & 0 \\ \frac{\lambda_{Bi}}{\lambda_{Po}-\lambda_{Rn}}\frac{\lambda_{Pb}}{\lambda_{Bi}-\lambda_{Rn}}\frac{\lambda_{Rn}}{\lambda_{Pb}-\lambda_{Rn}} & \frac{\lambda_{Bi}}{\lambda_{Po}-\lambda_{Pb}}\frac{\lambda_{Pb}}{\lambda_{Bi}-\lambda_{Pb}} & \frac{\lambda_{Bi}}{\lambda_{Po}-\lambda_{Bi}} & 1 \end{pmatrix}, \tag{A3}$$

$$M^{-1} = \begin{pmatrix} 1 & 0 & 0 & 0 \\ \frac{\lambda_{Rn}}{\lambda_{Rn}-\lambda_{Pb}} & 1 & 0 & 0 \\ \frac{\lambda_{Rn}}{\lambda_{Rn}-\lambda_{Bi}}\frac{\lambda_{Pb}}{\lambda_{Pb}-\lambda_{Bi}} & \frac{\lambda_{Pb}}{\lambda_{Pb}-\lambda_{Bi}} & 1 & 0 \\ \frac{\lambda_{Rn}}{\lambda_{Rn}-\lambda_{Po}}\frac{\lambda_{Pb}}{\lambda_{Pb}-\lambda_{Po}}\frac{\lambda_{Bi}}{\lambda_{Bi}-\lambda_{Po}} & \frac{\lambda_{Pb}}{\lambda_{Pb}-\lambda_{Po}}\frac{\lambda_{Bi}}{\lambda_{Bi}-\lambda_{Po}} & \frac{\lambda_{Bi}}{\lambda_{Bi}-\lambda_{Po}} & 1 \end{pmatrix}, \tag{A4}$$

where λ_{Rn}, λ_{Pb}, λ_{Bi} and λ_{Po} are the radioactive decay constants of ^{222}Rn, ^{210}Pb, ^{210}Bi and ^{210}Po, respectively.

References

1. Caltabiano, T.; Burton, M.; Giammanco, S.; Allard, P.; Bruno, N.; Mure, F.; Romano, R. Volcanic gas emissions from the summit craters and flanks of Mt. Etna, 1987–2000. In *Mt. Etna: Volcano Laboratory*; Wiley: Hoboken, NJ, USA, 2004; pp. 111–128.
2. Wallace, P.J. Volatiles in subduction zone magmas: Concentrations and fluxes based on melt inclusion and volcanic gas data. *J. Volcanol. Geotherm. Res.* **2005**, *140*, 217–240.
3. Shinohara, H. Excess degassing from volcanoes and its role on eruptive and intrusive activity. *Rev. Geophys.* **2008**, *46*, doi:10.1029/2007RG000244.
4. Allard, P. Endogenous magma degassing and storage at Mount Etna. *Geophys. Res. Lett.* **1997**, *24*, 2219–2222.
5. Bonaccorso, A.; Bonforte, A.; Currenti, G.; Negro, C.D.; Stefano, A.D.; Greco, F. Magma storage, eruptive activity and flank instability: Inferences from ground deformation and gravity changes during the 1993–2000 recharging of Mt. Etna volcano. *J. Volcanol. Geotherm. Res.* **2011**, *200*, 245–254.
6. Sharma, K.; Blake, S.; Self, S.; Krueger, A.J. SO2 emissions from basaltic eruptions, and the excess sulfur issue. *Geophys. Res. Lett.* **2004**, *31*, L13612, doi:10.1029/2004GL019688.
7. Steffke, A.M.; Harris, A.J.; Burton, M.; Caltabiano, T.; Salerno, G.G. Coupled use of COSPEC and satellite measurements to define the volumetric balance during effusive eruptions at Mt. Etna, Italy. *J. Volcanol. Geotherm. Res.* **2011**, *205*, 47–53.
8. Schiano, P.; Clocchiatti, R.; Ottolini, L.; Busa, T. Transition of Mount Etna lavas from a mantle-plume to an island-arc magmatic source. *Nature* **2001**, *412*, 900–904.
9. Kazahaya, K.; Shinohara, H.; Saito, G. Excessive degassing of Izu-Oshima volcano: Magma convection in a conduit. *Bull. Volcanol.* **1994**, *56*, 207–216.
10. Stevenson, D.S.; Blake, S. Modelling the dynamics and thermodynamics of volcanic degassing. *Bull. Volcanol.* **1998**, *60*, 307–317.
11. Beckett, F.; Burton, M.; Mader, H.; Phillips, J.; Polacci, M.; Rust, A.; Witham, F. Conduit convection driving persistent degassing at basaltic volcanoes. *J. Volcanol. Geotherm. Res.* **2014**, *283*, 19–35.
12. Ferlito, C.; Coltorti, M.; Lanzafame, G.; Giacomoni, P.P. The volatile flushing triggers eruptions at open conduit volcanoes: Evidence from Mount Etna volcano (Italy). *Lithos* **2014**, *184*, 447–455.
13. Moretti, R.; Métrich, N.; Arienzo, I.; Renzo, V.D.; Aiuppa, A.; Allard, P. Degassing vs. eruptive styles at Mt. Etna volcano (Sicily, Italy). Part I: Volatile stocking, gas fluxing, and the shift from low-energy to highly explosive basaltic eruptions. *Chem. Geol.* **2017**, doi:10.1016/j.chemgeo.2017.09.017.

14. Lambert, G.; Le Cloarec, M.F.; Ardouin, B.; Le Roulley, J.C. Volcanic emission of radionuclides and magma dynamics. *Earth Planet. Sci. Lett.* **1985**, *76*, 185–192.

15. Gauthier, P.J.; Le Cloarec, M.F.; Condomines, M. Degassing processes at Stromboli volcano inferred from short-lived disequilibria (^{210}Pb–^{210}Bi–^{210}Po) in volcanic gases. *J. Volcanol. Geotherm. Res.* **2000**, *102*, 1–19.

16. Lambert, G.; Bristeau, P.; Polian, G. Emission and enrichments of radon daughters from Etna volcano magma. *Geophys. Res. Lett.* **1976**, *3*, 724–726.

17. Pennisi, M.; Le Cloarec, M.F.; Lambert, G.; Le Roulley, J.C. Fractionation of metals in volcanic emissions. *Earth Planet. Sci. Lett.* **1988**, *88*, 284–288.

18. Le Cloarec, M.F.; Pennisi, M.; Ardouin, B.; Le Roulley, J.C.; Lambert, G. Relationship between gases and volcanic activity of Mount Etna in 1986. *J. Geophys. Res. Solid Earth* **1988**, *93*, 4477–4484.

19. Le Cloarec, M.F.; Pennisi, M. Radionuclides and sulfur content in Mount Etna plume in 1983–1995: New constraints on the magma feeding system. *J. Volcanol. Geotherm. Res.* **2001**, *108*, 141–155.

20. Allard, P.; Aiuppa, A.; Bani, P.; Métrich, N.; Bertagnini, A.; Gauthier, P.J.; Shinohara, H.; Sawyer, G.; Parello, F.; Bagnato, E.; et al. Prodigious emission rates and magma degassing budget of major, trace and radioactive volatile species from Ambrym basaltic volcano, Vanuatu island Arc. *J. Volcanol. Geotherm. Res.* **2016**, *322*, 119–143.

21. Gauthier, P.J.; Sigmarsson, O.; Gouhier, M.; Haddadi, B.; Moune, S. Elevated gas flux and trace metal degassing from the 2014–2015 fissure eruption at the Bárðarbunga volcanic system, Iceland. *J. Geophys. Res. Solid Earth* **2016**, *121*, 1610–1630.

22. Le Cloarec, M.F.; Lambert, G.; Le Guern, F.; Ardouin, B. Echanges de matériaux volatils entre phase solide, liquide et gazeuse au cours de l'éruption de l'Etna de 1983. *Comptes-Rendus des Séances de l'Académie des Sciences. Série 2, Mécanique-Physique, Chimie, Sciences de l'univers, Sciences de la terre* **1984**, *298*, 805–808. (In French)

23. Le Cloarec, M.F.; Lambert, G.; Ardouin, B. Isotopic enrichment of ^{210}Pb in gaseous emission from Mount Etna (Sicily). *Chem. Geol.* **1988**, *70*, 128.

24. Behncke, B.; Branca, S.; Corsaro, R.A.; Beni, E.D.; Miraglia, L.; Proietti, C. The 2011–2012 summit activity of Mount Etna: Birth, growth and products of the new SE crater. *J. Volcanol. Geotherm. Res.* **2014**, *270*, 10–21.

25. Geryes, T.; Monsanglant-Louvet, C. Determination of correction factors for alpha activity measurements in the environment (conditions of high dust loading). *Radiat. Prot. Dosim.* **2010**, *144*, 659–662.

26. Mather, T.A. Volcanoes and the environment: Lessons for understanding Earth's past and future from studies of present-day volcanic emissions. *J. Volcanol. Geotherm. Res.* **2015**, *304*, 160–179.

27. Gill, J.; Williams, R.; Bruland, K. Eruption of basalt and andesite lava degasses ^{222}Rn and ^{210}Po. *Geophys. Res. Lett.* **1985**, *12*, 17–20.

28. Reagan, M.; Tepley, F.J.; Gill, J.B.; Wortel, M.; Hartman, B. Rapid time scales of basalt to andesite differentiation at Anatahan volcano, Mariana Islands. *J. Volcanol. Geotherm. Res.* **2005**, *146*, 171–183.

29. Sigmarsson, O.; Condomines, M.; Gauthier, P.J. Excess ^{210}Po in 2010 Eyjafjallajökull tephra (Iceland): Evidence for pre-eruptive gas accumulation. *Earth Planet. Sci. Lett.* **2015**, *427*, 66–73.

30. Girard, G.; Reagan, M.K.; Sims, K.W.W.; Thornber, C.R.; Waters, C.L.; Phillips, E.H. ^{238}U–^{230}Th–^{226}Ra–^{210}Pb–^{210}Po Disequilibria Constraints on Magma Generation, Ascent, and Degassing during the Ongoing Eruption of Kīlauea. *J. Petrol.* **2017**, *58*, 1199–1226.

31. Gauthier, P.J.; Le Cloarec, M.F. Variability of alkali and heavy metal fluxes released by Mt. Etna volcano, Sicily, between 1991 and 1995. *J. Volcanol. Geotherm. Res.* **1998**, *81*, 311–326.

32. Aiuppa, A.; Dongarrà, G.; Valenza, M.; Federico, C.; Pecoraino, G. Degassing of trace volatile metals during the 2001 eruption of Etna. In *Volcanism and the Earth's Atmosphere*; Wiley: Hoboken, NJ, USA, 2003; pp. 41–54.

33. Kayzar, T.M.; Cooper, K.M.; Reagan, M.K.; Kent, A.J. Gas transport model for the magmatic system at Mount Pinatubo, Philippines: Insights from (^{210}Pb)/(^{226}Ra). *J. Volcanol. Geotherm. Res.* **2009**, *181*, 124–140.

34. Condomines, M.; Sigmarsson, O.; Gauthier, P. A simple model of ^{222}Rn accumulation leading to ^{210}Pb excesses in volcanic rocks. *Earth Planet. Sci. Lett.* **2010**, *293*, 331–338.

35. Berlo, K.; Turner, S. ^{210}Pb-^{226}Ra disequilibria in volcanic rocks. *Earth Planet. Sci. Lett.* **2010**, *296*, 155–164.

36. Reagan, M.; Turner, S.; Handley, H.; Turner, M.; Beier, C.; Caulfield, J.; Peate, D. ^{210}Pb-^{226}Ra disequilibria in young gas-laden magmas. *Sci. Rep.* **2017**, *7*, 45186.

37. Gauthier, P.J.; Condomines, M. ^{210}Pb–^{226}Ra radioactive disequilibria in recent lavas and radon degassing: Inferences on the magma chamber dynamics at Stromboli and Merapi volcanoes. *Earth Planet. Sci. Lett.* **1999**, *172*, 111–126.

38. Bateman, H. The solution of a system of differential equations occurring in the theory of radioactive transformations. *Proc. Camb. Philos. Soc.* **1910**, *15*, 423–427.

39. Moral, L.; Pacheco, A. Algebraic approach to the radioactive decay equations. *Am. J. Phys.* **2003**, *71*, 684–686.

40. Gauthier, P.J.; Condomines, M.; Hammouda, T. An experimental investigation of radon diffusion in an anhydrous andesitic melt at atmospheric pressure: Implications for radon degassing from erupting magmas. *Geochim. Cosmochim. Acta* **1999**, *63*, 645–656.

41. Métrich, N.; Allard, P.; Spilliaert, N.; Andronico, D.; Burton, M. 2001 flank eruption of the alkali- and volatile-rich primitive basalt responsible for Mount Etna's evolution in the last three decades. *Earth Planet. Sci. Lett.* **2004**, *228*, 1–17.

42. Spilliaert, N.; Métrich, N.; Allard, P. S–Cl–F degassing pattern of water-rich alkali basalt: Modelling and relationship with eruption styles on Mount Etna volcano. *Earth Planet. Sci. Lett.* **2006**, *248*, 772–786.

43. Sato, K.; Sato, J. Estimation of gas-releasing efficiency of erupting magma from ^{226}Ra–^{222}Rn disequilibrium. *Nature* **1977**, *266*, 439–440.

44. Sato, K.; Kaneoka, I.; Sato, J. Rare-gas releasing experiments and Rn degassing from erupting magma. *Geochem. J.* **1980**, *14*, 91–94.

45. Condomines, M.; Tanguy, J.C.; Michaud, V. Magma dynamics at Mt Etna: Constraints from U-Th-Ra-Pb radioactive disequilibria and Sr isotopes in historical lavas. *Earth Planet. Sci. Lett.* **1995**, *132*, 25–41.

46. Armienti, P.; Pareschi, M.; Innocenti, F.; Pompilio, M. Effects of magma storage and ascent on the kinetics of crystal growth. *Contrib. Mineral. Petrol.* **1994**, *115*, 402–414.

47. Allard, P.; Behncke, B.; D'Amico, S.; Neri, M.; Gambino, S. Mount Etna 1993–2005: Anatomy of an evolving eruptive cycle. *Earth-Sci. Rev.* **2006**, *78*, 85–114.

geosciences

MDPI

Article

Combining Spherical-Cap and Taylor Bubble Fluid Dynamics with Plume Measurements to Characterize Basaltic Degassing

Tom D. Pering [1,*] **and Andrew J. S. McGonigle** [1,2,3]

1   Department of Geography, University of Sheffield, Winter Street, Sheffield S10 2TN, UK;
    a.mcgonigle@sheffield.ac.uk
2   Istituto Nazionale di Geofisica e Vulcanologia, Sezione di Palermo, Via Ugo La Malfa, 90146 Palermo, Italy
3   School of Geosciences, The University of Sydney, Sydney, NSW 2006, Australia
*   Correspondence: t.pering@sheffield.ac.uk; Tel.: +44-1142-227961

Received: 11 October 2017; Accepted: 22 January 2018; Published: 26 January 2018

Abstract: Basaltic activity is the most common class of volcanism on Earth, characterized by magmas of sufficiently low viscosities such that bubbles can move independently of the melt. Following exsolution, spherical bubbles can then expand and/or coalesce to generate larger bubbles of spherical-cap or Taylor bubble (slug) morphologies. Puffing and strombolian explosive activity are driven by the bursting of these larger bubbles at the surface. Here, we present the first combined model classification of spherical-cap and Taylor bubble driven puffing and strombolian activity modes on volcanoes. Furthermore, we incorporate the possibility that neighboring bubbles might coalesce, leading to elevated strombolian explosivity. The model categorizes the behavior in terms of the temporal separation between the arrival of successive bubbles at the surface and bubble gas volume or length, with the output presented on visually-intuitive two-dimensional plots. The categorized behavior is grouped into the following regimes: puffing from (a) cap bubbles; and (b) non-overpressurized Taylor bubbles; and (c) Taylor bubble driven strombolian explosions. Each of these regimes is further subdivided into scenarios whereby inter-bubble interaction does/does not occur. The model performance is corroborated using field data from Stromboli (Aeolian Islands, Italy), Etna (Sicily, Italy), and Yasur (Vanuatu), representing one of the very first studies, focused on combining high temporal resolution degassing data with fluid dynamics as a means of deepening our understanding of the processes which drive basaltic volcanism.

Keywords: strombolian; puffing; Taylor bubble; gas slug; spherical-cap bubble; basaltic volcanism

1. Introduction

Basaltic volcanism is characterized by magmas of low viscosity, ranging from 10^1–10^4 Pa·s [1], which enable the free flow of gas bubbles within the melt, in contrast to the behavior of more viscous silicic systems [2]. In basaltic magmas spherical bubbles are generated following exsolution of gas from the melt [3]. These bubbles grow via diffusion, decompression-based expansion, or coalesce to form non-spherical bubbles, e.g., of spherical-cap morphology [4–6], which can transition into Taylor bubbles (also called gas slugs), which nearly span the conduit width, and are of a length greater than, or equal to, the conduit diameter (see Figure 1 for further details on the morphological characteristics of spherical-cap and Taylor bubbles) [4,7,8]. These distinct bubble morphologies give rise to a variety of potential classes of surface degassing activity, specifically, passive degassing of spherical bubbles [2]; puffing, from bursting of non-spherical bubbles or non-over-pressurized Taylor bubbles [9–11]; and explosions from over-pressurized Taylor bubbles [12–14]. The latter scenario is associated with strombolian volcanism, as manifested on the eponymous Stromboli volcano, e.g., [9,15], where the activity has been well characterized through measurements of the erupted gas masses, e.g., [11,16–18], and studies into the explosive dynamics, e.g., [19–21].

A number of other targets worldwide also exhibit strombolian volcanism, e.g., Yasur [22], Villarica [23], Etna [24], and Pacaya [25]. Similarly, puffing activity on volcanoes has been well studied using video, thermal imagery, and gas measurement approaches, particularly on Stromboli, e.g., [10,11,26,27].

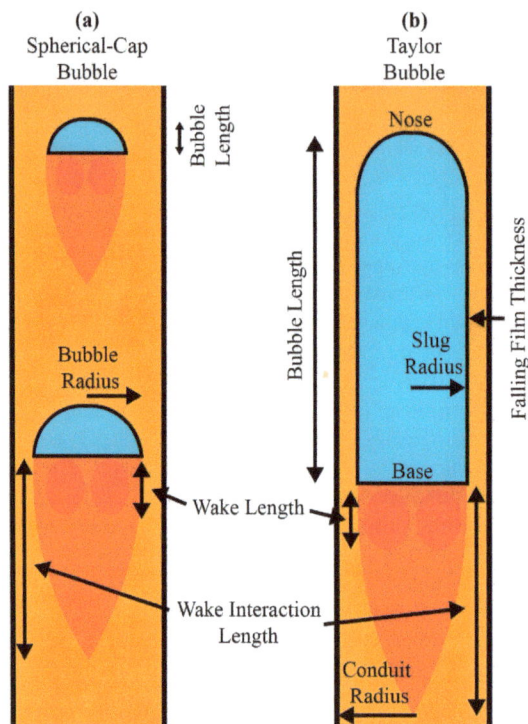

Figure 1. An illustration of (**a**) spherical-cap and (**b**) Taylor bubble morphologies including the bubble features relevant to the model described here. Any bubble falling within the wake length of the bubble ahead of it is considered liable for imminent coalescence. Any bubble beyond the interaction length would be considered to flow independently (e.g., in the single regions of the model outputs shown in Figures 2 and 3). Bubbles within the interaction length are affected by the leading bubble (e.g., falling within the rapid regions of Figures 2 and 3).

Hitherto, considerable attention has been devoted to the fluid dynamics of discrete aspects of slug flow in volcanoes, via mathematical, numerical, and laboratory modelling approaches. In particular, James et al. [13] investigated the evolution of Taylor bubble pressure and length during the ascent process. Furthermore, James et al. [28] developed a criterion to quantify the transition between puffing and explosive bursting, and Del Bello et al. [14] developed a static-pressure model for bursting Taylor bubbles. However, to date, there has been very little consideration of the fluid dynamics of spherical-cap bubbles in a volcanic scenario, bar the work of Bouche et al. [29]. Spherical-cap bubbles can be considered a transitionary morphology with characteristic shape (see Figure 1), prior to the formation of Taylor bubbles. These bubbles are characterized by a quasi-hemispherical nose and horizontal base and, unlike slugs, have lengths less than, or equal to, the conduit width, and have yet to develop a full falling film, as is the case with Taylor bubbles. There has also been very little attention devoted to resolving the implications of inter-bubble interactions within these volcanic bubble flow regimes. Recently, we have highlighted the importance of coalescence between multiple rising Taylor bubbles, in modulating the timing and intensity of high temporal resolution strombolian explosions, based on field observations on Mt. Etna [24], and laboratory experiments [30]; the potential importance of this phenomenon has also been highlighted by Gaudin et al. [26,27].

Here we present, for the first time, a combined model description of the fluid dynamics of puffing and strombolian volcanism driven by spherical-cap and Taylor bubbles. This has been achieved by: (a) bringing together prior model treatments of individual aspects of Taylor bubble flow from the volcanic and fluid dynamics literature; (b) considering, for the first time, spherical-cap bubbles in a volcanic scenario; and (c) including the previously little-considered (with notable exceptions, including [31,32]) possibility that bubbles might interact with one another to coalesce and generate larger, e.g., more explosive, masses. The model is also compared against degassing field data from Stromboli [11,17,18], Etna [24], and Yasur [33] volcanoes. This is one of the very first attempts to study volcanic degassing dynamics using a combination of modelling and gas flux time series [18], expedited by the advent of ≈1 Hz time resolution UV imaging of volcanic SO_2 fluxes, which enables the capture of rapid degassing phenomena in unprecedented detail [34–36]. This work is also one of only a few in recent years, focused on defining transitions between basaltic degassing modes, building on pioneering work performed in this area a number of decades ago, e.g., [37], that of Palma et al. [38] who identified the relationship between bubble bursting strength and the duration of the styles of basaltic volcanic activity relevant to this study, and the more recent work of Gaudin et al. [26], who categorized explosions based on bubble sizes and eruptive properties.

2. Modelling Transitions between Spherical, Non-Spherical and Taylor Bubble Flow Regimes

The model classifies strombolian and puffing degassing regimes as distinct areas on plots of inter-bubble burst spacing vs. bubble volume; this is illustrated schematically in Figure 2. In particular, the activity is categorized within the following classes: puffing from spherical-cap bubbles; and puffing from Taylor bubbles; and Taylor bubble-driven strombolian explosions. There is also a further subdivision of these classes into scenarios whereby the bubbles can/cannot interact with one another, i.e., "single" and "rapid" bursting regimes, and a region where bubble coalescence would inhibit the presence of independent bubbles. The model (available as an Excel spreadsheet) is contained within the supplementary material, with the underlying mathematics detailed below. In this section, transitions between bubble morphologies will be considered, i.e., the zonation with respect to bubble volume and length. In the following sections, the categorization in terms of inter-bubble spacing will be covered, specifically in terms of when inter-bubble interactions may occur.

Firstly, the degassing regimes are classified according to bubble volume (note that volume and length here are interchangeable using the formula for the volume of a cylinder, assuming a quasi-cylindrical geometry, and that data on bubble volume are often more readily acquirable than bubble lengths). Bubbles in basalts cease to act as spherical bubbles (e.g., following Stokes law) at low Reynolds numbers (Re_b), <0.3 [39], such that:

$$Re_b = \frac{\rho_m u_{sb} l}{\mu} \tag{1}$$

where ρ_m is the melt density, μ the melt viscosity, l the bubble length, and u_{sb} is the spherical bubble rise speed, from Stokes law:

$$u_{sb} = \frac{2(\rho_m - \rho_g)g\left(\frac{l}{2}\right)^2}{\mu} \tag{2}$$

where ρ_g is the density of the gas phase and g gravitational acceleration. The l at which a bubble ceases to be spherical can, therefore, be defined, enabling demarcation of the length (i.e., bubble volume) at which non-spherical bubbles form in Figure 2. Any bubbles smaller than this size would lead to passive degassing and effusive activity, for example from lava flows [40–42].

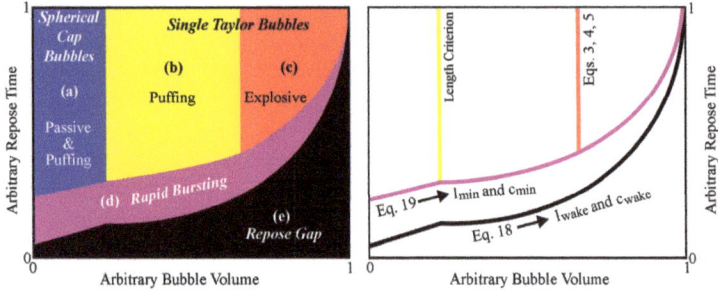

Figure 2. Left—an illustrative example of the zonation between activity classes associated with the model, plotted on arbitrary inter-bubble repose time vs. bubble volume axes. Within area (**a**) are spherical-cap bubbles, which produce passive activity or light puffing, (**b,c**) are Taylor bubble flow scenarios resulting in non-explosive and explosively-bursting scenarios, respectively. Area (**d**) represents cases where bubbles rise in sufficient proximity to one another to affect one another's fluid dynamics, while (**e**) corresponds to a region in which independent bubble bursting is unlikely due to coalescence between neighboring bubbles. Right—a further illustrative example with only defining lines between sections of the model. Important equations and bubble features are highlighted. Please see the text for full details on these.

The non-spherical bubble regime is dominated by spherical-cap and Taylor bubbles, particularly at targets such as Stromboli; therefore, our attention here will be focused on these bubble classes [5]. Spherical-cap bubbles burst passively at the magma surface or, as observed on Stromboli, can generate puffing, e.g., [10,11,19,26,27]. At larger bubble dimensions, spherical-cap bubbles transition into Taylor bubbles, when the bubble length exceeds that of the conduit diameter, e.g., [7]. Hence, areas are defined to the right of Figure 2 relating to Taylor bubble-driven activity involving puffing or strombolian explosions, depending on whether overpressure develops at the nose, e.g., [13]. This boundary has been defined by James et al. [28], who used the term, P^*_{slim}, which characterizes burst vigor, and is ≥ 1 for explosive gas release. Here, we adopt Equation (13) from [28] to define this transition, for a Taylor bubble reaching the magma surface at a pressure equal to surface atmospheric pressure, P_{surf}, giving:

$$P^*_{slim} = \frac{\sqrt{\rho_m g A' l P_{surf}}}{P_{surf}} \qquad (3)$$

where:

$$A' = 1 - \left(\frac{r_{TB}}{r_c}\right)^2 \qquad (4)$$

such that: r_{TB} is the Taylor bubble radius, which is the conduit radius, e.g., r_c minus λ', where λ' is the thickness of falling film surrounding the Taylor bubble. The falling film thickness is found, as per Llewellin et al. [43], from:

$$\lambda' = 0.204 + 0.123 \tanh(2.66 - 1.15 \log_{10} N_f) \qquad (5)$$

where the dimensionless inverse viscosity, N_f [7] is defined as:

$$N_f = \frac{\rho_m}{\mu} \sqrt{g(2r_c)^3} \qquad (6)$$

Hence, following Equation (3), the length, (and volume) at which slugs transition to explosive activity can be determined. The second element of the model is to consider the significance of inter-bubble spacing, in terms of determining whether bubbles rise independently of one another or whether interaction, which might lead to coalescence, could occur. This is considered in the following sections.

2.1. Interactions between Taylor Bubbles

The ascent velocity of a Taylor bubble base u_{TB} is determined using the Froude Number, e.g., [43,44]:

$$Fr = 0.34 \left[1 + \left(\frac{31.08}{N_f} \right)^{1.45} \right]^{-0.71} \tag{7}$$

such that:

$$u_{TB} = Fr \sqrt{2gr_c} \tag{8}$$

This speed is taken to be the average bulk rise velocity of a Taylor bubble [44] during ascent in a volcanic conduit, as expansion, which causes acceleration of the nose, occurs closer to the surface [13].

Other properties of Taylor bubbles are also important here, in particular, the wake length l_{wake}, and the wake interaction length l_{min}. The wake and wake interaction lengths are key features of a bubble during the coalescence process. Bubbles will begin to coalesce on meeting certain separation criteria. The wake length, which is typically around four times shorter than the wake interaction length, defines an area within which any trailing bubble will undergo near-instantaneous coalescence with the leading Taylor bubble, i.e., a rapid acceleration of the trailing bubble into the leading bubble as per [45]:

$$l_{wake} = 2r_c \left(0.30 + 1.22 \times 10^{-3} N_f \right) \tag{9}$$

The wake interaction length defines an area of fluidic disturbance beneath the Taylor bubble, within which any following bubble will be affected by the leading Taylor bubble through an increase in velocity, such that the trailing bubble will no longer act independently [46]:

$$l_{min} = 2r_c \left(1.46 + 4.75 \times 10^{-3} N_f \right) \tag{10}$$

2.2. Interactions between Spherical-Cap Bubbles

Spherical-cap bubbles can also be characterized in terms of Reynolds number, in this case appropriate to the length scale of these bubbles. This scale is determined by the equivalent diameter, d_e for bubbles of volume V_b [46]:

$$d_e = \left(\frac{6V_b}{\pi} \right)^{\frac{1}{3}} \tag{11}$$

Such that the Reynolds number in this case is defined as follows [46]:

$$Re = \frac{\rho_m d_e u}{\mu} \tag{12}$$

For u we use the ascent velocity of the Taylor bubble base (e.g., u_{TB}, to estimate the Re characteristics of the system using a known velocity including effects of pipe diameter). The bubble drag coefficient (C_d) is then calculated following [46]:

$$C_d = \left(2.67^{0.9} + \frac{16}{Re}^{0.9} \right)^{1.1} \tag{13}$$

This relationship is applicable for regimes with Morton numbers $>4 \times 10^{-3}$, a condition satisfied in volcanic scenarios, e.g., [14]. The spherical-cap bubble rise velocity (u_{CB}) can then be calculated using the following relationship, rearranged from Joseph [47], Equation (2.1):

$$u_{CB}^2 = \frac{\left(\frac{gd_e}{C_d}\right)}{4/3} \tag{14}$$

Coalescence between two spherical-cap bubbles occurs in the same manner as for Taylor bubbles [6,48]. To consider this, d_e is firstly converted to bubble diameter (d_b) using the constant 0.57 [4], which accounts for the non-spherical morphology of these bubbles:

$$d_b = \frac{d_e}{0.57} \tag{15}$$

The wake length of spherical-cap bubbles, c_{wake}, applicable for $Re < 200$ (e.g., appropriate to basaltic systems) is then determined as per Komasawa et al. [48]. Firstly, using bubble volume, V_b, we can calculate the volume of the wake [48]:

$$V_w = V_b Re^{0.66} \tag{16}$$

which is then used to determine the wake length, c_{wake} from the following relationship:

$$V_w = \left(\frac{\pi d_w^2}{4}\right) \times (c_{wake} - l_{CB}) \tag{17}$$

where d_w is the maximum wake diameter, which is taken as d_b, and l_{CB} is the length of the cap bubble (taken as $d_b/2$) [48]. For spherical-cap interaction length (c_{min}), there is no available prior modelling literature to refer to here, hence, we take this to be four times greater than c_{wake}, as this is the approximate scaling between the wake and wake interaction lengths in the Taylor bubble case, although, as spherical-cap bubble volume decreases, the influence of the interaction length will also decrease.

2.3. Bubble Interactions

Using the theory presented in Section 2.1 and in Section 2.2 a temporal inter-bubble separation, t_{min}, can be defined as a function of bubble volume (i.e., length) below which it would be highly improbable for an independent trailing bubble to burst at the surface. In this case, any following bubble would be travelling within the leading bubble's wake, hence, would be liable for imminent coalescence; t_{min} was, therefore, taken to be equal to the rise time of the trailing bubble (i.e., the ascent velocity of the spherical-cap or Taylor bubble) through a column of liquid of thickness equal to that of the leading wake length (l_{wake} or c_{wake}) plus the height of fluid (l_{film}), arising from complete drainage of the film surrounding the leading bubble, following each burst. The latter was constrained, in the case of Taylor bubbles, from the film volume, as a function of the slug volume, i.e., height, with knowledge of the film thickness, as per Equation (5). For spherical-cap bubbles, this was constrained from the bubble volume and conduit volume around the bubble (i.e., applying the formula for the volume of a cylinder, for given conduit and/or bubble radii):

$$t_{min} = \frac{l_{film} + l_{wake}}{u_{TB}} \text{ or } t_{min} = \frac{l_{film} + c_{wake}}{u_{CB}} \tag{18}$$

This, therefore, defines a region on Figure 2e where bubble bursts are significantly less likely to occur, i.e., within this region, all trailing bubbles are likely to undergo near-instantaneous coalescence with the leading bubble. This area is termed the "repose gap" region, following the terminology of Pering et al. [24], who, in a study of rapid bursting events on Mt. Etna, noted an absence of large mass, low repose time events, hypothesizing that this behavior was due to coalescence; this region is also discussed in a modelling framework in Pering et al. [30].

We can also define the time, $t_{transition}$, below which adjacent rising bubbles cannot be considered to behave independently (e.g., bubbles are located within the wake areas of those ahead of them), such that:

$$t_{transition} = \frac{l_{film} + l_{min}}{u_{TB}} \text{ or } t_{transition} = \frac{l_{film} + c_{min}}{u_{CB}} \tag{19}$$

These durations correspond to the time taken for a Taylor or spherical-cap bubble to rise through a column of melt of height equal to that of the drained film plus the wake interaction length, l_{min} or c_{min}, of the leading bubble (Figure 2), and are based on the lengths or volumes of bubble in question. Hence, this line subdivides the non-spherical bubble degassing classes in Figure 2, into categories where the bubbles can/cannot be considered as rising independently of one another, respectively. Note that a single area is used to denote non-independent (i.e., rapidly-bursting) Taylor and spherical-cap bubbles (Figure 2d), without segregation between explosive and non-explosive cases for Taylor bubbles, as complexities associated with pressure differences in this regime could lead to cases where the P_{slim}^* parameter would not apply.

3. Model Application to Target Volcanoes and Comparison with Field Data

The model was initiated with a range of Taylor and spherical-cap bubble volumes relevant to the volcano scale, e.g., with corresponding bubble lengths ranging from the centimeter scale for spherical-cap bubbles, to the order of meters for Taylor bubbles. We apply the model for conditions appropriate to three target volcanoes: Stromboli, Etna, and Yasur, where field data concerning puffing/strombolian explosive behavior are available with inter-event repose intervals on the order of seconds to minutes. The model outputs, with field data overlain, are shown in Figure 3. During bubble ascent, the overlying viscous magma acts to retard expansion, i.e., creating a gas overpressure. To account for this phenomenon we applied the model of Del Bello et al. [14] for Taylor bubble data for Etna and Stromboli, which provides estimates of gas overpressure at burst and the resulting bubble lengths as a function of magmatic and conduit parameters, from which burst volumes can be extracted.

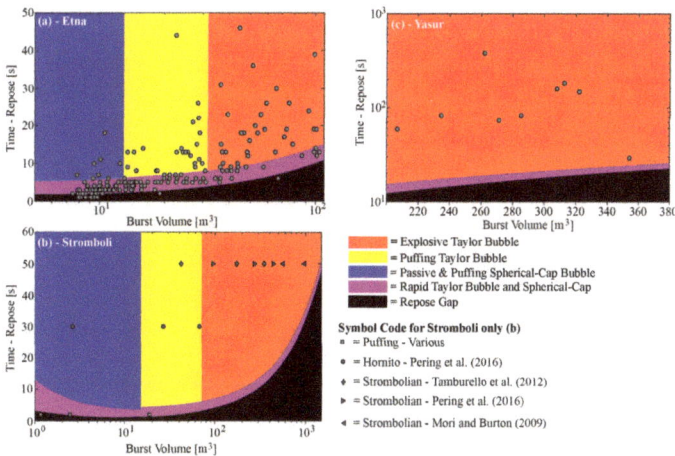

Figure 3. Outputs from the model run with input conduit and fluid dynamic parameters appropriate to: (**a**) Mt. Etna; (**b**) Stromboli; and (**c**) and Yasur volcanoes. Overlain on the plots are data points derived from field measurements on these targets, for Etna from Pering et al. [24], and Yasur from Kremers et al. [33]. In the Stromboli case, a number of literature sources are referred to, as detailed above. In each case a single repose interval is applied, which is a minimum for the observed activity. Figure 3b also just shows the maximum, mean, and minimum burst volumes from each of the Stromboli papers to simplify the graphic.

The schematic representation of the model in Figure 2 necessarily applies sharp definitions to the boundaries between defined degassing areas. In reality, there will be a degree of fuzziness around these, as the model provides a mathematical simplification of the 'real world' conditions in volcanic conduits.

Firstly, we considered the case of Mt. Etna (Figure 3a). Here we took UV camera field data from Pering et al. [24], captured during a period of very rapid bubble-bursting activity (modal inter-burst period ≈ 4 s) observed at the Bocca Nuova crater. Here, observed masses were converted to in-conduit volumes, using a pressure value of ≈65 kPa (i.e., for Mt. Etna's summit craters' altitude ≈3300 m) within the Del Bello et al. [14] model. The parameters applied within our model were: magma density of 2600 kg·m^{-3}, viscosity of 2000 Pa·s, and conduit radius of 1.5 m, e.g., [49]. In general, the field data clearly fall above the repose gap area, affirming the model suggestion from Pering et al. [24] that independently-bursting, large-volume, low repose time events would be improbable, due to inter-bubble coalescence in the conduit. The majority of bursts (62%) fall within the rapid bursting Taylor and spherical-cap bubble area, with 15% contained in the single Taylor bubble explosive area, 12% in the single Taylor bubble puffing area, and most of the remainder in the single puffing area, indicative of activity spanning the strombolian explosive-puffing spectrum. In the case of this rapid bursting scenario, the model points towards bubble interaction playing a key role in the fluid dynamics, as previously suggested by Pering et al. [24].

Secondly, for Stromboli (Figure 3b), we ran the model with density, viscosity, and conduit radius values of 2700 kg·m^{-3} [50,51], 300 Pa·s [52], and 1.5 m, respectively [53,54]. Figure 3b also includes a range of field data points, based on literature-derived main vent burst volumes [11,17,18]. These data generally fall within the single explosive Taylor bubble region, in line with the classical strombolian activity associated with this target. However, the very smallest bursts from the Tamburello et al. [11] dataset fall within the single Taylor bubble puffing region, capturing the spectrum of activity exhibited at the volcano. For the specifically-described 'puffing' events from Tamburello et al. [11], all the data points are located away from the rapid Taylor bubble bursting area. Note that in all the above cases, minimum repose times from the literature have been assigned, e.g., 50 s from Ripepe et al. [10] for puffing. Hence, even for these rather extreme prescriptions of inter-burst temporal resolution, the model points towards clearly independent bubble flow behavior.

In addition, Pering et al. [18] reported on non-explosive puffing events, from a hornito adjacent to the southeast crater, with a minimum repose interval of 30 s. These data fall within the single puffing areas for Taylor and spherical-cap bubbles. Even smaller decimeter-sized bubbles have also been associated with puffing activity from smaller vent openings at Stromboli with repose intervals of ≈0.5–2 s [9,10]. When plotted on Figure 3b, these events mostly lie in the rapid puffing area of the model, in view of their rather smaller inter-event durations than those puffing events reported in Pering et al. [18].

Thirdly, the model was run for the Yasur volcano (Figure 3c). Here, Kremers et al. [33] reported on infrasonic observations of Taylor bubble bursting, quoting slug lengths and inter-event intervals for a number of events. Here, we applied magma density and viscosity values of 2600 kg·m^{-3} and 1000 Pa·s, with a conduit radius of 1.5 m [22] within the model. In particular, we converted the Kremers et al. [33] data to the volume (see Table 1) using the formula for the volume of a cylinder (the infrasound derived length data in Kremers et al. [33] already account for pressure and viscous effects), and plot against repose time in Figure 3c. In this case all data fall within the single explosive Taylor bubble region, in line with the existent strombolian activity, and indicating independent bubble flow, well outside the repose gap region.

Table 1. Slug length and repose time data associated with rapid strombolian activity in the Yasur volcano (Kremers et al. [33] Table 2, note, only bursts with defined repose times are taken and this is then taken as the time between bursts) in addition to calculated explosive gas volumes.

Length (m)	Volume (m^3)	Repose (s)
118.5	285	82
174.9	354	29
138.5	308	160
68.6	207	59
109.7	271	73
149.5	322	149
142.1	313	185
102.6	262	380
83.7	235	82

4. Discussion and Limitations

Here, we present the first unified fluid dynamic treatment of spherical-cap and Taylor bubble-driven puffing and strombolian explosive activity in basaltic volcanism, the most ubiquitous class of activity on Earth. This involves concatenation of discrete modelled aspects of Taylor bubble flow in volcanic scenarios, the consideration of spherical-cap bubble fluid dynamics in volcanology (building on the considerations of [29]), and incorporation of the possibility of inter-bubble interaction, which has been little considered, hitherto. We also compared the model against field data from Etna, Stromboli, and Yasur volcanoes, resulting in the field data falling, as would be expected within the areal zonation of activity regimes, e.g., in respect of whether strombolian activity or puffing was manifested. In particular, the general absence of data points in the repose gap region of the plots, affirmed the expectation that bubble-coalescence would mitigate against independent bubble bursting in this area. A further model success is the seamless flow between the c_{min} and l_{min} traces, i.e., the transition between Taylor and spherical-cap regions. Hence, the mathematical treatments presented here appear not to break down close to this regime shift. The flow between c_{wake} and l_{wake} for the Etna data (Figure 3a) demonstrates similar seamlessness, however, there is a slight mismatch for the Stromboli data (Figure 3b), highlighting a need for further study into interaction lengths of spherical-cap bubbles. In this treatment, we assume that rapid strombolian activity, such as that observed at Etna, are driven by trains of fully-formed gas slugs, with associated fluid dynamic features, for the duration of ascent from depth prior to burst.

Whilst a few data fell slightly within the repose gap region of Figure 3a,b, this is likely explicable by the following: (a) The repose gap region refers to an area where bursting is improbable, rather than impossible, and is based on the assumption that a bubble cannot exist independently within the wake of another bubble; in reality such a bubble would have a very short, yet finite, lifetime which could account for the slight overlap of some data into the repose gap region, and is commensurate with the close proximity of all such events to the transition line; (b) for the Etna data, all such events are within a second of the repose gap line, which is likely at least partially a result of the margin of error of the inter-event durations, given the quantized camera acquisition time stamp and finite exposure times (100 s of ms); (c) for the Stromboli data, the very smallest puffing events from Ripepe et al. [10] fall within the repose gap, potentially indicating that this activity is associated with somewhat different magmatic and rheological properties than those reported in the literature for the bulk conduit conditions of Stromboli, and assigned to the model; alternately, this could indicate that our assumption that c_{min} is four times c_{wake}, based on the behaviour of Taylor bubbles, could be an overestimate; and (d) a final issue is that the model does not include complexities associated with the flow of bubble trains. Here, all bubble ascent velocities have been based on models associated with the flow of single bubbles in vertical conduits. In fact, in bubble trains, the rise velocities are higher than in the single bubble case [55–57], and there is also the issue of near-surface expansion of trailing bubbles, e.g., [13]. Both of these effects will act to lower the c_{min}, c_{wake}, l_{min}, and l_{wake} traces in Figures 2 and 3, plausibly also accounting for the very few data points that fell within the repose areas of the Etna and Stromboli plots. Future model development could take into consideration these effects, in addition to inclusion of more complex and realistic conduit conditions, e.g., [12], such as inclination [58,59] and formation of viscous caps at the top of the conduit [26,60], both of which are possibilities on Stromboli.

5. Conclusions

Here, we present the first unified model treatment of cap bubble and slug-based puffing and strombolian explosive degassing behavior in volcanoes. This model illustrates the exciting new scientific frontiers expedited by the recent advent of high speed imaging of volcanic gas plumes, such that models for subterranean fluid dynamics can be corroborated with surface degassing observations [61] in far more detail than previously possible with then-available temporally-coarser degassing data. Indeed, this work is one of the very first to exploit this opportunity, following from Pering et al. [18,24]. In particular, this framework offers the possibility of diagnosing underground fluid dynamic, conduit, or magmatic conditions, based on surface observations of burst masses and timings. Future work could focus on augmenting this combined UV camera to model a development framework with contemporaneous in situ gas composition data [62,63], and infrasound measurements [25]. In addition, future work will focus on validating the model using numerical and laboratory models, building on the recent work of Pering et al. [30].

This work is also focused on defining fluid dynamic transitions between disparate basaltic degassing classes, highlighting the key role played by inter-bubble separation and coalescence during such activity. There has been very little work in this area since seminal research, e.g., [37] a number of decades ago. This new capacity to develop models, informed by high time resolution UV camera observations offers exciting promise to provide step change advances in this field, extending to a wider range of basaltic styles, e.g., covering Hawaiian activity. In particular, at basaltic volcanoes such as Mt. Etna, where activity styles can transition rapidly between puffing and passive degassing through single strombolian explosions to more rapid bursting events and, finally, to Hawaiian lava fountaining [64,65], the model could be of utility in eruption forecasting.

Supplementary Materials: The following are available online at www.mdpi.com/xxx/s1, Table S1: Basaltic degassing model.

Acknowledgments: Andrew J. S. McGonigle acknowledges a Leverhulme Trust Research Fellowship (RF-2016-580), funding from the Rolex Institute, EPSRC GCRF institutional quick spend funds, a Google Faculty Research Award and NERC grant (NE/M021084/1). Tom D. Pering acknowledges the support of a NERC studentship (NE/K500914/1), the Royal Society (RG170226), the University of Sheffield, and ESRC Impact Acceleration funding. We thank three anonymous reviewers for their comments, which have improved the quality of the manuscript.

Author Contributions: Tom D. Pering designed and implemented the described model. Tom D. Pering and Andrew J. S. McGonigle wrote the manuscript.

Conflicts of Interest: The authors declare no conflict of interest.

Notation and Greek Letters

This section contains the notation and Greek letters used throughout this manuscript (listed in appearance order). Units used, where applicable, are included in brackets.

Re_b	Bubble Reynolds number
ρ_m	Magma density (kg·m^{-3})
u_{sb}	Ascent velocity of a spherical bubble (m·s^{-1})
l	Bubble length (m)
μ	Magma viscosity (Pa·s)

ρ_g	Gas density (kg·m^{-3})
g	Gravitational acceleration (m·s^{-2})
P^*_{slim}	Dimensionless burst vigour
A'	Ratio of bubble radius to pipe radius
P_{surf}	Atmospheric pressure at the surface (Pa)
r_{TB}	Taylor bubble radius (m)
r_c	Conduit radius (m)
λ'	Falling film thickness (m)
N_f	Dimensionless inverse viscosity
Fr	Froude number
u_{TB}	Taylor bubble base ascent velocity (m·s^{-1})
l_{wake}	Taylor bubble wake length (m)
l_{min}	Taylor bubble interaction length (m)
d_e	Equivalent diameter (m)
V_b	Bubble volume (m^3)
Re	Reynolds number
C_d	Bubble drag coefficient
u_{CB}	Spherical cap bubble base ascent velocity (m·s^{-1})
d_b	Bubble diameter (m)
V_w	Volume of spherical cap bubble wake (m^3)
c_{wake}	Spherical cap bubble wake length (m)
l_{CB}	Cap bubble length (m)
c_{min}	Spherical cap bubble interaction length (m)
t_{min}	Minimum repose time (s)
$t_{transition}$	Transition time (s)

References

1. Shaw, H.R.; Wright, T.L.; Peck, D.L.; Okamura, R.R. The viscosity of basaltic magma; an analysis of field measruements in Makaophui lava lake, Hawaii. *Am. J. Sci.* **1968**, *266*, 225–264. [CrossRef]
2. Parfitt, E.A. A discussion of the mechanisms of explosive basaltic eruptions. *J. Volcanol. Geotherm. Res.* **2004**, *134*, 77–107. [CrossRef]
3. Sparks, R.S.J. The dynamics of bubble formation and growth in magmas: A review and analysis. *J. Volcanl. Geotherm. Res.* **1978**, *3*, 1–37. [CrossRef]
4. Davies, R.M.; Taylor, G.I. The mechanics of large bubbles rising through extended liquids and through liquids in tubes. *Proc. R. Soc. Lond. A* **1950**, *200*, 375–390. [CrossRef]
5. Wegener, P.P.; Parlange, J. Spherical-cap bubbles. *Annu. Rev. Fluid Mech.* **1973**, *5*, 79–100. [CrossRef]
6. Bhaga, D.; Weber, M.E. Bubbles in viscous liquids: Shapes, wakes, and velocities. *J. Fluid Mech.* **1981**, *105*, 61–85. [CrossRef]
7. Wallis, G.B. *One-Dimensional Two-Phase Flow*; McGraw-Hill: New York, NY, USA, 1969.
8. Morgado, A.O.; Miranda, J.M.; Araújo, J.D.P.; Campos, J.B.L.M. Review on vertical gas-liquid slug flow. *Int. J. Multiph. Flow* **2016**, *85*, 348–368. [CrossRef]
9. Ripepe, M.; Gordeev, E. Gas bubble dynamics model for shallow volcanic tremor at Stromboli. *J. Geophys. Res.* **1999**, *104*, 10639–10654. [CrossRef]
10. Ripepe, M.; Harris, A.J.L.; Carniel, R. Thermal, seismic and infrasonic evidences of variable degassing rates at Stromboli volcano. *J. Volcanol. Geotherm. Res.* **2002**, *118*, 285–297. [CrossRef]
11. Tamburello, G.; Aiuppa, A.; Kantzas, E.P.; McGonigle, A.J.S.; Ripepe, M. Passive vs. active degassing modes at an open-vent volcano (Stromboli, Italy). *Earth Planet. Sci. Lett.* **2012**, *359–360*, 106–116. [CrossRef]
12. Seyfried, R.; Freundt, A. Experiments on conduit flow and eruption behaviour of basaltic volcanic eruptions. *J. Geophys. Res.* **2000**, *105*, 23727–23740. [CrossRef]
13. James, M.R.; Lane, S.J.; Corder, S.B. Modelling the rapid near-surface expansion of gas slugs in low-viscosity magmas. *Geol. Soc. Spec. Publ.* **2008**, *307*, 147–167. [CrossRef]

14. Del Bello, E.; Llewellin, E.W.; Taddeucci, J.; Scarlato, P.; Lane, S.J. An analytical model for gas overpressure in slug-drive explosions: Insights into Strombolian volcanic eruptions. *J. Geophys. Res. Solid Earth* **2012**, *117*, B02206. [CrossRef]

15. Chouet, B.; Hamisevi, N.; McGetchi, T.R. Photoballistics of volcanic jet activity at Stromboli, Italy. *J. Geophys. Res.* **1974**, *79*, 4961–4976. [CrossRef]

16. McGonigle, A.J.S.; Aiuppa, A.; Ripepe, M.; Kantzas, E.P.; Tamburello, G. Spectroscopic capture of 1 Hz volcanic SO_2 fluxes and integration with volcano geophysical data. *Geophys. Res. Lett.* **2009**, *36*. [CrossRef]

17. Mori, T.; Burton, M. Quantification of the gas mass emitted during single explosions on Stromboli with the SO_2 imaging camera. *J. Volcanol. Geotherm. Res.* **2009**, *188*, 395–400. [CrossRef]

18. Pering, T.D.; McGonigle, A.J.S.; James, M.R.; Tamburello, G.; Aiuppa, A.; Delle Donne, D.; Ripepe, M. Conduit dynamics and post-explosion degassing on Stromboli: A combined UV camera and numerical modelling treatment. *Geophys. Res. Lett.* **2016**, *43*, 5009–5016. [CrossRef] [PubMed]

19. Taddeucci, J.; Scarlato, P.; Capponi, A.; Del Bello, E.; Cimarelli, C.; Palladino, D.M.; Kueppers, U. High-speed imaging of Strombolian explosions: The ejection velocity of pyroclasts. *Geophys. Res. Lett.* **2012**, *39*, L02301. [CrossRef]

20. Taddeucci, J.; Alatorre-Ibarguengoitia, M.A.; Palladino, D.M.; Scarlato, P.; Camaldo, C. High-speed imaging of Strombolian eruptions: Gas-pyroclast dynamics in initial volcanic jets. *Geophys. Res. Lett.* **2015**, *42*, 6253–6260. [CrossRef]

21. Delle Donne, D.; Ripepe, M.; Lacanna, G.; Tamburello, G.; Bitetto, M.; Aiuppa, A. Gas mass derived by infrasound and UV cameras: Implications for mass flow rate. *J. Volcanol. Geotherm. Res.* **2016**, *325*, 169–178. [CrossRef]

22. Kremers, S.; Lavallée, Y.; Hanson, J.; Hess, K.-U.; Chevrel, O.; Wassermann, J.; Dingwell, D.B. Shallow magma-mingling-driven Strombolian eruptions at Mt. Yasur volcano, Vanuatu. *Geophys. Res. Lett.* **2012**, *39*, L21304. [CrossRef]

23. Shinohara, H.; Witter, J.B. Volcanic gases emitted during mild Strombolian activity of Villarrica volcano, Chile. *Geophys. Res. Lett.* **2005**, *32*, L20308. [CrossRef]

24. Pering, T.D.; Tamburello, G.; McGonigle, A.J.S.; Aiuppa, A.; James, M.R.; Lane, S.J.; Sciotto, M.; Cannata, A.; Patane, D. Dynamics of mild strombolian activity on Mt. Etna. *J. Volcanol. Geotherm. Res.* **2015**, *300*, 103–111. [CrossRef]

25. Dalton, M.P.; Waite, G.P.; Watson, I.M.; Nadeau, P.A. Multiparameter quantification of gas release during weak Strombolian eruptions at Pacaya volcano, Guatemala. *Geophys. Res. Lett.* **2010**, *37*, L09303. [CrossRef]

26. Gaudin, D.; Taddeucci, J.; Scarlato, P.; Del Bello, E.; Ricci, T.; Orr, T.; Houghton, B.; Harris, A.; Rao, S.; Bucci, A. Integrating puffing and explosions in a general scheme for Strombolian-style activity. *J. Geophys. Res. Solid Earth* **2017**, *122*, 1860–1875. [CrossRef]

27. Gaudin, D.; Taddeucci, J.; Scarlato, P.; Harris, A.; Bombrun, M.; Del Bello, E.; Ricci, T. Characteristics of puffing activity revealed by ground-based, thermal infrared imaging: The example of Stromboli Volcano (Italy). *Bull. Volcanol.* **2017**, *79*. [CrossRef]

28. James, M.R.; Lane, S.J.; Wilson, L.; Corder, S.B. Degassing at low magma-viscosity volcanoes: Quantifying the transition between passive bubble-burst and Strombolian eruption. *J. Volcanol. Geotherm. Res.* **2009**, *180*, 81–88. [CrossRef]

29. Bouche, E.; Vergniolle, S.; Staudacher, T.; Nercessian, A.; Delmont, J.-C.; Frogneux, M.; Cartault, F.; Le Pichon, A. The role of large bubbles detected from acoustic measurements on the dynamics of Erta 'Ale lava lake (Ethiopia). *Earth Planet. Sci. Lett.* **2010**, *295*, 37–48. [CrossRef]

30. Pering, T.D.; McGonigle, A.J.S.; James, M.R.; Capponi, A.; Lane, S.J.; Tamburello, G.; Aiuppa, A. The dynamics of slug trains in volcanic conduits: Evidence for expansion driven slug coalescence. *J. Volcanol. Geotherm. Res.* **2017**, *348*, 26–35. [CrossRef]

31. Llewellin, E.; Del bello, E.; Lane, S.J.; Capponi, A.; Mathias, S.; Taddeucci, J. Cyclicity in slug-driven basaltic eruptions: Insights from large-scale analogue experiments. In Proceedings of the EGU General Assembly; 2013, Vienna, Austria, 7–12 April 2013.

32. Llewellin, E.W.; Burton, M.R.; Mader, H.M.; Polacci, M. Conduit speed limit promotes formation of explosive 'super slugs'. In Proceedings of the AGU Fall Meeting 2014, San Francisco, CA, USA, 15–19 December 2014.

33. Kremers, S.; Wassermann, J.; Meier, K.; Pelties, C.; van Driel, M.; Vasseur, J.; Hort, M. Inverting the source mechanism of Strombolian explosions at Mt. Yasur, Vanuatu, using a multi-parameter dataset. *J. Volcanol. Geotherm. Res.* **2013**, *262*, 104–122. [CrossRef]

34. Mori, T.; Burton, M. The SO_2 camera: A simple, fast and cheap method for ground-based imaging of SO_2 in volcanic plumes. *Geophys. Res. Lett.* **2006**, *33*, L24804. [CrossRef]

35. Bluth, G.J.S.; Shannon, J.M.; Watson, I.M.; Prata, A.J.; Realmuto, V.J. Development of an ultra-violet digital camera for volcanic SO_2 imaging. *J. Volcanol. Geotherm. Res.* **2007**, *161*, 47–56. [CrossRef]

36. Kantzas, E.P.; McGonigle, A.J.S.; Tamburello, G.; Aiuppa, A.; Bryant, R.G. Protocols for UV camera volcanic SO_2 measurements. *J. Volcanol. Geotherm. Res.* **2010**, *194*, 55–60. [CrossRef]

37. Parfitt, E.A.; Wilson, L. Explosive volcanic eruptions—IX. The transition between Hawaiian-style lava fountaining and Strombolian explosive activity. *Geophys. J. Int.* **1995**, *121*, 226–232. [CrossRef]

38. Palma, J.L.; Calder, E.S.; Basualto, D.; Blake, S.; Rothery, D.A. Correlations between SO_2 flux, seismicity, and outgassing activity at the open vent of Villarrica volcano, Chile. *J. Geophys. Res. Solid Earth* **2008**, *113*, B10201. [CrossRef]

39. James, M.R.; Lane, S.J.; Houghton, B.F. Unsteady Explosive Activity: Strombolian Eruptions. In *Modeling Volcanic Processes: The Physics and Mathematics of Volcanism*; Fagents, S.A., Gregg, T.K.P., Lopes, R.M.C., Eds.; Cambridge University Press: Cambridge, UK, 2013; pp. 107–129.

40. Manga, M. Waves of bubbles in basaltic magmas and lavas. *J. Geophys. Res.* **1996**, *101*, 17457–17465. [CrossRef]

41. Herd, R.A.; Pinkerton, H. Bubble coalescence in basaltic lava: Its impact on the evolution of bubble populations. *J. Volcanol. Geotherm. Res.* **1997**, *75*, 137–157. [CrossRef]

42. Harris, A.J.L.; Dehn, J.; Calvari, S. Lava effusion rate definition and measurement: A review. *Bull. Volcanol.* **2007**, *70*. [CrossRef]

43. Llewellin, E.W.; Del Bello, E.; Taddeucci, J.; Scarlato, P.; Lane, S.J. The thickness of the falling film of liquid around a Taylor bubble. *Proc. R. Soc. A* **2012**, *468*. [CrossRef]

44. Viana, F.; Pardo, R.; Yánez, R.; Trallero, J.L.; Joseph, D.D. Universal correlation for the rise velocity of long gas bubbles in round pipes. *J. Fluid Mech.* **2003**, *494*, 379–398. [CrossRef]

45. Campos, J.B.L.M.; Guedes de Carvalho, J.R.F.G. An experimental study of the wake of gas slugs rising in liquids. *J. Fluid Mech.* **1988**, *196*, 27–37. [CrossRef]

46. Pinto, A.M.F.R.; Campos, J.B.L.M. Coalescence of two gas slugs rising in a vertical column of liquid. *Chem. Eng. Sci.* **1996**, *51*, 45–54. [CrossRef]

47. Joseph, D.D. Rise velocity of a spherical cap bubble. *J. Fluid Mech.* **2003**, *488*, 213–223. [CrossRef]

48. Komasawa, I.; Otake, T.; Kamojima, M. Wake behaviour and its effect on interaction between spherical-cap bubbles. *J. Chem. Eng. Japan* **1980**, *13*, 103–109. [CrossRef]

49. Corsaro, R.A.; Pompilio, M. *Dynamics of Magmas at Mount Etna, in Mt. Etna: Volcano Laboratory*; Bonaccorso, A., Calvari, S., Coltelli, M., del Negro, C., Falsaperla, S., Eds.; American Geophysical Union: Washington, DC, USA, 2004.

50. Vergniolle, S.; Brandeis, G. Strombolian explosions: 1. A large bubble breaking at the surface of a lava column as a source of sound. *J. Geophys. Res.* **1996**, *101*, 20433–20447. [CrossRef]

51. Métrich, N.; Bertagnini, A.; Landi, P.; Rosi, M. Crystallization driven by decompression and water loss at Stromboli volcano (Aeolian Islands, Italy). *J. Petrol.* **2001**, *42*, 1471–1490. [CrossRef]

52. Vergniolle, S.; Brandeis, G.; Mareschal, J.C. Strombolian explosions 2. Eruption dynamics determined from acoustic measurements. *J. Geophys. Res.* **1996**, *101*, 20449–20466. [CrossRef]

53. Harris, A.J.L.; Stevenson, D.S. Thermal observations of degassing open conduits and fumaroles at Stromboli and Vulcano using remotely sensed data. *J. Volcanol. Geotherm. Res.* **1997**, *76*, 175–198. [CrossRef]

54. Delle Donne, D.; Ripepe, M. High-frame rate thermal imagery of Strombolian explosions: Implications for explosive and infrasonic source dynamics. *J. Geophys. Res. Solid Earth* **2012**, *117*, B09206. [CrossRef]

55. Krishna, R.; Urseanu, M.I.; van Baten, J.M.; Ellenberger, J. Rise velocity of a swarm of large gas bubbles in liquid. *Chem. Eng. Sci.* **1999**, *54*, 171–183. [CrossRef]

56. Mayor, T.S.; Pinto, A.M.F.R.; Campos, J.B.L.M. Vertical slug flow in laminar regime in the liquid and turbulent regime in the bubble wake—Comparison with fully turbulent and fully laminar regimes. *Chem. Eng. Sci.* **2008**, *63*, 3614–3631. [CrossRef]

57. Mayor, T.S.; Pinto, A.M.F.R.; Campos, J.B.L.M. On the gas expansion and gas hold-up in vertical slugging columns—A simulation study. *Chem. Eng. Prog.* **2008**, *47*, 799–815. [CrossRef]

58. Chouet, B.; Dawson, P.; Ohminato, T.; Martini, M.; Saccorotti, G.; Giudicpietro, F.; De Luca, G.; Milana, G.; Scarpa, R. Source mechanisms of explosions at Stromboli Volcano, Italy, determined from moment-tensor inversions of very-long period data. *J. Geophys. Res.* **2003**, *108*, ESE 7-1–ESE 7-25. [CrossRef]

59. James, M.R.; Lane, S.J.; Chouet, B.; Gilbert, J.S. Pressure changes associated with the ascent and bursting of gas slugs in liquid-filled vertical and inclined conduits. *J. Volcanol. Geotherm. Res.* **2004**, *129*, 61–82. [CrossRef]

60. Capponi, A.; James, M.R.; Lane, S.J. Gas slug ascent in a stratified magma: Implications of lfow organisation and instability for Strombolian eruption dynamics. *Earth Planet. Sci. Lett.* **2016**, *435*, 159–170. [CrossRef]

61. McGonigle, A.J.S.; Pering, T.D.; Wilkes, T.C.; Tamburello, G.; D'Aleo, R.; Bitetto, M.; Aiuppa, A. Ultraviolet imaging of volcanic plumes: A new paradigm in volcanology. *Geosciences* **2017**, *7*, 68. [CrossRef]

62. Aiuppa, A.; Federico, C.; Paonita, A.; Giudice, G.; Valenza, M. Chemical mapping of a fumarolic field: La Fossa Crater, Vulcano Island (Aeolian Islands, Italy). *Geophys. Res. Lett.* **2005**, *32*, L13309. [CrossRef]

63. Shinohara, H. A new technique to estimate volcanic gas composition: Plume measurements with a portable multi-sensor system. *J. Volcano Geotherm. Res.* **2005**, *143*, 319–333. [CrossRef]

64. Allard, P.; Burton, M.; Muré, F. Spectroscopic evidence for a lava fountain driven by previously accumulated magmatic gas. *Nature* **2005**, *433*, 407–410. [CrossRef] [PubMed]

65. Aiuppa, A.; Moretti, R.; Federico, C.; Giudice, G.; Gurrieri, S.; Liuzzo, M.; Papale, P.; Shinohara, H.; Valenza, M. Forecasting Etna eruptions by real-time observation of volcanic gas composition. *Geology* **2007**, *35*, 1115–1118. [CrossRef]

geosciences

MDPI

Article

Assessment of the Combined Sensitivity of Nadir TIR Satellite Observations to Volcanic SO$_2$ and Sulphate Aerosols after a Moderate Stratospheric Eruption

Henda Guermazi [1,2,*], Pasquale Sellitto [2], Mohamed Moncef Serbaji [1], Bernard Legras [2] and Farhat Rekhiss [1]

[1]   National School of Engineers of Sfax, Water, Energy and Environment Laboratory L3E, University of Sfax, B.P 1173, 3038 Sfax, Tunisia; moncef.serbaji@fss.rnu.tn (M.M.S.); farhat.rekhiss@enis.tn (F.R.)
[2]   Laboratoire de Météorologie Dynamique, Institut Pierre Simon Laplace, Ecole Normale Supérieure, PSL Research University, Ecole Polytechnique, Université Paris-Saclay, Sorbonne Universités, UPMC Université Paris 6, CNRS, 24 rue Lhomond, 75005 Paris, France; psellitto@lmd.ens.fr (P.S.); legras@lmd.ens.fr (B.L.)
*   Correspondence: hguermazi@lmd.ens.fr; Tel.: +216-5258-4050

Received: 20 July 2017; Accepted: 24 August 2017; Published: 13 September 2017

Abstract: Monitoring gaseous and particulate volcanic emissions with remote observations is of particular importance for climate studies, air quality and natural risk assessment. The concurrent impact of the simultaneous presence of sulphur dioxide (SO$_2$) emissions and the subsequently formed secondary sulphate aerosols (SSA) on the thermal infraRed (TIR) satellite observations is not yet well quantified. In this paper, we present the first assessment of the combined sensitivity of pseudo-observations from three TIR satellite instruments (the Infrared Atmospheric Sounding Interferometer (IASI), the MODerate resolution Imaging Spectro radiometer (MODIS) and the Spinning Enhanced Visible and InfraRed Imager (SEVIRI)) to these two volcanic effluents, following an idealized moderate stratospheric eruption. Direct radiative transfer calculations have been performed using the 4A (Automatized Atmospheric Absorption Atlas) radiative transfer model during short-term atmospheric sulphur cycle evolution. The results show that the mutual effect of the volcanic SO$_2$ and SSA on the TIR outgoing radiation is obvious after three to five days from the eruption. Therefore, retrieval efforts of SO$_2$ concentration should consider the progressively formed SSA and vice-versa. This result is also confirmed by estimating the information content of the TIR pseudo-observations to the bi-dimensional retrieved vector formed by the total masses of sulphur dioxide and sulphate aerosols. We find that it is important to be careful when attempting to quantify SO$_2$ burdens in aged volcanic plumes using broad-band instruments like SEVIRI and MODIS as these retrievals present high uncertainties. For IASI, the total errors are smaller and the two parameters can be retrieved as independent quantities.

Keywords: satellite remote sensing; volcanic emissions; SO$_2$; SSA; radiative transfer

1. Introduction

Volcanic eruptions are a major natural source of various trace gases and aerosols types that can perturb the atmospheric composition (e.g., [1,2]) and the Earth's radiative budget (e.g., [3]). These effluents, injected into the stratosphere, can produce atmospheric impacts on a relatively long time-scale [4]. About 7.5–10.0 Tg·S·year^{-1} of sulphur dioxide (SO$_2$) of volcanic origin are globally released to the atmosphere [5]. This contributes the third most abundant gas releases from volcanic activity, after water vapour and carbon dioxide. Sulphur dioxide is a precursor of secondary sulphate aerosol (SSA). These particles are efficient scatterers for short wave (solar) radiation, which can result in a global cooling of the climate system following massive [3] to moderate stratospheric eruptions [6]. In addition, SSA can absorb long wave radiation, which can result in a local warming [7]. Secondary sulphate aerosols can also promote the destruction of the stratospheric ozone by heterogeneous chemistry [8] and modify the occurrence and optical properties of clouds systems [9].

Once SO_2 is released to the atmosphere, its evolution to form aerosols particles follows two mechanisms [10]. The first mechanism represents the aqueous oxidation of SO_2 to sulphuric acid $H_2SO_{4(aq)}$ in a dilute water droplet, according to the following series of equations:

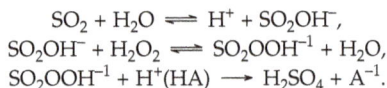

$$SO_2 + H_2O \rightleftharpoons H^+ + SO_2OH^-,$$
$$SO_2OH^- + H_2O_2 \rightleftharpoons SO_2OOH^{-1} + H_2O,$$
$$SO_2OOH^{-1} + H^+(HA) \longrightarrow H_2SO_4 + A^{-1}.$$

A is the equilibrium of SO_2 between the gas and aqueous phase. The aqueous phase oxidation occurs mostly in the troposphere where 60% to 80% of the tropospheric SO_2 are removed by this process [11]. The second mechanism is the gas phase oxidation taking place in both the troposphere and the stratosphere and involves three steps: SO_2 reacts rapidly with OH to form HSO_3, which reacts with O_2 to form SO_3. The latter reacts with H_2O to form sulfuric acid H_2SO_4 with bimolecular rate constant of 9×10^{-13} $cm^3 \cdot molecules^{-1} \cdot s^{-1}$ [12] represented by the third equation in the following series:

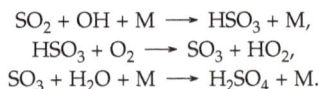

$$SO_2 + OH + M \longrightarrow HSO_3 + M,$$
$$HSO_3 + O_2 \longrightarrow SO_3 + HO_2,$$
$$SO_3 + H_2O + M \longrightarrow H_2SO_4 + M.$$

M represents an inert species N_2 or O_2 necessary for the energetic of the reaction. The lifetime of SO_2 depends on the plume altitude. It is longer in the stratosphere where the concentration of hydroxyl is relatively small [13]. It also depends on different physical processes, like the dry deposition, and the scavenging by cloud or rainwater droplets [14]. Starting from the gaseous H_2SO_4 formed with this mechanism, SSA are formed by homogeneous nucleation [15]. In the stratosphere, they are generally formed of about 75% H_2SO_4 and 25% H_2O [4] and have small deposition rates ensuring long lifetimes (of the order of months to years [16]). Sulphate aerosols formed in the troposphere are depleted by precipitation and have, therefore, shorter life times, varying from days to weeks [17]. Sulphate aerosols can also be directly emitted from the volcanic vents [18], in which case they are called primary sulphate aerosols. The average conversion rate of SO_2 to stratospheric aerosols represents an e-folding time of 30 to 40 days [16,19]. It is generally assumed that all the SO_2 emissions are converted to SSA and other sink processes are negligible.

Satellite measurements are well established tools to detect volcanic eruption and characterize the emissions, specifically for volcanoes that are not monitored by ground measurements. Their contribution is crucial, e.g., for aviation hazard mitigation. Both InfraRed (IR) and UltraViolet (UV) sensors provide near-real time measurements of SO_2 (see a review of the capabilities of the satellite instruments available at present in [20]). Aerosol remote sensing in the IR channels has received more and more attention in recent years. The imaginary part of the refractive indices of many aerosols compositions has a strong spectral variability in this domain, thus giving access to specific information on both their distribution and their composition [21,22]. Sulphur dioxide and SSA have spectral signatures in the same infrared window between 700 and 1400 cm^{-1}. This region includes two absorption bands of SO_2 centered at about 1150 and 1370 cm^{-1} [23] and also two distinctive SSA absorption features localized at 905 and 1170 cm^{-1} [24].

Monitoring the aforementioned atmospheric sulphur cycles using remote sensing is essential in order to better understand the inherent processes, and to estimate the SSA impact on the radiative transfer. In addition, usual methods to derive information on volcanic SO_2 emissions neglect the impact of subsequently formed SSA on SO_2 retrievals. This is particularly critical when using satellite data in the infrared spectral region, as both SO_2 and SSA have spectral signatures in the same band, namely between 700 and 1400 cm^{-1} [22,25,26]. In this paper, we present the first sensitivity analysis of the SO$_2$ and SSA mutual interference on pseudo-observations of three Thermal InfraRed (TIR) satellites instruments, the Infrared Atmospheric Sounding Interferometer (IASI), the MODerate resolution Imaging Spectro radiometer (MODIS) and the Spinning Enhanced Visible and InfraRed Imager (SEVIRI) , after an idealized moderate stratospheric volcanic eruption. We also assess the information content of these pseudo-observations to SO$_2$ and SSA total masses.

The paper is structured as follows: in Section 2, we present the data and methods used for this study. The results are presented and discussed in Section 3. Finally, we summarize our findings in Section 4.

2. Data and Methods

2.1. Satellite Data

In the present study, we consider three prototypical TIR nadir satellite instruments: IASI, MODIS and SEVIRI. These three instruments present different technical characteristics and advantages to observe SO$_2$ and SSA. SEVIRI has a high temporal resolution. MODIS has a high spatial resolution. Both instruments have a limited spectral resolution. On the contrary, IASI is characterized by a high spectral resolution. The MODIS and SEVIRI infrared channels, in the spectral range 700–1400 cm^{-1} (sensitive region to SO$_2$ and SSA absorption bands, as mentioned in Section 1), used in this work, are detailed in Table 1.

Table 1. MODIS and SEVIRI TIR bands in the spectral range 700–1400 cm^{-1} and their spectral characteristics. (* = not used in this work because of the interaction with the strong ozone absorption band at 9.7 µm (e.g., [26])).

Instrument	Channel	Central Wavenumber (cm^{-1})	Central Wavelength (µm)	Minimum Wavelength (µm)	Maximum Wavelength (µm)
SEVIRI	IR8.7	1149.42	8.70	8.30	9.10
	IR9.7 *	1035.19	9.66	9.38	9.94
	IR10.8	925.93	10.80	9.80	11.80
	IR12.0	833.33	12.00	11.00	13.00
	IR13.4	746.27	13.40	12.40	14.40
MODIS	28	1365.18	7.32	7.17	7.47
	29	1169.60	8.55	8.40	8.70
	30 *	1027.75	9.73	9.58	9.88
	31	909.62	11.03	10.78	11.28
	32	831.95	12.02	11.77	12.27
	33	749.91	13.34	13.18	13.48
	34	733.13	13.64	13.48	13.78
	35	717.36	13.94	13.78	14.08
	36	702.25	14.24	14.08	14.38

2.1.1. IASI

The Infrared Atmospheric Sounding Interferometer (IASI) is on-board MetOp-A and MetOp-B, polar orbiting meteorological satellites launched by EUMETSAT (European Organisation for the Exploitation of Meteorological Satellites) in October 2006 and September 2012, respectively. These sun-synchronous satellites perform measurements at an altitude of around 817 km and crosses the equator twice daily at 9:30 a.m. and at 9:30 p.m. local time in a descending and an ascending node, respectively. The characteristics of the IASI instrument are detailed by Clerbaux et al. [27]. The IASI instrument is a Fourier Transform Spectrometer that measures the infrared radiation emitted from the Earth in the range of 3.4–15.5 µm corresponding to 645–2760 cm^{-1}. This important spectral coverage allows the retrieval of temperature and water vapour profiles, and contains absorption bands of many atmospheric gases, like carbon dioxide, ozone, methane and others. The spectral resolution, in our spectral region of interest, is 0.5 cm^{-1}, after apodisation. IASI observations have been extensively used to monitor volcanic SO_2 amounts (e.g., [23,28]). Volcanic SSA have been also recently studied (e.g., [22,25]).

2.1.2. MODIS

The MODerate resolution Imaging Spectroradiometer (MODIS) is part of the National Aeronautics and Space Administration (NASA) Earth Observing System (EOS). It provides long-term global observation of the Earth's land, ocean and atmospheric properties. The MODIS instrument is designed to achieve a trade-off of relatively high spectral, spatial and temporal resolution, with a priority on spatial resolution and imaging capabilities. MODIS observes the Earth with a 2330 km swath, from a polar orbit approximately 700 km above the surface and ±55° views scan. MODIS is flying on two NASA satellites, Terra and Aqua, which are polar-orbiting sun-synchronous platforms. Terra and Aqua were launched on 18 December 1999 and 4 May 2002, respectively. The Terra orbit passes from North to South and crosses the equator at about 10:30 a.m., while Aqua has an ascending orbit and crosses the equator at about 1:30 p.m. This instrument acquires data at 36 spectral bands (0.4–14.4 μm), with 29 spectral bands (bands 8–36) are located in the middle and long wave TIR spectral regions. In these bands, the spatial resolution is 1 km. MODIS observations have been extensively used to monitor volcanic SO_2 and ash amounts (e.g., [24,29,30]), but no SSA inversion algorithm is available at present, to our knowledge.

2.1.3. SEVIRI

The Spinning Enhanced Visible and Infrared Imager (SEVIRI), the main sensor of the Meteosat Second Generation (MSG) geostationary satellite, orbits the Earth at an approximate altitude of 36,000 km with a period of 24 h and a nadir point of approximately 3° W, over the equator. The instrument is a line-by-line scanning radiometer, which provides image data in four Visible and Near-InfraRed (VNIR) channels and eight IR channels. The spatial resolution of IR channels is 3 km. The key feature of this imaging instrument is the repeat cycle of 15 min. Despite their limited spectral resolution, SEVIRI measurements have been used to quantifying volcanic SO_2 [31] and detecting subsequent SSA formation [26].

2.2. Stratospheric Volcanic Sulphur Cycle

To quantitatively study the concurrent impact of volcanic SO_2 emission and subsequently formed SSA, we introduce, in this section, a chemical/micro-physical simplified model of SO_2 to SSA formation. In this work, we focus on stratospheric eruptions. The model developed here is based on the model introduced by Miles et al. [32].

We assume that all the consumed SO_2 gaseous emissions oxidise to form sulphuric acid, and that this chemical process is the sole sink for these emissions. The oxidation phase, and then the time evolution of volcanic SO_2 burdens, is controlled by Equation (1) (e.g., [13,33]), where $M_{SO_2}(t)$ is the SO_2 mass at a time t, $M_{SO_2}(t_0)$ is the total SO_2 mass loading injected by the volcano (the day of eruption, $t = 0$) and a is the e-folding time for this process. In our study, we fix SO_2 e-folding time to 3.10^{-7} s^{-1} (about 38 days lifetime), as observed by Oppenheime et al. (e.g.,[13]), and suggested as a typical value for stratospheric sulphur cycles:

$$M_{SO_2}(t) = M_{SO_2}(t_0)e^{-at}. \tag{1}$$

The oxidised SO_2 forms gaseous sulphuric acid (H_2SO_4). Starting from the amount of gaseous H_2SO_4 at time t obtained with Equation (1), the micro-physical processes leading to the formation of SSA are represented with Equation (2). In this equation, $M_{SSA}(t)$ is the time-resolved SSA effective mass volume concentration and b is the e-folding time of gaseous to particulate H_2SO_4 conversion (the other quantities have been introduced before). The factor b describes different processes going from nucleation, condensation to coagulation and is assumed here as relative to a lifetime of approximately three months, which is typical of stratospheric volcanic sulphur cycle [32]. In the present study, the loss rate of SSA (different physical processes, like gravitational settling, evaporation and wet removal) are not taken into account, having an e-folding time of the order of one year in the stratosphere (e.g., [32,34]). Our study is targeted on a short-term evolution (the first few days after the eruptive event). We suppose the formed SSA particles as binary solution systems formed of 75% H_2SO_4 and 25% H_2O. The increase in mass for this binary-solution transformation (oxidation and then nucleation/hydration) is 2.04 [32]. Thus, the initial mass of SO_2 is approximately doubled upon the aerosol formation:

$$M_{SSA}(t) = \frac{2.04abM_{SO_2}(t_0)}{b-a}(\frac{1-e^{-at}}{a} + \frac{e^{-bt}-1}{b}).$$ (2)

From the SSA mass $M_{SSA}(t)$ calculated with Equation (2), we derive the number concentration of sulphate aerosols using a logarithmic size distribution. We fix a mean radius r_m of 0.2 μm, a standard deviation σ_r of 1.86 and a sulphuric acid mixing ratio of 75% H_2SO_4 and 25% H_2O. These are representative values of SSA distributions in the upper troposphere lower stratosphere (UTLS) [35] and have been previously used by Sellitto et al. (e.g., [24]). The number concentration of aerosol distributions N_0 are calculated using Equation (3). The three parameters (N_0, r_m and sulfuric acid mixing ratio) of the aerosol size distribution are used as inputs to calculate the optical properties of aerosols (extinction, absorption and scattering coefficients and phase function) using a Mie code. These radiative calculations are described in Section 2.3:

$$M_{SSA}(t) = \frac{4}{3}\pi\rho(c)re^3 Ne.$$ (3)

In our subsequent radiative simulations, we considered an initial mass volume concentration of SO_2 injected in the stratosphere at three different plume altitudes: 18.5, 20.0 and 21.3 km. We sample four different time intervals during evolution since the eruption: 1, 3, 5 and 10 days. The evolution over time of SO_2 and SSA masses, as well as of SSA number concentration, as a function of the injection height, are summarized in Table 2.

Table 2. SO_2 mass concentration, secondary sulphate aerosols (SSA) mass concentration and particle distribution properties for the investigated time intervals and injection altitudes.

Altitude (km)	Time (days)	M_{SO_2} (g/m^3)	M_{SSA} (g/m^3)	N_0 (particles cm^{-3})	r_m (μm)	H_2SO_4 Mixing Ratio (%)
	0	1.91×10^{-4}				
	1	1.85×10^{-4}	4.98×10^{-8}	0.21		
18.5	3	1.76×10^{-4}	4.37×10^{-7}	1.88		
	5	1.67×10^{-4}	1.18×10^{-6}	5.11	0.2	75
	10	1.47×10^{-4}	4.48×10^{-6}	19.28		
	0	1.91×10^{-4}				
	1	1.85×10^{-4}	5.74×10^{-8}	0.25		
20.0	3	1.76×10^{-4}	5.05×10^{-7}	2.17		
	5	1.67×10^{-4}	1.37×10^{-6}	5.89	0.2	75
	10	1.47×10^{-4}	5.17×10^{-6}	22.24		
	0	1.91×10^{-4}				
	1	1.85×10^{-4}	6.22×10^{-8}	0.26		
21.3	3	1.76×10^{-4}	5.47×10^{-7}	2.35		
	5	1.67×10^{-4}	1.48×10^{-6}	6.39	0.2	75
	10	1.47×10^{-4}	5.60×10^{-6}	24.10		

2.3. Radiative Transfer Simulations

The IASI, MODIS and SEVIRI pseudo-observations are obtained using the radiative transfer model 4A (Automatized Atmospheric Absorption Atlas OPerational) [36]. This model is a line-by-line radiative transfer model, developed by the Laboratoire de Météorologie Dynamique and the NOVELTIS company [37] with the support of CNES (Centre National d'Études Spatiales), to allow fast forward radiative transfer calculations in the IR spectral region, using optical thickness databases, called *Atlases*. Spectra are computed at high resolution and can be convolved with various types of instrument Relative Spectral Response (RSR) functions. We simulate radiances in the range of 700–1400 cm^{-1}, with zero viewing zenith angle and with spectral resolution of 0.50 cm^{-1}, in order to fit IASI high-spectral resolution observations. The TIR calculations are subsequently convolved with MODIS and SEVIRI RSRs, to produce corresponding pseudo-observations. We consider as input for the radiative transfer calculation a typical tropical atmosphere in terms of temperature, pressure and trace gases profiles. Following the considerations of Section 2.2, a fixed SO$_2$ amount of 10 Dobson Units (DU) is injected at different altitudes (18.5, 20.0 and 21.3 km) to test the impact of volcanic effluents injections at different lower-stratospheric altitudes. The amount of 10 DU has been selected to simulate a moderate stratospheric eruption, like the recent Nabro, Sarichev or Kasatochi eruptions (see, e.g., [6]). The SO$_2$ mass concentration has been re-calculated for each injection altitude, due to the slightly different layer thicknesses in 4A (ranging between 1.3 and 1.5 km, in this altitude range). No vertical diffusion or vertical plume structure has been simulated and the SO$_2$ perturbation has been produced at one single layer each time. During the first day of the volcanic eruption ($t = 0$), only SO$_2$ perturbations are present. Subsequent SSA formation (and SO$_2$ loss) were considered at $t = 1$, $t = 3$, $t = 5$ and $t = 10$. The aerosol optical parameters (extinction coefficient, single scattering albedo and asymmetry parameters) described above for each layer are required as inputs, when aerosols are considered. These parameters are calculated using a Mie code, using the time- and altitude-dependent particle size distribution introduced in Section 2.2 and the H$_2$SO$_4$ mixing ratio-dependent refractive index of Bierman et al. [38], taken from the GEISA (Gestion et Etude des Information Spectroscopiques) spectroscopic database. The real and imaginary part of the refractive indices have been taken for a reference temperature of 215 K and for H$_2$SO$_4$ mixing ratio of 75%. The Mie scattering routines are obtained from the Earth Observation Data Group of the Department of Physics of Oxford University and they are run in IDL (Interactive Data Language). In order to solve the Radiative Transfer Equation (RTE) for the scattering aerosol contribution, we use the DIScrete ORdinaTe (DISORT) algorithm [39]. A baseline run is performed for a clear atmosphere (in the absence of both volcanic SO$_2$ and SSA), to compare with the time-dependent volcanically-perturbed simulations.

3. Results and Discussion

3.1. The SSA Spectral Extinction Coefficient Variability during Short-Term Plume Evolution

We first analyse the variability of the spectral extinction coefficient of SSA layers during the plume evolution, following the chemical/micro-physical model of Section 2.2. The evolution of the spectral extinction coefficient is illustrated in Figure 1, between 600 and 1400 cm^{-1}, for an initial volcanic injection at 20.0 km. For each time interval, the aerosol extinction generally increases with the wavenumber, as discussed by Sellitto and Legras [22]. A minimum extinction between 650 and 800 cm^{-1} and a maximum extinction at about 1170 cm^{-1} are found. A secondary maximum is also found around 905 cm^{-1}. This behaviour is principally attributed to the absorption features of the undissociated H$_2$SO$_4$ in the aerosol droplets ([22] and references therein). The extinction of the SSA layer starts to be particularly important (higher than 0.001 km^{-1} at 1170 cm^{-1}) after about five days. The extinction at 10 days is about 40 times larger than at one day after the eruption, due to the steep increase of particles number concentration from volcanic SO$_2$ conversion. According to Table 2, our chemical/micro-physical model generates SSA number concentrations of 0.2 and 22.2 particles cm^{-3}, at 1 and 10 days, respectively.

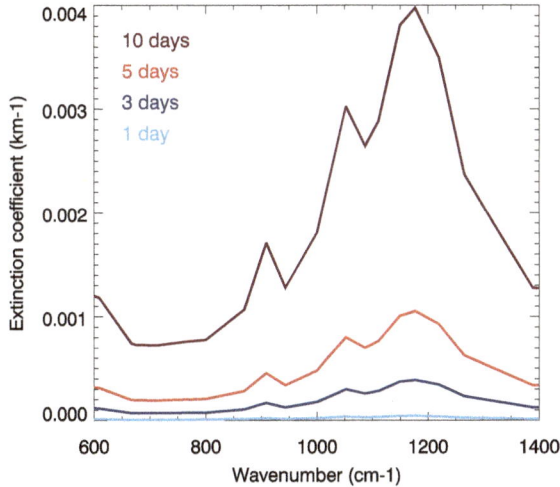

Figure 1. SSA spectral extinction coefficients variability for the investigated time intervals since eruption (one day: sky blue; three days: dark blue; five days: light red; ten days: dark red). The considered SO_2 injection altitude is 20.0 km

3.2. Brightness Temperature Pseudo-Observations Variability

In this section, we discuss the concurrent impact on the outgoing IR radiation of volcanic SO_2 and the subsequently formed SSA. In order to quantify these impacts, we use the SO_2 varying concentrations and the aerosol extinction coefficients as inputs for the forward modeling with 4A. These volcanic effluents are modelled as a single layer and introduced in a typical tropical atmosphere. The mutual SSA and SO_2 interference is studied using three prototype instrumental models, to produce SEVIRI, MODIS and IASI pseudo-observations. To get insights into the altitude dependent concurrent radiative impacts, three injection altitudes have been tested, 18.5, 20.0 and 21.3 km (near and above the tropical tropopause; the levels are fixed in the 4A Radiative Transfer Model (RTM)). The modelled plume is then extending vertically over a thickness of about 1.5 km.

Figure 2 shows the synthetic spectra for IASI BT (Brightness Temperature) plume signatures (BT pseudo-observations with SO_2 mixing ratio and SSA layer extinction, as simulated at each time interval, minus the pseudo-observation for the reference clear atmosphere), considering a plume injection altitude of 20.0 km. At the time of eruption, the residual spectrum is characterised by the SO_2-only spectral absorption in the range 1100–1200 cm^{-1}, with the largest absorption at 1150 and 1160 cm^{-1} (e.g., [23]). A second, stronger absorption feature is visible in the range 1300–1400 cm^{-1}. This absorption band presents competitive interference with water vapour [31] and is discarded from our subsequent analyses. Therefore, we consider only the spectral range between 700 and 1300 cm^{-1}. Throughout the whole evolution, spectral signatures are also influenced by the presence of other interfering species, like the ozone (strong absorption band at about 1030 cm^{-1}). For this reason, the instrumental channels affected by the strong ozone absorption are also discarded from our subsequent analyses (Channels IR8.7 for SEVIRI and 29 for MODIS). Then, for time intervals of one day, three days and five days after the eruption, the residual spectral is still markedly characterized by the SO_2 absorption, but the whole-band signature of SSA gradually appears, including the maximum extinction at about 905 and 1150 cm^{-1}. Starting from day 5, the spectral signature of the combined SSA and SO_2 is particularly apparent. The plume residual signature at about 1150–1200 cm^{-1} for day 10 is approximately two times greater than for day 5 (about −1.0 K), reaching a value of about −2.0 K. This evolution is consistent with the SO_2 depletion and SSA formation over time. These analyses confirm that the residual signature of SSA in IASI-like instruments becomes more and more important as SO_2 converts to SSAs, and the effect of SO_2 is overestimated due to the formation of these particles. As a matter of fact, starting from day 5, the SSA extinction is dominant with respect to SO_2 absorption.

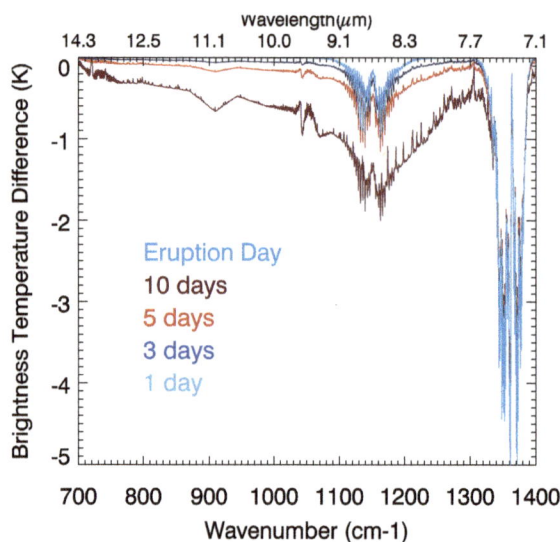

Figure 2. Plume residual IASI pseudo-observations at four different time intervals (one day: sky blue; three days: dark blue; five days: light red; ten days: dark red) since eruption (reference pseudo-observation at the eruption in light blue). An injection altitude of 20.0 km is considered here.

3.3. The Impact of the Plume Altitude

In order to investigate the influence of the initial volcanic SO_2 injection altitude on the synthetic TIR observations, in Figure 3, we show the SEVIRI (Figure 3a), MODIS (Figure 3b) and IASI (Figure 3c) pseudo-observations for plumes at three different altitudes levels: 18.5, 20.0, and 21.3 km. The BT residual for the three instruments have similar spectral behaviour, though with lower spectral resolution for SEVIRI and MODIS with respect to IASI. The maximum signature of IASI pseudo-observations at 1100–1200 cm^{-1} is translated with a maximum signature at band 29 for MODIS and at band IR8.7 for SEVIRI. At this region, the signatures of SEVIRI and MODIS are less strong than the one of IASI. This is due to the internal convolution, during 4A RTM post-processing, with SEVIRI and MODIS RSR. Accurate information on one specific chemical species can get lost due to this band averaging. The main difference of MODIS with respect to SEVIRI is the larger number of exploitable bands (eight versus five), which, in turn, provides a better information content (as discussed later). Comparing the different curves, we notice very small differences, in general smaller than about 0.1 K, as a function of the altitude. Stronger signatures are associated with higher altitudes. This difference is more and more pronounced as the conversion to SSA progresses, and is stronger after 10 days from the eruption. Thus, the higher the plume layer, the greater its impact on the BT signature. This is a reasonable result because, when the aerosol layer is more distant from the satellite platform, the absorption of radiation by overlying gases (H_2O and O_3) partially hides the signature and reaches the satellite with more attenuation. However, it must be stressed that a BT difference of 0.1 K is generally under the radiometric noise of the three satellite instruments explored in this work and then the altitude information is hardly inferable from these differences.

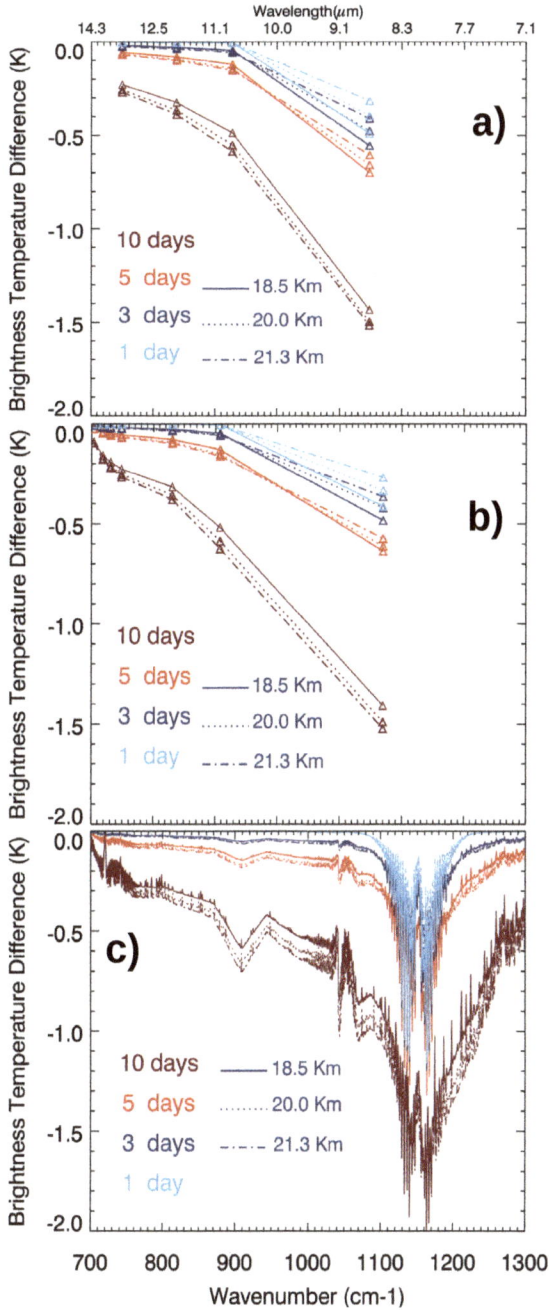

Figure 3. (a) SEVIRI; (b) MODIS and (c) IASI BT residual pseudo-observations at four different time intervals (one day: sky blue; three days: dark blue; five days: light red; ten days: dark red) and for three plume altitudes (18.5 km: solid lines; 20.0 km: dotted lines; 21.3 km: dot-dashed lines).

For the analysed case, the atmospheric sulphur cycle evolution has an important and complex impact on the TIR pseudo-observation. The effect induced by SO_2 is enhanced by the presence of sulphate aerosols. Retrieval efforts of SO_2 concentrations should consider the interference with surely coexisting SSAs and vice-versa.

3.4. Information Content of SEVIRI, MODIS and IASI Pseudo-Observations

To more accurately quantify the sensitivity of TIR pseudo-observations to SO_2 and SSA and their mutual interference shown in the previous section, here we aim to estimate the information content of these pseudo-observations on the retrieved parameters (SO_2 and SSA masses). In particular, we evaluate this information content using Rodgers theory [40]. We set-up our ideal retrieval, defining a time-dependent parameter vector $\mathbf{x}(t) = [M_{SO_2}(t), M_{SSA}(t)]$.

We first calculate the Jacobian matrix, representing the sensitivity of the spectral pseudo-observations to parameter vectors elements (in our case the two masses, M_{SO_2} and M_{SSA}). The Jacobian matrix elements, whose analytic expression is in Equation (4), are the partial derivatives of the BT measurement, at each wavenumber (index i), with respect to the retrieved parameters of the state vector (index j). From a numerical calculation perspective, for each investigated time interval t, we took a mean value of SO_2 and SSA masses ($\bar{\mathbf{x}}(t)$). The IASI, MODIS and SEVIRI spectra were then simulated with the same atmospheric and instrumental set-up described in the Section 2.3, considering small positive and negative variation of, alternatively, M_{SO_2} and M_{SSA}:

$$\mathbf{K}_j^i(t) = \left. \frac{\partial F_i(\mathbf{x}(t))}{\partial x_j(t)} \right|_{\bar{\mathbf{x}}(t)}. \tag{4}$$

$\mathbf{K}(t)$ contains two lines that represent the weighting functions with respect to the two parameters M_{SO_2} and M_{SSA}. The weighting functions of IASI, MODIS and SEVIRI pseudo-observations, as well as their temporal variability, are shown in Figure 4. Only time intervals $t = 1$ day and $t = 10$ days are shown, for a plume altitude of 20.0 km. The two time intervals represent the two extremes and so intermediate time intervals have, correspondingly, an intermediate behaviour between these two extremes. The sensitivity of TIR pseudo-observations to SSA, in terms of the weighting function, is about one order of magnitude higher that the sensitivity to SO_2. Values as high as 0.50 K·μg^{-1}·m^{-3} are found for SSA in the range of maximum sensitivity (1100–1200 cm^{-1}) for the three instruments. Approximately in the same spectral region, the weighting function values for SO_2 are about ten times smaller (about 0.05 K·μg^{-1}·m^{-3}). This is a strong indication that the SSA layer, even after a few days from the eruption, when the conversion of SO_2 has generated only a small amount of SSA in terms of its mass, is significantly more active from a radiative point of view than SO_2. From another perspective, the sensitivity of TIR observations to SO_2 can be dramatically hampered by SSA formation, even after a few days from the eruption event. Another spectral region with strong sensitivity to SSA is found around 900 cm^{-1} (IASI) and band 31 (MODIS), due to the secondary absorption feature of undissociated H_2SO_4 in SSA droplets. The MODIS band 31 is very well adapted to catch this feature, being nicely centred around the peak at 905 cm^{-1} (please refer to Figure 1 to identify this peak). On the contrary, no similar band is available for SEVIRI, which hampers the exploitation of this information. As discussed before, we avoid in the following the region with ozone absorption interference, identified in Figure 4 by white crosshatches.

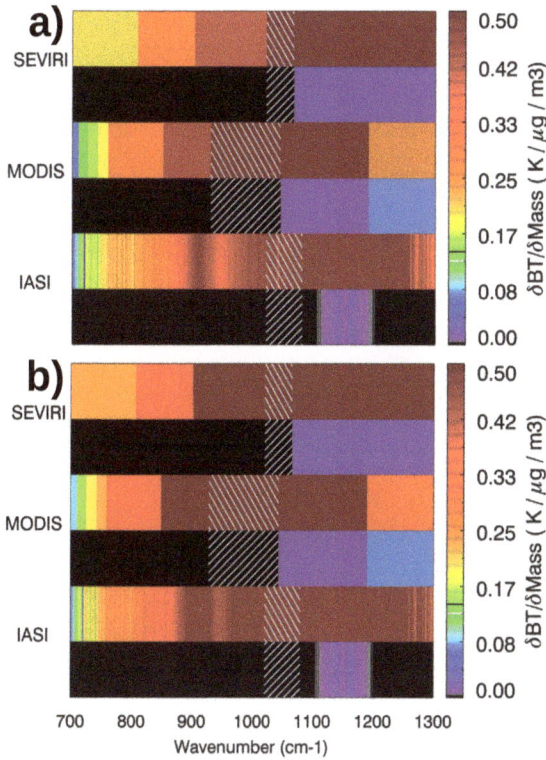

Figure 4. Jacobian matrices for SEVIRI, MODIS and IASI pseudo-observations, at time intervals (**a**) $t = 1$ day and (**b**) $t = 10$ days. In the figure, the two line Jacobian matrix lines (the weighting functions), for each instrument and time interval, are ordered as follows: line 1 is the weighting function with respect to M_{SSA} and line 2 is the weighting function with respect to M_{SO_2}. The ozone absorption region is identified by white crosshatches and excluded.

We further characterize the information content of the synthetic observations, with the averaging kernel matrix **A** (Equation (5)), which represents the sensitivity of the retrieved state to the true state. In Equation (5), S_ϵ and S_a represent, respectively, the measurement error covariance matrix and the a priori covariance matrix. The measurement error covariance matrix S_ϵ, representing the radiometric noise, is assumed diagonal with each diagonal element representing the Noise Equivalent Brightness Temperature (NEBT) in each spectral band, for each instrument. For IASI, the NEBT is taken as equal to 0.2 K for all wavenumbers. For MODIS, the NEBT is taken equal to 0.05 K for channels 29, 31, 32, to 0.25 K for channels 33, 34 and 35 and to 0.35 K for channel 36 [41]. For SEVIRI, NEBT values of 0.80, 0.94, 0.93 and 0.74 K are taken for channels IR8.7, IR10.8, IR12.0 and IR13.4 [42]. For our case study, we consider an a priori covariance matrix with two diagonal elements equal to 50% of the modelled SO_2 mass and 100% of the modelled SSA mass, at each time interval t. This reflects the fact that a priori SSA information is hardly available, from both satellite and ground-based observations, and atmospheric chemical/micro-physical modelling. On the contrary, constraints on SO_2 mixing ratios, e.g., at the time of eruption, can be derived from satellite (e.g., [23,29]) or, in the case of well instrumented volcanoes, from ground-based observations (e.g., [43]). We also added an additional water vapour uncertainty S_i, due to the known interference with spectrally-ubiquitous water vapour absorption lines. These uncertainties are smaller than 0.2 K in our spectral region of interest [22,44]. Consequently, this value has been chosen as a conservation water vapour-related error in the subsequent calculations.

$$\mathbf{A} = \left(\mathbf{K}^T \mathbf{S}_e^{-1} \mathbf{K} + \mathbf{S}_a^{-1} \right)^{-1} \mathbf{K}^T \mathbf{S}_e^{-1} \mathbf{K}. \tag{5}$$

Starting from the averaging kernel matrix, two important diagnostics of the sensitivity of the retrieval can be derived: the number of degrees of freedom (DOF) and total error. The DOF is the trace of $\mathbf{A}$ and quantifies the number of independent parameters that can be retrieved from the measurements. In Table 3, the DOFs for our simulated retrieval, for all the investigated time intervals and the three instrument configurations, are summarized. In our case, 2 being the dimensionality of the retrieved state vector ($[M_{SO_2}, M_{SSA}]$), the DOFs can take values between 0.0 (no retrievable independent parameters) and 2.0 (two perfectly independent retrievable parameters). The DOF for IASI pseudo-observations is about 2.0, which points at the fact that the two parameters can be retrieved independently. For MODIS and SEVIRI, the DOF values are around 1.0 (between 0.99 and 1.09 for MODIS and between 0.88 and 0.92 for SEVIRI). This confirms that SO$_2$ and SSA are strongly inter-dependent and to discriminate the individual radiative contributions of the two effluents is virtually impossible without high-spectral resolution observations.

Table 3. DOFs of SEVIRI, MODIS and IASI pseudo-observations to the retrieval of the state vector $[M_{SO_2}, M_{SSA}]$, at different time intervals since the eruption event.

Time (days)	1	3	5	10
SEVIRI	0.92	0.92	0.88	0.94
MODIS	1.05	1.09	0.99	1.05
IASI	1.99	1.99	1.98	1.99

The theoretical estimation of the total uncertainties, using the Rodgers theory, is also very important to assess the expected quality of the retrievals. The total error covariance matrix $\mathbf{S}_x$ (see Equation (6)) is generally expressed in terms of the smoothing error $\mathbf{S}_{sm}$ and the radiometric noise $\mathbf{S}_m$ [40]. As mentioned before, we consider an additional error component $\mathbf{S}_i$ that takes into account the interference of water vapour absorption:

$$\begin{aligned}
\mathbf{S}_x = \mathbf{S}_{sm} + \mathbf{S}_m + \mathbf{S}_g + \mathbf{S}_i &= (\mathbf{A} - I)\mathbf{S}_a(\mathbf{A} - I)^T + \\
(\mathbf{K}^T \mathbf{S}_e^{-1} \mathbf{K} + \mathbf{S}_a^{-1})^{-1} \mathbf{K}^T \mathbf{S}_e^{-1} \mathbf{K} (\mathbf{K}^T \mathbf{S}_e^{-1} \mathbf{K} + \mathbf{S}_a^{-1})^{-1} + \\
(\mathbf{K}^T \mathbf{S}_i^{-1} \mathbf{K} + \mathbf{S}_a^{-1})^{-1} \mathbf{K}^T \mathbf{S}_i^{-1} \mathbf{K} (\mathbf{K}^T \mathbf{S}_i^{-1} \mathbf{K} + \mathbf{S}_a^{-1})^{-1}.
\end{aligned} \tag{6}$$

The total error estimations for the two retrieved quantities (the diagonal elements of $\mathbf{S}_x$) are summarized in Table 4. For MODIS and SEVIRI, SO$_2$ retrieval uncertainties are found around 50%. On the contrary, SO$_2$ retrieval uncertainties with IASI observations are smaller than 7.0%. As for the SSA, uncertainties of SEVIRI observations reach values as high as about 25% to 35%. MODIS shows lower uncertainties, smaller than 10%. This is mostly due to the smaller radiometric noise and the contribution of band 31, centred around the distinct and peculiar absorption peak of SSAs around 900 cm^{-1}. Theoretical uncertainties are definitely smaller for IASI observations, smaller than 1%. To summarize, broad-band instruments like SEVIRI and MODIS cannot gain quantitative observations of both SO$_2$ and SSA mass burden as independent informations. After just one day from the eruption event, the SO$_2$ can only be observed with significant (around 50%) uncertainties, while the radiatively predominant SSAs, even if in small amounts, can be quantified in terms of their mass, with reasonable uncertainties. These results thus suggest that care must be taken when attempting to quantitatively observe SO$_2$ burdens from volcanic eruptions with broad-band instruments. These observations are accurate only under a few hours from the eruptive event and aged plumes, containing a mixture of SO$_2$ and SSAs (and possibly other gaseous and particulate volcanic effluents) are difficult to characterize in terms of SO$_2$-only information. This is mainly due to the concurrent spectral features of these two effluents, basically in the same band, centred around 1200 cm^{-1}. As for IASI-like high-spectral-resolution sounders, the two effluents are in principle independently retrievable with limited uncertainties on both.

Table 4. Total error (%) of SEVIRI, MODIS and IASI pseudo-observations to the retrieval of the state vector [M_{SO_2}, M_{SSA}], at different time intervals since the eruption event.

Instrument	Parameters	Time (Days)			
		1	3	5	10
SEVIRI	SSA	28.67	27.55	34.70	24.16
	SO$_2$	49.98	49.98	49.99	49.99
MODIS	SSA	7.94	7.95	8.76	7.61
	SO$_2$	48.57	47.57	49.93	48.56
IASI	SSA	0.28	0.27	0.56	0.23
	SO$_2$	3.51	2.64	6.88	3.41

4. Conclusions

The present study gives a detailed analysis of IASI, MODIS and SEVIRI TIR pseudo-observations and their information content on SO$_2$ and SSA, at different time intervals and altitudes following an ideal moderate stratospheric eruption (injection of 10 DU of SO$_2$ around 20 km of altitude, in a tropical atmosphere). The mutual interference of SSA and SO$_2$ on the outgoing TIR radiation and then on the observed BT signal observed by satellite instruments has never been studied in these terms, to our knowledge. Our analyses demonstrate that, despite the relatively small amount of SSA formed (in terms of their total mass), the combined effect of the volcanic SO$_2$ and SSA on the TIR pseudo-observations is apparent after 3 to 5 days from the eruption and is very important after 10 days. In effect, the maximum spectral signature of the two volcanic effluents locate approximately in the same region (about 1100–1200 cm^{-1}, linked to a vibrational mode of both SO$_2$ and the undissociated H$_2$SO$_4$ present in SSA droplets). These results are quantitatively confirmed by assessing the information content of the TIR pseudo-observations to a test bi-dimensional state vector of retrieved parameters, constituted by the SO$_2$ and SSA masses. The sensitivity of TIR pseudo-observations to SSA, in terms of the Jacobian matrix, is about one order magnitude bigger than the sensitivity to SO$_2$ (0.50 versus 0.05 K·µg^{-1}·m^{-3} in the range of maximum sensitivity around 1100–1200 cm^{-1}, for the three instruments). For the broad-band instruments like MODIS or SEVIRI, the information content of TIR pseudo-observations to SO$_2$ and SSA mass burdens are strongly inter-dependent (DOF around 1.0) and to discriminate the individual radiative contributions of the two effluents is virtually impossible without the high spectral resolution of IASI-like instruments (DOF of about 2.0). The theoretical uncertainties for MODIS and SEVIRI are about 50% for the SO$_2$, 10% (SEVIRI) and 25% to 35% (MODIS) for the SSA. IASI- related uncertainties are, on the contrary, smaller than 7.0%, for the SO$_2$, and lower than 1%, for the SSA. This demonstrates that the high-spectral-resolution observations of IASI-like instruments allows, in principle, to quantitatively observe these two volcanic effluents as independent, and low-uncertainties parameters are found, through the analysed 10-days short-term evolution. On the contrary, broad-band instruments like SEVIRI and MODIS cannot gain quantitative observations of both SO$_2$ and SSA mass burden as independent pieces of information.

Further analyses using new generation TIR instruments such as Himawari, Visible Infrared Imaging Radiometer Suite (VIIRS) and IASI-NG (New Generation) are needed to characterize future possibilities in complex volcanic plume combined monitoring. In addition, studies regarding the case of tropospheric eruptions, where atmospheric processes are potentially more complex, are still ongoing.

Acknowledgments: We would like to acknowledge NOVELTIS (L'innovation au service de la Protection du Vivant) and Alain Chédin for supporting us with the radiative transfer model 4A. The optical parameters of sulphate aerosol layers used in this work are obtained with the Interactive Data Language (IDL) Mie scattering routines developed by the Earth Observation Data Group of the Department of Physics of Oxford University, and available via the following website: http://eodg.atm.ox.ac.uk/MIE/. This project has been partially supported by the EU 7th Framework Program under Grant No. 603557 (StratoClim). The two anonymous reviewers are gratefully acknowledged.

Author Contributions: Henda Guermazi and Pasquale Sellitto conceived this study and wrote the paper. Henda Guermazi performed the simulation. Pasquale Sellitto supervised all activities. All authors revised and approved the manuscript.

Conflicts of Interest: The authors declare no conflict of interest.

Abbreviations

The following abbreviations are used in this manuscript:

BT	Brightness Temperature
CNES	Centre National D'études Spatiales
DISORT	DIScrete ORdinaTe algorithm
DOF	Degrees of Freedom
DU	Dobson Unit
EOS	Earth Observing System
EUMETSAT	EUropean organisation for the exploitation of METeorological SATellite
H$_2$SO$_4$	Sulphuric acid
IASI	Infrared Atmospheric Sounding Interferometer
IR	InfraRed
IDL	Interactive Data Language
GEISA	GEstion des Informations Spectroscopiques Atmosphériques
MetOP	METeorological OPerational
MODIS	MODerate resolution Imaging Spectroradiometer
MSG	Meteosat Second Generation
NASA	National Aeronautics and Space Administration
NEBT	Noise Equivalent Brightness Temperature
OH	Hydroxyl
O$_2$	Oxygen
RSR	Relative Spectral Response
RTM	Radiative Transfer Model
SEVIRI	Spinning Enhanced Visible and Infrared Imager
SO$_2$	Sulphur Dioxide
SSA	Secondary Sulphate Aerosol
TIR	Thermal InfraRed
UTLS	Upper Troposphere Lower Stratosphere
UV	UltraViolet
VIIRS	visible Infrared Imager Radiometer Suite
VNIR	Visible Near InfraRed
4A	Automatized Atmospheric Absorption Atlas

References

1. Graf, H.F.; Langmann, B.; Feichter, J. The contribution of Earth degassing to the atmospheric sulfur budget. *Chem. Geol.* **1998**, *147*, 131–145.
2. Sellitto, P.; Zanetel, C.; di Sarra, A.; Salerno, G.; Tapparo, A.; Meloni, D.; Pace, G.; Caltabiano, T.; Briole, P.; Legras, B. The impact of Mount Etna sulfur emissions on the atmospheric composition and aerosol properties in the central Mediterranean: A statistical analysis over the period 2000–2013 based on observations and Lagrangian modelling. *Atmos. Environ.* **2017**, *148*, 77–88.
3. Robock, A. Volcanic eruptions and climate. *Rev. Geophys.* **2000**, *38*, 191–219.
4. Grainger, D.G.; Highwood, E.J. Changes in stratospheric composition, chemistry, radiation and climate caused by volcanic eruptions. *Geol. Soc.* **2013**, *213*, 329–347.
5. Halmer, M.; Schmincke, H.U.; Graf, H.F. The annual volcanic gas input into the atmosphere, in particular into the stratosphere: A global data set for the past 100 years. *J. Volcanol. Geotherm. Res.* **2002**, *115*, 511–528.
6. Ridley, D.A.; Solomon, S.; Barnes, J.E.; Burlakov, V.D.; Deshler, T.; Dolgii, S.I.; Herber, A.B.; Nagai, T.; Neely, R.R.; Nevzorov, A.V.; et al. Total volcanic stratospheric aerosol optical depths and implications for global climate change. *Geophys. Res. Lett.* **2014**, *41*, 7763–7769.

7. Stenchikov, G.L.; Kirchner, I.; Robock, A.; Graf, H.F.; Antuña, J.C.; Grainger, R.G.; Lambert, A.; Thomason, L. Radiative forcing from the 1991 Mount Pinatubo volcanic eruption. *J. Geophys. Res. Atmos.* **1998**, *103*, 13837–13857.

8. Solomon, S.; Portmann, R.W.; Garcia, R.R.; Thomason, L.W.; Poole, L.R.; McCormick, M.P. The role of aerosol variations in anthropogenic ozone depletion at northern midlatitudes. *J. Geophys. Res. Atmos.* **1996**, *101*, 6713–6727.

9. Malavelle, F.F.; Haywood, J.M.; Jones, A.; Gettelman, A.; Clarisse, L.; Bauduin, S.; Allan, R.P.; Karset, I.H.H.; Kristjánsson, J.E.; Oreopoulos, L.; et al. Strong constraints on aerosol-cloud interactions from volcanic eruptions. *Nature* **2017**, *546*, 485–491.

10. Seinfeld, J.; Pandis, S. *Atmospheric Chemistry and Physics: From Air Pollution to Climate Change*; John Wiley: New York, NY, USA, 2012.

11. Doeringer, D.; Eldering, A.; Boone, C.D.; González Abad, G.; Bernath, P.F. Observation of sulfate aerosols and SO_2 from the Sarychev volcanic eruption using data from the Atmospheric Chemistry Experiment (ACE). *J. Geophys. Res. Atmos.* **2012**, *117*, D03203.

12. Castleman, A.W., Jr.; Davis, R.D.; Tang, I.N.; Ball, J.A. Heterogeneous processes and the chemistry of aerosol formation in the upper atmosphere. In Proceedings of the Fourth Conference on the Climatic Assessment Program DOT-TSC-OST-75-38, Department of Transportation, Cambridge, MD, USA, 4–7 February 1975.

13. Oppenheimer, C.; Francis, P.; Stix, J. Depletion rates of sulfur dioxide in tropospheric volcanic plumes. *Geophys. Res. Lett.* **1998**, *25*, 2671–2674.

14. Pruppacher, J.; Klett, J. *Microphyscics of Clouds and Precipitation, Atmospheric and Oceanic Sciences Library*, 2nd ed.; Kluwer Academic Publishers: Norwell, MA, USA, 2004; p. 954.

15. Steel, H.; Hamill, P. Effects of temperature and humidity on the grouwth and optical properties of sulfuric acid water droplets in the stratosphere. *J. Aerosol Sci.* **1981**, *12*, 517–528.

16. Bluth, G.J.S.; William, I.; Rose, I.E.S.; Krueger, A.J. Stratospheric Loading of Sulfur from Explosive Volcanic Eruptions. *J. Geol.* **1997**, *1105*, 671–683.

17. Stevenson, D.S.; Johnson, C.E.; Highwood, E.J.; Gauci, V.; Collins, W.J.; Derwent, R.G. Atmospheric impact of the 1783–1784 Laki eruption: Part I Chemistry modelling. *Atmos. Chem. Phys.* **2003**, *3*, 487–507.

18. Allen, A.G.; Oppenheimer, C.; Ferm, M.; Baxter, P.J.; Horrocks, L.A.; Galle, B.; McGonigle, A.J.S.; Duffell, H.J. Primary sulfate aerosol and associated emissions from Masaya Volcano, Nicaragua. *J. Geophys. Res. Atmosp.* **2002**, *107*, ACH 5-1–ACH 5-8.

19. Coffey, M.T. Observations of the impact of volcanic activity on stratospheric chemistry. *J. Geophys. Res. Atmos.* **1996**, *101*, 6767–6780.

20. Carn, S.; Clarisse, L.; Prata, A. Multi-decadal satellite measurements of global volcanic degassing. *J. Volcanol. Geotherm. Res.* **2016**, *311*, 99–134.

21. Clarisse, L.; Coheur, P.F.; Prata, F.; Hadji-Lazaro, J.; Hurtmans, D.; Clerbaux, C. A unified approach to infrared aerosol remote sensing and type specification. *Atmos. Chem. Phys.* **2013**, *13*, 2195–2221.

22. Sellitto, P.; Legras, B. Sensitivity of thermal infrared nadir instruments to the chemical and microphysical properties of UTLS secondary sulfate aerosols. *Atmos. Meas. Tech.* **2016**, *9*, 115–132.

23. Carboni, E.; Grainger, R.; Walker, J.; Dudhia, A.; Siddans, R. A new scheme for sulphur dioxide retrieval from IASI measurements: Application to the Eyjafjallajokull eruption of April and May 2010. *Atmos. Chem. Phys.* **2012**, *12*, 11417–11434.

24. Sellitto, P.; di Sarra, A.; Corradini, S.; Boichu, M.; Herbin, H.; Dubuisson, P.; Sèze, G.; Meloni, D.; Monteleone, F.; Merucci, L.; et al. Synergistic use of Lagrangian dispersion and radiative transfer modelling with satellite and surface remote sensing measurements for the investigation of volcanic plumes: The Mount Etna eruption of 25–27 October 2013. *Atmos. Chem. Phys.* **2016**, *16*, 6841–6861.

25. Karagulian, F.; Clarisse, L.; Clerbaux, C.; Prata, A.J.; Hurtmans, D.; Coheur, P.F. Detection of volcanic SO_2, ash, and H_2SO_4 using the Infrared Atmospheric Sounding Interferometer (IASI). *J. Geophys. Res. Atmos.* **2010**, *115*, D00L02.

26. Sellitto, P.; Sèze, G.; Legras, B. Secondary sulphate aerosols and cirrus clouds detection with SEVIRI during Nabro volcano eruption. *Int. J. Remote Sens.* **2017**, *38*, 5657–5672.

27. Clerbaux, C.; Coheur, P.F.; Clarisse, L.; Hadji-Lazaro, J.; Hurtmans, D.; Turquety, S.; Bowman, K.; Worden, H.; Carn, S.A. Measurements of SO_2 profiles in volcanic plumes from the NASA Tropospheric Emission Spectrometer (TES). *Geophys. Res. Lett.* **2008**, *35*, L22807.

28. Carboni, E.; Grainger, R.G.; Mather, T.A.; Pyle, D.M.; Thomas, G.E.; Siddans, R.; Smith, A.J.A.; Dudhia, A.; Koukouli, M.E.; Balis, D. The vertical distribution of volcanic SO_2 plumes measured by IASI. *Atmos. Chem. Phys.* **2016**, *16*, 4343–4367.

29. Corradini, S.; Merucci, L.; Prata, A.J. Retrieval of SO_2 from thermal infrared satellite measurements: Correction procedures for the effects of volcanic ash. *Atmos. Meas. Tech.* **2009**, *2*, 177–191.

30. Dubuisson, P.; Herbin, H.; Minvielle, F.; Compiègne, M.; Thieuleux, F.; Parol, F.; Pelon, J. Remote sensing of volcanic ash plumes from thermal infrared: A case study analysis from SEVIRI, MODIS and IASI instruments. *Atmos. Meas. Tech.* **2014**, *7*, 359–371.

31. Prata, A.J.; Kerkmann, J. Simultaneous retrieval of volcanic ash and SO_2 using MSG-SEVIRI measurements. *Geophys. Res. Lett.* **2007**, *34*, L05813.

32. Miles, G.M.; Grainger, R.G.; Highwood, E.J. The significance of volcanic eruption strength and frequency for climate. *Q. J. R. Meteorol. Soc.* **2004**, *130*, 2361–2376.

33. McCormick, B.T.; Herzog, M.; Yang, J.; Edmonds, M.; Mather, T.A.; Carn, S.A.; Hidalgo, S.; Langmann, B. A comparison of satellite- and ground-based measurements of SO_2 emissions from Tungurahua volcano, Ecuador. *J. Geophys. Res. Atmos.* **2014**, *119*, 4264–4285.

34. Lambert, A.; Grainger, R.G.; Rodgers, C.D.; Taylor, F.W.; Mergenthaler, J.L.; Kumer, J.B.; Massie, S.T. Global evolution of the Mt. Pinatubo volcanic aerosols observed by the infrared limb-sounding instruments CLAES and ISAMS on the Upper Atmosphere Research Satellite. *J. Geophys. Res. Atmos.* **1997**, *102*, 1495–1512.

35. SPARC. *SPARC Assessment of Stratospheric Aerosol Properties (ASAP)*; Technical Report No. 4, WCRP-124, WMO/TD-No. 1295; SPARC offices: Paris, French; Toronto, ON, Canada, 2006.

36. Scott, N.; Chedin, A. A fast line-by-line method for atmospheric absorption computations: The Automatized Atmospheric Absorption Atlas. *J. Appl. Meteorol.* **1981**, *20*, 802–812.

37. Noveltis. L'innovation au Service de la Protection du Vivant. Available online: http://www.noveltis.com/ (accessed on 13 September 2017). (In French)

38. Biermann, U.M.; Luo, B.P.; Peter, T. Absorption Spectra and Optical Constants of Binary and Ternary Solutions of H_2SO_4, HNO_3, and H_2O in the Mid Infrared at Atmospheric Temperatures. *J. Phys. Chem. A* **2000**, *104*, 783–793.

39. Stamnes, K.; Tsay, S.C.; Wiscombe, W.; Jayaweera, K. Numerically stable algorithm for discrete-ordinate-method radiative transfer in multiple scattering and emitting layered media. *Appl. Opt.* **1988**, *27*, 2502–2509.

40. Rodgers, C.D. *Inverse Methods for Atmospheric Sounding: Theory and Practice*; Series on Atmospheric Oceanic and Planetary Physics; World Scientific: London, UK, 2000; Volume 2, pp. 43–64.

41. Wan, Z. Estimate of noise and systematic error in early thermal infrared data of the Moderate Resolution Imaging Spectroradiometer (MODIS). *Remote Sens. Environ.* **2002**, *80*, 47–54.

42. EUMETSAT. Typical Radiometric Accuracy and Noise for MSG-1/2. Available online: http://www.eumetsat.int/website/wcm/idc/idcplg?IdcService=GET_FILE&dDocName=PDF_TYP_307RADIOMET_ACC_MSG-1-2&RevisionSelectionMethod=LatestReleased&Rendition=Web (accessed on 26 February 2007).

43. Salerno, G.; Burton, M.; Oppenheimer, C.; Caltabiano, T.; Randazzo, D.; Bruno, N.; Longo, V. Three-years of SO_2 flux measurements of Mt. Etna using an automated UV scanner array: Comparison with conventional traverses and uncertainties in flux retrieval. *J. Volcanol. Geotherm. Res.* **2009**, *183*, 76–83.

44. Pougatchev, N.; August, T.; Calbet, X.; Hultberg, T.; Oduleye, O.; Schlüssel, P.; Stiller, B.; Germain, K.S.; Bingham, G. IASI temperature and water vapor retrievals—Error assessment and validation. *Atmos. Chem. Phys.* **2009**, *9*, 6453–6458.

geosciences

MDPI

Article

Nonlinear Spectral Unmixing for the Characterisation of Volcanic Surface Deposit and Airborne Plumes from Remote Sensing Imagery

Giorgio A. Licciardi [1,2,*], Pasquale Sellitto [3], Alessandro Piscini [4] and Jocelyn Chanussot [2,5]

[1] Research Consortium Hypatia, 00133 Rome, Italy
[2] GIPSA-Lab, Institut Polytechnique de Grenoble, 38000 Grenoble, France;
jocelyn.chanussot@gipsa-lab.grenoble-inp.fr
[3] Laboratoire de Météorologie Dynamique, Institut Pierre Simon Laplace, École Normale Supérieure,
75231 Paris, France; psellitto@lmd.ens.fr
[4] Istituto Nazionale di Geofisica e Vulcanologia, 00143 Rome, Italy; alessandro.piscini@ingv.it
[5] Faculty of Electrical and Computer Engineering, University of Iceland, 101 Reykjavik, Iceland
[*] Correspondence: giorgio.licciardi@consorzioipazia.it; Tel.: +39-06-8567696

Received: 29 March 2017; Accepted: 20 June 2017; Published: 23 June 2017

Abstract: In image processing, it is commonly assumed that the model ruling spectral mixture in a given hyperspectral pixel is linear. However, in many real life cases, the different objects and materials determining the observed spectral signatures overlap in the same scene, resulting in nonlinear mixture. This is particularly evident in volcanoes-related imagery, where both airborne plumes of effluents and surface deposit of volcanic ejecta can be mixed in the same observation line of sight. To tackle this intrinsic complexity, in this paper, we perform a pilot test using Nonlinear Principal Component Analysis (NLPCA) as a nonlinear transformation, that projects a hyperspectral image onto a reduced-dimensionality feature space. The use of NLPCA is twofold: (1) it is used to reduce the dimensionality of the original spectral data and (2) it performs a linearization of the information, thus allowing the effective use of successive linear approaches for spectral unmixing. The proposed method has been tested on two different hyperspectral datasets, dealing with active volcanoes at the time of the observation. The dimensionality of the spectroscopic problem is reduced of up to 95% (ratio of the elements of compressed nonlinear vectors and initial spectral inputs), by the use of NLPCA. The selective use of an atmospheric correction pre-processing is applied, demonstrating how individual plume and volcanic surface deposit components can be discriminated, paving the way to future application of this method.

Keywords: nonlinear spectral unmixing; nonlinear PCA; volcanic plumes; hyperspectral remote sensing

1. Introduction

Volcanoes can inject a great amount of gaseous and particulate effluents to the atmosphere, like water vapour, sulphur dioxide and ash, both at background degassing conditions and during explosive eruptions [1]. Depending on the volcanic activity and the emission fluxes, these effluents can organize as volcanic clouds, which interact with solar and terrestrial radiation, thus affecting the observed spectra in the remotely observed images. The main types of volcanic aerosols are mineral ash, directly emitted by the eruption and secondary sulphate aerosols (SSA), which are produced in-plume by oxidation and hydration of SO_2 emission. Ash can be detected by thermal infrared (TIR) observations using absorption signatures between 8 and 12 µm (1250 and 833 cm^{-1}), typically centred around 10 µm (1000 cm^{-1}) (e.g., [2,3]). Sulphur dioxide emissions are measured both with TIR (e.g., [4–6]) and ultraviolet/visible (UV/VIS) satellite instruments [7]. Recently, targeted sensitivity analyses have shown that also SSA have a typical spectral signature in the TIR spectral range that can be used to detect and characterise this volcanic particulate product [8,9].

Apart from quantitative and semi-quantitative observation of these emitted species using high spectral resolution (sounder) instruments, the interaction of the radiation with the volcanic plumes allow the detection and tracking of the volcanic plume and the identification of the species therein, by multi- and hyper-spectral (imaging) observations (e.g., [10–14]). Recently, neural network (NN) approaches have been tried in order to quantify volcanic ash and SO_2 using both multi-spectral and hyper-spectral data [15–17]. In such cases, unmixing the volcanic cloud spectral signature from other spectral features, as those arising from other atmospheric components in the line of sight of the instruments or from the surface, is vital to rule out these different contributions [3,18].

Hyper-spectral imaging sensors normally record scenes in which numerous interacting objects and material substances, both at the surface and in the overlying atmosphere, contribute to the spectrum measured from a single pixel, by their interaction with the atmospheric radiation recorded by the sensor. Given such mixed pixels, the process of identification of the individual constituent materials in the mixture (endmembers), as well as the proportions in which they appear (abundances), is commonly referred to as spectral unmixing. In remote sensing images, usually the endmembers correspond to the spectral response of macroscopic materials present in the scene, such as surface water, soil, human structures (like buildings) and dominating atmospheric features (like thick meteorological or aerosols clouds) [19].

In recent literature, unmixing techniques are characterized as linear or nonlinear processes [20]. Linear mixtures are dominant when the incident light interacts with objects composed of one individual material before reaching the sensor (different physical/micro-physical properties but the same chemical/mineralogical composition). Thus, light reflected from different materials are mixed in the sensor itself, with minimal interference [21–23]. On the other hand, nonlinear mixing occurs as the result of physical interactions between light scattered from multiple materials, with different chemical/mineralogical composition. These interactions can be sub-divided into *multi-layered* and *microscopic* interactions. Multi-layered interactions occur when light, reflected from one individual material, interacts with other individual and distinct objects before reaching the sensor. Microscopic mixing occurs when two or more materials are physically mixed and this mixture interacts with radiation. In the case of the multi-layered mixing, the first order terms are sufficient to describe the mixture leading to the bilinear model [24,25]. On the other hand, microscopic interactions require an extremely complex physical modelling of the mixture and the interaction with radiation. For these reasons, only approximation are presently proposed in literature [26,27]. However, techniques developed for the processing of internal mixtures are inefficient in the multiple interaction scenario (and vice versa). Moreover, these models are based on the assumption that the mixtures are produced by different materials having Lambertian surfaces. In many real cases, the light that interacts with different materials does not produce in an isotropically distributed radiance. All these effects result in nonlinear mixing.

A more flexible solution can be achieved by using machine learning approaches, such as NN [28–30]. These algorithms can learn nonlinear correlations in a supervised fashion based on a collection of examples (training dataset). However, the universal approximation properties of NN [31] assumes that the training dataset covers all the possible physical scenarios and interactions between radiation and fixed materials. Practically, a very large training dataset is necessary to approach the convergence of NN properties in several possible scenarios, given the problem under investigation. This large training dataset is not always available. A possible alternative is to project the hyper-spectral image into a linearised feature space by means of a nonlinear transformation. In this way, any kind of linear unmixing method can be effectively applied to these linear features. Based on this idea, in this paper, we propose the use of the nonlinear Principal Component Analysis (NLPCA) for the projection of the hyper-spectral image into a feature space [32]. The obtained features are then used as input to a linear unmixing algorithm for the identification of the endmembers. Finally, the abundance estimation is performed solving the constrained least square problem.

The remainder of the paper is organized as follows: in Section 2, the the proposed approach is introduced, while in Section 3, experimental methodology as well as an in-deep analysis of the results are provided. Finally, Section 4 gives some conclusion remarks.

2. Nonlinear Spectral Unmixing

The general framework of spectral unmixing, either linear or nonlinear, is composed of three main processing steps: (1) dimensionality reduction; (2) endmember identification and (3) abundance estimation, as depicted in Figure 1. In cases where the surface characterisation is the sole target of the analysis, the observed radiances can be converted to surface reflectances by means of an atmospheric correction, aimed at compensating for the atmospheric attenuation and scattering. Performing an accurate atmospheric correction is often an arduous task, due to coarse information on the atmospheric composition. Moreover, applying a correction for the atmospheric contribution may have a negative impact on the spectral unmixing when the task of the analysis is to find endmembers present in the atmosphere. The three processing steps are discussed in the following.

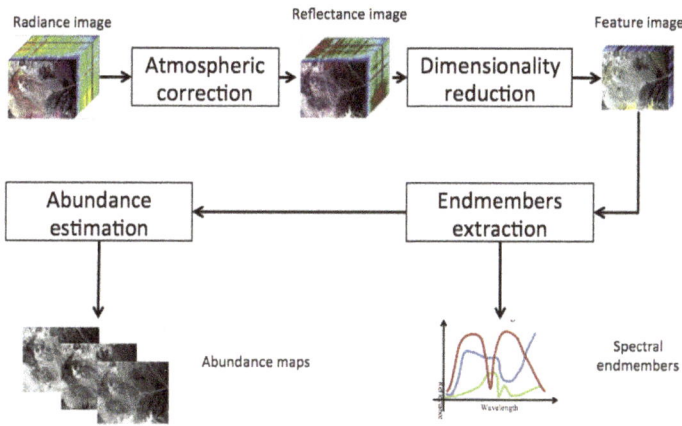

Figure 1. Schematic diagram describing the complete processing for hyperspectral unmixing.

As for step 1 (dimensionality reduction), since hyperspectral images are composed by hundreds of extremely correlated bands, it is possible, and indeed beneficial, to reduce the effective dimension of the input data by use of decorrelation approaches. This processing step projects the image into a reduced dimensionality feature space. The algorithm performance in terms of computation time, complexity and performances are, then, generally improved [33]. However, a careful selection of the dimensionality of the feature space is essential because this choice limits the number of possible final endmembers. As for step 2 (endmembers identification), there exist several approaches to rule out endmembers in a given image scene, which fall into three main groups. Geometrical approaches are based on the hypothesis that linearly mixed vectors are in a simplex set. Statistical approaches use parameter estimation techniques to determine the endmember. Sparse regression approaches formulate unmixing as a linear sparse regression problem. As for step 3 (abundances estimation), given the hyperspectral image and the endmembers identified in the previous step, the abundance of each endmember is quantified solving a constrained optimization problem, which minimizes the residual between the observed spectral vectors and the linear space spanned by the inferred spectral signatures. Usually, in linear unmixing, the fractional abundances are constrained to be nonnegative and to sum to one. While atmospheric correction and dimensionality reduction are optional, the endmembers' determination and abundances' estimation steps are central to any unmixing approaches. Often, steps 2 and 3 are implemented simultaneously [34]. In the case of nonlinear mixtures, the use of linear approaches may result in the false identification of non-physical endmembers and in most cases cannot detect the whole of endmembers actually present in the scene. In the specific case of nonlinear approaches, the endmembers' determination and abundances' estimation steps require a prior detailed knowledge of the possible physical interactions between materials, which is not always available. In order to override this limitation, in this paper, we propose a novel approach that, instead of deriving an extremely complex nonlinear model, linearizes the input data by means of a preliminary nonlinear transformation. In particular, we propose to use the NLPCA to project the original hyperspectral image into a linearized feature space. The advantage of the use of NLPCA is twofold. On the one hand, it performs a linearization of the original information. On the other hand, it reduces the data dimensionality. The NLPCA, thanks to its nonlinear functions basis, permits obtaining a set of features that do not present nonlinear correlations, thus allowing the subsequent use of linear approaches for the endmember determination and inversion phases. As for the endmember identification, we proposed the use of the N-FINDR algorithm [35]. This choice is based on the main advantage of the N-FINDR, which is able to automatically detect endmembers without a priori information. This characteristic is extremely valuable in the case of NLPCA, where the obtained features may not always have a physical interpretation. The abundance estimation step will be then carried out by using the SUnSAL algorithm [36] for the solution of the constrained least square problem. In the following, a detailed description of the NLPCA approach, as well as the N-FINDR and SUnSAL algorithms are reported.

2.1. Nonlinear Principal Component Analysis

In this work, NNPCA, commonly referred to as a nonlinear generalization of the PCA techniques, is performed by an AutoAssociative Neural Network (AANN) or auto-encoder [32]. A standard auto-encoder is a conventional feedforward neural network, having a symmetrical three layer topology, where input and output layers have the same number of nodes, and a hidden layer, usually referred to as bottleneck, of smaller dimension than either input/output layers. Both input and output layers have sigmoidal activation functions. The auto-encoder is trained to perform identity mapping, meaning that the input has to be equal to the output [37]. This means that, after a successful training phase, the fewer nodes in the bottleneck layer, than in the input/output, represent (in fact encode) the information of the inputs in a smaller dimensionality space. Due to the nonlinear nature of NNs, this information compression is obtained with nonlinear combinations of the inputs. In other words, data compression caused by the network bottleneck forces hidden units to represent significant features in the data, removing redundancies. The smaller-dimensionality information vector can, then, be used as input for the subsequent processing.

Different from standard auto-encoders, the topology of a nonlinear AANN uses by default three hidden layers, including the internal bottleneck layer of smaller dimension than either input or output layers (Figure 2). In order to understand why three hidden layers are necessary to obtain a nonlinear representation of the data, it is useful to consider the nonlinear AANN as a combination of two successive neural networks or functional mappings, namely *coding* and *decoding sub-networks*, as depicted in Figure 2.

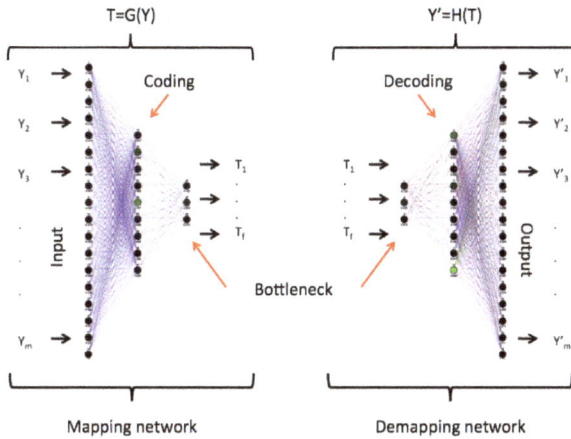

Figure 2. General structure of the Autoassociative Neural Network (AANN). The AANN can be seen as a concatenation of two simple networks. The first network project the original data Y into a lower dimensionality subspace T by means of the transformation function G. The second network reproject the compressed data T back into the original space by means of the transformation function H.

The first sub-network represents the encoding or extraction function:

$$T = G(Y), \tag{1}$$

which projects the original m-dimensional data X onto a lower dimensional subspace defined by the activations of the units in the central hidden layer (bottleneck), producing a smaller-dimensionality (say n-dimensional) vector **T** by means of the mapping F. Similarly, the second sub-network defines an arbitrary functional mapping:

$$Y' = H(T), \tag{2}$$

which projects from the smaller-dimensionality feature space back onto the original m-dimensional space, by means of the mapping funtion H. The ability of any NN to fit arbitrary nonlinear functions depends on the presence of a hidden layer with nonlinear nodes. In [31], it was shown that any nonlinear function can be approximated by a superposition of a set of $\sigma(x)$ transformations that are continuous, bounded and monotonically increasing functions, with $\sigma(x) \rightarrow 1$ as $x \rightarrow +\infty$ and $\sigma(x) \rightarrow 0$ as $x \rightarrow -\infty$. This property is often called *universal fitting* [38], and is a generalisation of the Weierstrass theorem (which applies to polynomial functions). Thus, a network lacking a hidden layer is only capable of producing linear combinations of the inputs, given linear nodes in the output layer. In the same way, a network lacking a hidden layer but including nonlinear activation functions in the output layer is only capable of approximating multi-variable sigmoidal functions. Similarly, a NN with linear nodes in the hidden layer will return linear combinations of the inputs. From these considerations, it can be affirmed that the NLPCA can be implemented by two NNs approximating the nonlinear functions G and H. The NN producing the G mapping has as an input layer of m nodes followed by the hidden layer (often called *mapping layer*) with $m_1 > n$ nodes and sigmoidal transfer functions (to assure the *universal fitting* property of NNs). The output layer of this subnet contains $n < m$ nodes and for this reason it is often called *bottleneck*. The second NN (also called *decoding subnet*) producing the H mapping, has an input layer with n nodes, followed by the hidden layer (often called *demapping layer*) with $m_2 > n$ nodes and sigmoidal transfer functions. The output layer yields the reconstructed data and thus contains m nodes.

In order to correctly define the two mapping functions G and H, a supervised training of the two NNs is required. This needs, in principle, a complete knowledge of the relations between the input and output spaces, which is only an ideal case. From a practical point of view, this means that while the inputs of the coding subnet is known, the output is unknown. Conversely, the input of the decoding layer is unknown while its output is known (i.e., the input, as the coding+decoding NN are aimed at approximating an identity mapping). Therefore, a direct supervised training of the two sub-networks is not feasible. However, one can observe that combining in series the two ANNs is equivalent to define a composite function $Y' = H(G(Y))$ that links the original data $\mathbf{Y}$ with its reconstruction version Y'. Thus, practically, the combined network is trained to produce the identity mapping. This means that the parameters of the network representing Y' are optimized so that the reconstructed outputs match the inputs as closely as possible. The training aimed at learning the identity mapping has been called *self-supervised backpropagation* or *autoassociation*, leading to the definition of these NNs as AutoAssociative NNs (AANNs).

For AANNs, the training phase is an iterative process and is completed when the following sum of squared errors is minimized:

$$E = \sum_{p=1}^{n} \sum_{i=1}^{m} (\widehat{\mathbf{y}}_i - \mathbf{y}_i)_p{}^2. \tag{3}$$

In Equation (3), $\widehat{\mathbf{y}}_i$ and $\mathbf{y}_i$ are the calculated and target output vectors of the AANN, for a training set of p examples.

Once the AANN is trained, it is possible to use extract the *coding sub-network* only, to project the original data into a lower dimensional space given by the bottleneck layer. Thus, the f NLPCA can be obtained from the bottleneck. From a topological point of view, it can be noted that, while the number of nodes in the bottleneck layer defines the features subspace, the nodes in the coding and decoding layers are related to the complexity of the mapping and demapping functions. As it can be seen in Figure 3, the NLPCA defines a set of nonlinear functions able to describe the nonlinear correlations between input variables. The result is a linearised feature space. However, one of the main difficulties in designing the AANN relies in the selection of the correct number of nodes in the three hidden layers, since the mapping functions as well as the subspace dimension strongly depend on them.

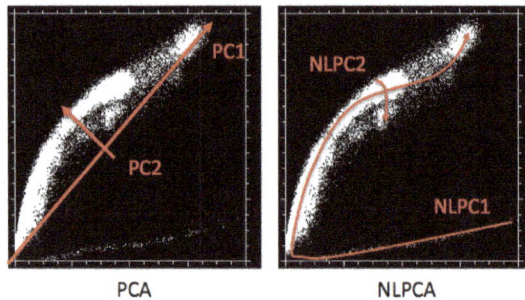

Figure 3. Example of a nonlinear dataset being mapped into feature space by means of linear and nonlinear PCA.

The best NN topology can be retrieved by using a simple heuristic grid search algorithm that varies recursively the number of nodes of the hidden layers and evaluated the value of the Means Square Error (MSE) error [39]. Then, the topology presenting the smallest error is selected. However, without starting assumptions, this approach can be extremely time consuming and different optimal solutions should be found. Starting from p that represents the number of samples in the training set, a separate constraint is imposed by each output node, so that the total number of the possible adjustable parameters (weights and biases for all network connections and nodes, respectively) must be less than

$p \times m$. Moreover, m, M_1, M_2 and n being the number of nodes of input/output, coding, decoding and bottleneck layers, and analysing the structure of the AANN used here, it can be found that the number of adjustable parameters is $(M_1 + M_2)(m + n + 1) + m + n$ that implies the following inequality:

$$M_1 + M_2 \ll \frac{m(n-1) - n}{m + n + 1}. \tag{4}$$

The aim of a dimensionality reduction method is to reduce the original spectral dimension into a lower dimensional space. This can be translated into the AANN structure as a condition on n, i.e., $n \ll m, p$. Then, Equation (4) becomes:

$$M_1 + M_2 \ll n. \tag{5}$$

Assuming a balanced structure of the AANN, M_1 and M_2 should have the same dimensions $(M_1 \sim M_2 = M)$, we have:

$$2M \ll n. \tag{6}$$

It is worth noting that Equation (6) is effective only if the number of mapping/demapping nodes M is greater then the number of nodes in the bottleneck layer n. Otherwise, there will not be enough data to effectively extract n NLPCs. It is also worth underlining that, since the output has to simply replicate the input, there is no need to have a specific a priori knowledge for the learning phase implementation. This implies that the AANN training can be performed in a fully automatic way and that all pixels in the image can be used as training samples for this task. Practically, this has actually been the technique adopted in this paper.

2.2. Endmember Extraction and Abundance Estimation

After the dimensionality reduction phase been accomplished, the subsequent endmember extraction and abundance estimation phases need to be accomplished. Among the several algorithms developed for automatic or semiautomatic extraction of spectral endmembers, the N-FINDR algorithm attempts to automatically find the simplex of maximum volume that can be inscribed within the hyperspectral data set [35]. The N-FINDR implementation is firstly initialized by randomly selecting a set of q endmembers $\{E_1, E_2, ..., E_q\}$, where $q \leq n + 1$, and n corresponds to the dimension of the feature space. Then, the volume of the simplex defined by the current set of endmembers is derived by:

$$V(E_1, E_2, ..., E_q) = \frac{\left| det \begin{bmatrix} 1 & 1 & \cdots & 1 \\ (E_1 & E_2 & \cdots & E_q) \end{bmatrix} \right|}{(q-1)!}. \tag{7}$$

Then, for each pixel vector $X(i, j)$ of the input hyperspectral data, the volume V is recalculated by testing the pixel in the first endmembers position:

$$\begin{aligned} V\left(X(1,1), E_2, ..., E_q\right) \\ V\left(X(1,2), E_2, ..., E_q\right) \\ \vdots \\ V\left(X(r,c), E_2, ..., E_q\right), \end{aligned} \tag{8}$$

where r and c represent the number of rows and columns of the image. If one of the volumes calculated in Equation (8) is greater than $V(E_1, E_2, ..., E_q)$, then E_1 is replaced with the pixel corresponding the the maximum volume, and a new set of endmembers is produced. The same procedure is then carried out iteratively by testing the volumes in the other endmembers positions, retaining the combinations corresponding to the maximum volumes. The processing ends when all the pixels in the input data have been tested in each endmember position.

Since in the proposed approach the input data corresponds to the features derived by the NLPCA, the obtained endmembers are represented in the feature space. In order to obtain the equivalent endmembers in the spectral domain, the decoding sub-network of the NLPCA is used.

Once the endmembers are obtained, the fractional abundances estimated by minimizing the total squared error, under the constraints of non-negativity and/or the sum to one:

$$min_x \left(\tfrac{1}{2}\right) \left\| Ea - S \right\|_2^2,$$
$$x \geq 0,$$
$$1^T x = 1,$$

(9)

where $E \in \Re^q$ denote the matrix containing the q endmembers, $a \in \Re^k$ the fractional abundance vector, and $S \in \Re^k$ the observed mixed pixel. In order to solve the optimization problems, we used the SUnSAL algorithm, which is an instance of the Constrained Split Augmented Lagrangian Shrinkage Algorithm (C-SALSA) methodology to effectively solve a large number of constrained least-squares problems sharing the same matrix system in [36]. Figure 4 depicts the complete schematic of the proposed approach.

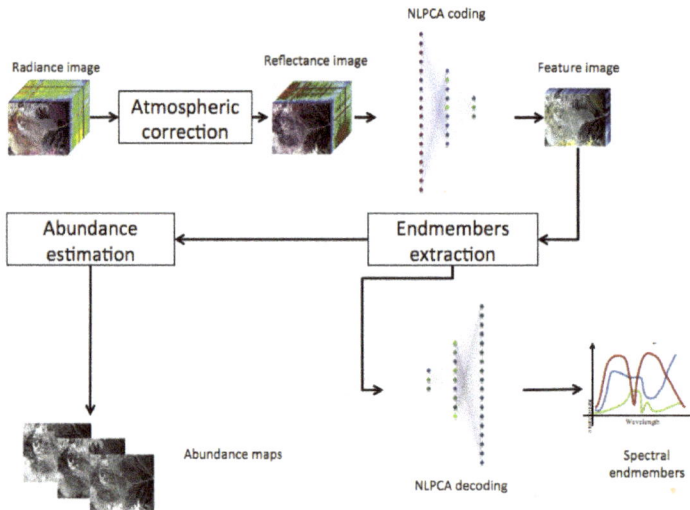

Figure 4. Schematic diagram describing the proposed processing for hyperspectral unmixing.

3. Experimental Results

In this section, two real measurements datasets have been considered to test the proposed technique. Both radiance and reflectance data, i.e., without and with atmospheric correction, are used in order to analyze the effect of these two options in our approach. Differently from simulated data, real images are, generally, strongly conditioned by a great deal of additional circumstances, such as differences in illumination through the scene, angle of view as well as multiple scattering effects. These factors have been reported to influence the endmember selection in linear unmixing approaches [34]. As for the assessment of the effectiveness of the proposed technique, a qualitative analysis has been carried out by considering comparisons with ground truth fractional abundance maps and spectral library. In the case of abundance maps' ground truth, the effectiveness of the technique can be assessed by estimating the abundances of the endmembers in the scene and comparing the obtained values with reference fractions. In using ground truth spectral library, the quality of the endmembers is evaluated by comparing them with some reference spectral signatures using spectral similarity criteria. In order to further appreciate the effectiveness of the proposed method in handling nonlinear mixtures, the obtained results have been compared with those obtained applying the endmembers extraction and abundance estimation phases, directly to the hyperspectral image and to the linear features obtained through the use of standard PCA.

3.1. Campi Flegrei

On a first experiment, we applied the proposed technique to a hyperspectral image acquired with the Hyperion satellite-borne sensor, onboard the EO-1 satellite. Hyperion acquires 220 hyperspectral bands in the Visible/near-infrared, from 0.4 to 2.5 μm. However, only 155 bands have been retained from the original dataset, discarding the most noisy and the bands without relevant information for this application [40]. The considered Hyperion image has been acquired in 2008 over the Campi Flegrei (CF) area, northwest of Naples, Italy. In this experiment, we focused on the caldera area, comprising more than 24 craters and volcanic edifices and presenting effusive gaseous manifestations, in particular in the Solfatara crater.

The CF region is located in the Campanian plain, which is a NW-SE trending Plio-Quaternary extensional basin bordered by carbonate platforms. The CF caldera has been interested by volcanism and hydrothermal activity for thousands of years [41]. The main structures of CF consist of two nested calderas formed after the Campanian Ignimbrite and the Neapolitan Yellow Tuff eruptions [42]. The outer is the Campania Ignimbrite (CI) caldera, dated 39 ky, while the inner is the Neapolitan Yellow Tuff (NYT) caldera, dated 15 ky [43]. Currently, the surroundings of Solfatara crater are the area with the strongest geothermal emission in CF. Large quantities of volcanic-hydrothermal CO_2 are released through soil diffuse emission. The gas emissions from this area consists in about 5000 t day^{-1} of a CO_2/H_2O mixture. Its power is 100 MW, which is 10 times higher than the conductive heat flux over the whole caldera surface [44]. Since the image has been acquired from satellite, the atmospheric contribution has a relevant role. In this first example, an atmospheric correction is applied. Figure 5 reports the ground truth of the Hyperion image.

The Hyperion image after the atmospheric correction has been processed to mitigate the striping effect. The corrected image has been then projected in the feature space by means of the NLPCA. In particular, we found that the optimal topology for the AANN was 155 input/output nodes, 80 nodes in the coding/decoding layers and nine nodes in the bottleneck layers. Then, the dimensionality of the hyperspectral input vector is the reduced by about the 95% by the nonlinear compression of the implemented AANN.

Once the AANN is trained, the N-FINDR algorithm has been applied to the nine nonlinear principal components obtained from the bottleneck layer of the AANN and 10 endmembers have been detected. The obtained endmembers have been used as input to the SUnSAL algorithm in order to estimate the fractional abundances of each endmember (Figure 6). Finally, the 10 endmembers (EM, in the following) have been processed through the decoding sub-network of the AANN in order to retrieve their spectral signatures reported in Figure 7.

	Water
	Dense manmade
	Sparse manmade
	Highway
	Low vegetation
	High vegetation
	Proximal volcanic deposits

Figure 5. Hyperion image: ground truth.

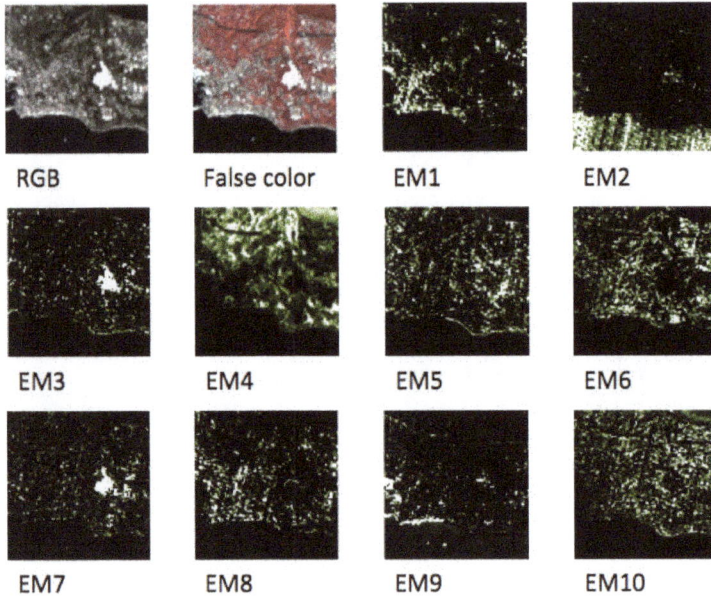

Figure 6. Hyperion image: fractional abundances obtained from the NLPCA derived data corresponding to the obtained endmembers (EM1, EM2,...,EM10).

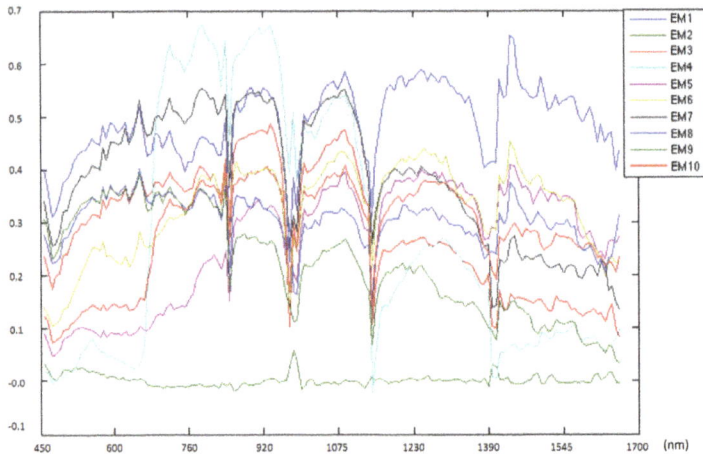

Figure 7. Spectral signatures of the endmembers obtained with the proposed method for the Hyperion image.

From the obtained endmembers, it can noted that EM1, EM6 and EM8 present similar spectral signatures, which, according to the available ground truth, represent different extents of man-made structures. Then, a new endmember representing man-made structures is obtained by summing up these three endmembers (Figure 8). In a similar way, both EM4 and EM10 could be associated to different types of vegetations. Finally, EM2 and EM5 clearly refer to water and bare soil, respectively. The remaining three endmembers EM3, EM7 and EM9 are the most interesting from a volcanic products deposit detection point of view. Proximal volcanic material is collectively identified by EM3 and 7, while distal material, in the coastline area and more northwestwards, by EM9. The spectral signatures of EM3, 7 and 9, while each one different with respect to the others, present a similar lower reflectance (higher absorption) in the range between about 950 nm and 1000 nm. This identifies similar material, while in morphological (surface structure, rugosity, macroscopic properties of the deposit) and, probably, detailed mineralogical composition. Further studies are ongoing, starting with this pilot analysis, to detect the precise composition of the structures identified with these endmembers, using their retrieved spectral signatures.

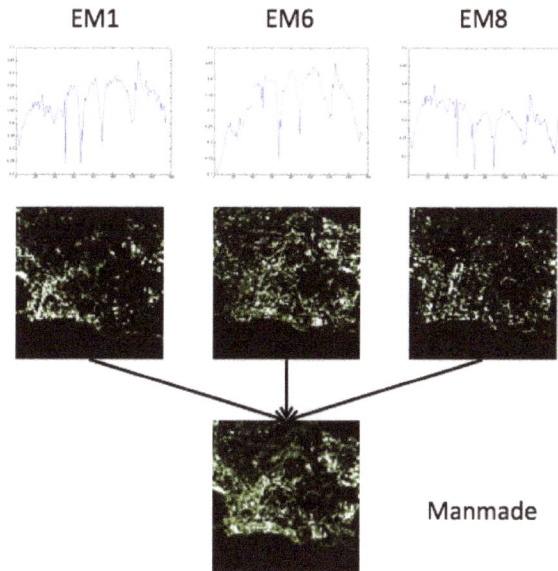

Figure 8. EM1, EM6 and EM8 combined to form a new endmember representing manmade surfaces.

For sake of comparison, the endmember extraction (N-FINDR) and abundance estimation (SUnSAL) phases have been applied also to the original image and the low-dimensionality image obtained through the use of linear PCA, respectively. In particular, for the PCA, there have been selected nine linear principal components in order to have a comparison with the NLPCA. The obtained abundance maps are reported in Figures 9 and 10. In a first analysis, it can be noted that, in both cases, the noise have a strong impact on the abundance maps. As for the abundance maps obtained with the original image, EM1, EM4 and EM7 can be associated to manmade surfaces. EM3 seems to be associated to water surfaces while EM2 can be associated to low reflective surfaces. EM5 and EM6 are related to vegetation while EM9 could be associated to bare soil. EM8 is associated to both volcanic deposits and also dense vegetation. Finally, EM10 is strongly influenced by noise and cannot be associated to any material. On a more deep analysis, it can be noted that volcanic products deposit cannot be identified in one endmember but are present in both EM7 and EM8. A similar result can be obtained analyzing the abundances derived from the nine linear PCs. However, the amount of noise affecting the abundance maps is higher. In both cases, the volcanic products deposit are not completely identified.

The main result of this first experiment is that, using an atmospheric correction routine, the volcanic plume features have been apparently removed and the characterisation of the surface volcanic material deposit is then possible as a lesser complex spectroscopic interpretation problem, and with a limited output space dimensionality.

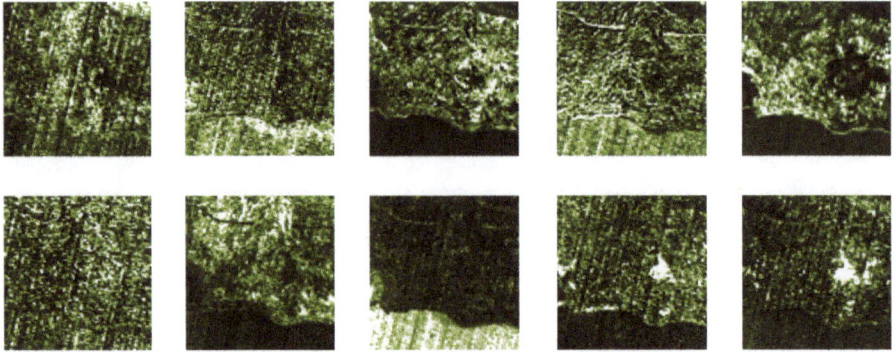

Figure 9. Fractional abundances obtained from the original Hyperion image.

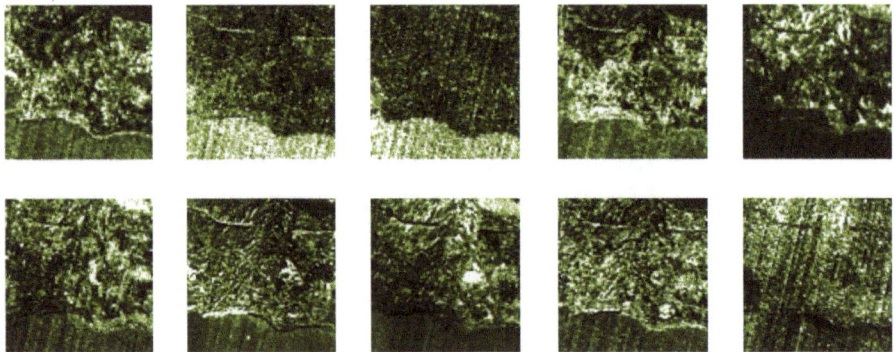

Figure 10. Hyperion image: fractional abundances obtained from the PCA derived image.

3.2. Kilauea Volcano

A second experiment has been carried out using a hyperspectral image obtained with the AVIRIS (Airborne Visible/Infrared Imaging Spectrometer) sensor. AVIRIS is an airborne optical sensor delivering calibrated images of the upwelling spectral radiance in 224 contiguous spectral bands with wavelengths from 0.4 to 2.5 μm. The image used in this case-study has been acquired in 2007 over the Puu Oo crater, in the eastern rift zone of the Kilauea volcano of the Hawaiian Islands Figure 11. The volcanic activity is usually characterized by emissions of both ash and sulphur dioxide. This image was taken during moderate activity of the volcano.

In this case, an atmospheric correction is not applied. By opposition with respect to the previous case, this processing chain allows to actually identify plume features by means of the endmember extraction.

As done for the previous experiment, the AVIRIS radiance image has been projected in the feature space by means of NLPCA. In this case, the optimal topology is found to have 224 nodes in the input/output layers, 110 nodes in the coding/decoding layers and 10 nodes in the bottleneck layer. Then, as for the CF case-study, also in this case, the dimensionality of the hyperspectral input vector is reduced by about the 95% by the nonlinear compression of the implemented AANN. Subsequently, seven endmembers and the corresponding fractional abundance maps have been identified through the use of the N-FINDR and SUnSAL algorithms (Figures 12 and 13).

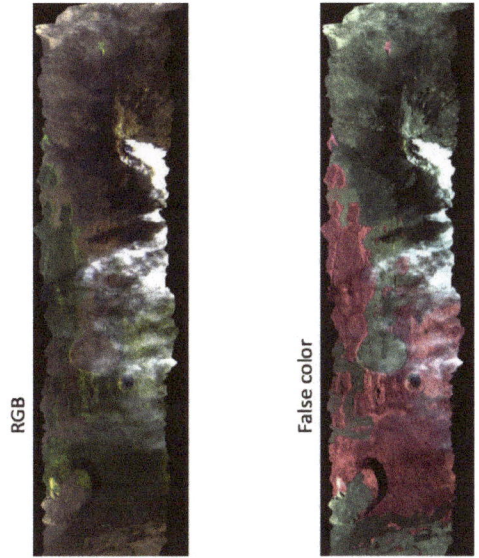

Figure 11. Color and false color representations of the AVIRIS image.

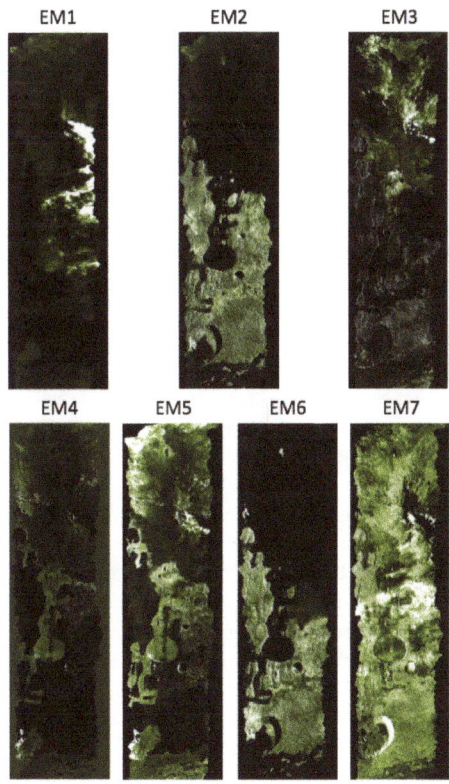

Figure 12. Abundance maps obtained from the AVIRIS image using the NLPCA approach.

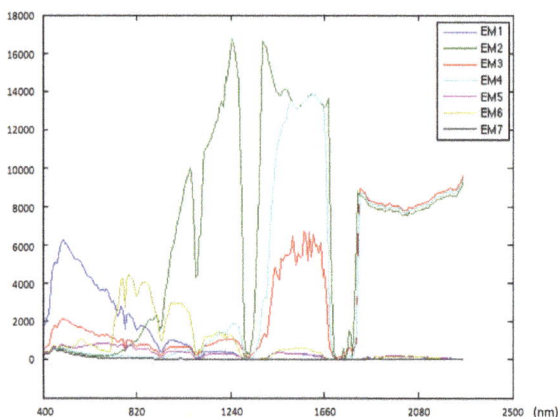

Figure 13. Spectral signatures of the endmembers obtained with the proposed method for the AVIRIS image.

Surface features have been identified, as for the previous case. Thanks to comparison with ground truth, endmembers EM2 and 6 are associated with different types of vegetation and EM4 is associated to bare soil, i.e., basalt. Surface volcanic deposit are also identified, as for EM3 that is associated with tephra deposits. Differently with respect to the CF case, volcanic plume features here are also identified, as they are not screened out by the atmospheric correction carried out for the previous case. Endmember EM1 can be associated with water vapour emissions and the subsequent condensation to liquid clouds of volcanic origin. It tracks very well the plume geometry, as water vapour is a major volcanic effluent and is very apparent in the extracted image using only EM1. Looking at the spectral signature of this endmember, the rapid decrease of the reflectivity between the visible range (maximum at about 500 nm) and the near infrared, point at scattering processes by relatively small particles, as freshly nucleated water droplets following the emission by the volcano. Endmember EM5, while presenting several surface structures, including volcanic deposit, as for EM3, still holds a certain representativity of the plume object, in particular in the near range, as visible from Figure 12. The spectral signature of EM5 has a larger spectral structure than EM1, in the visible range, and so it can be argued that it carries information about larger ash particles. This hypothesis is supported, in conjunction with these spectroscopic/scattering considerations on the spectral signature of this endmember, by the fact that: (a) the plume identification of EM5 is confined to the near-crater area, thus suggesting coarse particles with quick sedimentation, and (b) the simultaneous identification of surface volcanic deposit features, probably with mineralogical composition consistent with the plume-ash discussed here. Further studies are required to attempt finer attribution of these mixed plume/surface endmembers' identification to specific volcanic products, even if these first results are encouraging towards the application of the methodology. Finally, EM7 is associated to the different illumination paths (shadows), and then carry very limited information on atmospheric or surface features.

Similarly to the previous experiment, a comparison with the results of the spectral unmixing (N-FINDR + SUnSAL), applied to the original and the PCA derived images, respectively, has been carried out. According to the abundance maps reported in Figures 14 and 15, it can be noted that, similarly to the NLPCA case, in both cases, there are endmembers related to different vegetation surfaces, bare soil and water vapour. However, the main difference relies in the way the volcanic materials are detected. In particular, with linear approaches, it is not possible to detect all the volcanic materials present in the scene, as it was possible with the NLPCA approach.

Figure 14. Abundance maps obtained from the original AVIRIS image.

Figure 15. AVIRIS image: abundance maps obtained from seven linear PCs.

4. Conclusions

In this paper, a nonlinear spectral unmixing procedure is tested, using AANN, to separate the various components of remote sensing imagery of volcanoes and their products. The problem of separating background, surface deposits of volcanic ejecta and airborne plumes of volcanic effluents, basing on their individual spectral signatures, from a given remote sensing image, is tackled. The method is tested on two different cases: (1) Campi Flegrei area (deposits and fumaroles) and (2) Kilauea volcano (deposits and plume for a moderate eruption). The dimensionality of the spectroscopic problem is reduced of up to 95%, using narrow-bottleneck AANNs, thus removing the redundancies of the hyperspectral input data. An atmospheric correction procedure is applied to the first case but not to the second. The Campi Flegrei case-study is very suitable to try the complete removal of the plume interference, due to the limited effluents concentration for this case. The atmospheric correction has enabled the access to the identification of volcanic and background features for this case. On the contrary, the thicker plume for the Kilauea case-study made this case very well adapted to investigate both plume and surface volcanic products. Then, an atmospheric correction is not performed to retain the airborne plume information for the subsequent analyses. For Kilauea, the method allows the identification partial separation of different plume (e.g., water vapour/condensed droplets, ash) and surface tephra deposits. While more analyses and validation efforts, using ground truth or complementary remote observations, are needed, this pilot study demonstrate the potential of this methodology to tackle different unmixing problems linked to the identification of volcanic products from remote-sensing imagery. This is also confirmed by comparing the results obtained with the proposed method with those obtained using classical linear approaches. However, it is important to state that the obtained endmembers are extreme points only in the feature space but they might not be in the original observation space. For this reason, the obtained abundances may not lead to objective quantification of the endmembers. Finally, if not correctly designed, the dimensionality reduction using the NLPCA approach, may lead to endmembers that have physical meaning but do not exist in the real world. This is the case of features representing the illumination or in some cases sensor artifacts.

Acknowledgments: The work of Pasquale Sellitto has been partially funded by the FP7 Med-SuV and the INGV SMED projects. The work of Giorgio Licciardi and Jocelyn Chanussot has been funded by the ANR ASTRID (project APHYPIS) under grant ANR-16-ASTR-0027-01.

Author Contributions: G.A.L. conceived the experiments; G.A.L., A.P., P.S. and J.C. analysed the data and discussed the results; G.A.L. and P.S. wrote the paper.

Conflicts of Interest: The authors declare no conflict of interest.

References

1. Von Glasow, R. Atmospheric chemistry in volcanic plumes. *Proc. Natl. Acad. Sci. USA* **2010**, *107*, 6594–6599.
2. Prata, A.J. Infrared radiative transfer calculations for volcanic ash clouds. *Geophys. Res. Lett.* **1989**, *16*, 1293–1296.
3. Gangale, G.; Prata, A.J.; Clarisse, L. On the infrared spectral signature of volcanic ash. *Remote Sens. Environ.* **2010**, *114*, 414–425.
4. Karagulian, F.; Clarisse, L.; Clerbaux, C.; Prata, A.J.; Hurtmans, D.; Coheur, P.F. Detection of volcanic SO_2, ash, and H_2SO_4 using the Infrared Atmospheric Sounding Interferometer (IASI). *J. Geophys. Res. Atmos.* **2010**, *115*, 1–10.
5. Clarisse, L.; Hurtmans, D.; Clerbaux, C.; Hadji-Lazaro, J.; Ngadi, Y.; Coheur, P.F. Retrieval of sulphur dioxide from the infrared atmospheric sounding interferometer (IASI). *Atmos. Meas. Tech.* **2012**, *5*, 581–594.
6. Carboni, E.; Grainger, R.; Walker, J.; Dudhia, A.; Siddans, R. A new scheme for sulphur dioxide retrieval from IASI measurements: Application to the Eyjafjallajokull eruption of April and May 2010. *Atmos. Chem. Phys.* **2012**, *12*, 11417–11434.
7. Li, C.; Krotkov, N.A.; Carn, S.; Zhang, Y.; Spurr, R.J.D.; Joiner, J. New-generation NASA Aura Ozone Monitoring Instrument (OMI) volcanic SO_2 dataset: Algorithm description, initial results, and continuation with the Suomi-NPP Ozone Mapping and Profiler Suite (OMPS). *Atmos. Meas. Tech.* **2017**, *10*, 445–458.

8. Sellitto, P.; Legras, B. Sensitivity of thermal infrared nadir instruments to the chemical and microphysical properties of UTLS secondary sulfate aerosols. *Atmos. Meas. Tech.* **2016**, *9*, 115–132.

9. Sellitto, P.; Sèze, G.; Legras, B. Secondary sulphate aerosols and cirrus clouds detection with SEVIRI during Nabro volcano eruption. In Proceedings of the EGU General Assembly 2016, Vienna, Austria, 17–22 April 2016. (In typesetting for International Journal of Remote Sensing)

10. Prata, A.J. Observations of volcanic ash clouds in the 10–12 μm window using AVHRR/2 data. *Int. J. Remote Sens.* **1989**, *10*, 751–761.

11. Prata, A.J.; Kerkmann, J. Simultaneous retrieval of volcanic ash and SO_2 using MSG-SEVIRI measurements. *Geophys. Res. Lett.* **2007**, *34*, L05813.

12. Corradini, S.; Merucci, L.; Prata, A.J.; Piscini, A. Volcanic ash and SO_2 in the 2008 Kasatochi eruption: Retrievals comparison from different IR satellite sensors. *J. Geophys. Res. Atmos.* **2010**, *115*, doi:10.1029/2009JD013634.

13. Campion, R.; Salerno, G.G.; Coheur, P.F.; Hurtmans, D.; Clarisse, L.; Kazahaya, K.; Burton, M.; Caltabiano, T.; Clerbaux, C.; Bernard, A. Measuring volcanic degassing of SO_2 in the lower troposphere with ASTER band ratios. *J. Volcanol. Geotherm. Res.* **2010**, *194*, 42–54.

14. Rix, M.; Valks, P.; Hao, N.; van Geffen, J.; Clerbaux, C.; Clarisse, L.; Coheur, P.F.; Erbertseder, T.; Zimmer, W.; Emmadi, S. Satellite Monitoring of Volcanic Sulfur Dioxide Emissions for Early Warning of Volcanic Hazards. *IEEE J. Sel. Top. Appl. Earth Obs. Remote Sens.* **2009**, *2*, 196–206.

15. Picchiani, M.; Chini, M.; Corradini, S.; Merucci, L.; Sellitto, P.; Del Frate, F.; Stramondo, S. Volcanic ash detection and retrievals using MODIS data by means of neural networks. *Atmos. Meas. Tech.* **2011**, *4*, 2619–2631.

16. Piscini, A.; Carboni, E.; Del Frate, F.; Grainger, R.G. Identifying volcanic endmembers in hyperspectral images using spectral unmixing. In Proceedings of the SPIE 9242, Remote Sensing of Clouds and the Atmosphere XIX, and Optics in Atmospheric Propagation and Adaptive Systems XVII, Amsterdam, The Netherlands, 22 September 2014.

17. Piscini, A.; Picchiani, M.; Chini, M.; Corradini, S.; Merucci, L.; Del Frate, F.; Stramondo, S. A neural network approach for the simultaneous retrieval of volcanic ash parameters and SO_2 using MODIS data. *Atmos. Meas. Tech.* **2014**, *7*, 4023–4047.

18. Piscini, A.; Carboni, E.; Del Frate, F.; Grainger, R.G. Simultaneous retrieval of volcanic sulphur dioxide and plume height from hyperspectral data using artificial neural networks. *Geophys. J. Int.* **2014**, *198*, 697.

19. Keshava, N. A Survey of Spectral Unmixing Algorithms. *MIT Linc. Lab. J.* **2003**, *14*, 55–78.

20. Keshava, N.; Mustard, J.F. Spectral Unmixing. *IEEE Signal Process. Mag.* **2002**, *19*, 44–57.

21. Singer, R.B.; McCord, T.B. Mars: Large scale mixing of bright and dark surface materials and implications for analysis of spectral reflectance. In Proceedings of the Lunar and Planetary Science Conference, Houston, TX, USA, 19–23 March 1979; pp. 1835–1848.

22. Hapke, B. Bidirection reflectance spectroscopy. I. Theory. *J. Geophys. Res.* **1981**, *86*, 3039–3054.

23. Clark, R.N.; Roush, T.L. Reflectance spectroscopy: Quantitative analysis techniques for remote sensing applications. *J. Geophys. Res.* **1984**, *89*, 6329–6340.

24. Fan, W.; Hu, B.; Miller, J.; Li, M. Comparative study between a new nonlinear model and common linear model for analyzing laboratory simulated-forest hyperspectral data. *Int. J. Remote Sens.* **2009**, *30*, 2951–2962.

25. Altmann, Y.; Dobigeon, N.; Tourneret, J.Y. Bilinear models for nonlinear unmixing of hyperspectral images. In Proceedings of the 2011 3rd Workshop on Hyperspectral Image Signal Processing: Evolution in Remote Sensing (WHISPERS), Lisbon, Portugal, 6–9 June 2011.

26. Kulbelka, P.; Munk, F. Reflection characteristics of paints. *Z. Tech. Phys.* **1931**, *12*, 593–601.

27. Shkuratov, Y.; Starukhina, L.; Hoffmann, H.; Arnold, G. A model of spectral albedo of particulate surfaces: Implications for optical properties of the Moon. *Icarus* **1999**, *137*, 235–246.

28. Plaza, J.; Plaza, A. Spectral mixture analysis of hyperspectral scenes using intelligently selected training samples. *IEEE Geosci. Remote Sens. Lett.* **2010**, *7*, 371–375.

29. Altmann, Y.; Dobigeon, N.; McLaughlin, S.; Tourneret, J.Y. Nonlinear unmixing of hyperspectral images using radial basis functions and orthogonal least squares. In Proceedings of the IEEE International Conference on Geoscience and Remote Sensing (IGARSS), Vancouver, BC, Canada, 24–29 July 2011; pp. 1151–1154.

30. Licciardi, G.; Frate, F.D. Pixel Unmixing in Hyperspectral Data by Means of Neural Networks. *IEEE Trans. Geosci. Remote Sens.* **2011**, *49*, 4163–4172.

31. Cybenko, G. Approximations by superpositions of sigmoidal functions. *Math. Control Signals Syst.* **1989**, *2*, 303–314.

32. Kramer, M.A. Nonlinear principal component analysis using autoassociative neural networks. *AIChE J.* **1991**, *37*, 233–243.

33. Sellitto, P. Artificial neural networks for spectral sensitivity analysis to optimize inversion algorithms for satellite-based earth observation: Sulfate aerosol observations with high-resolution thermal infrared sounders. In *Sensitivity Analysis in Earth Observation Modelling*; Petropoulos, G.P., Srivastava, P.K., Eds.; Elsevier: Amsterdam, The Netherlands, 2017; pp. 161–175.

34. Bioucas-Dias, J.M.; Plaza, A.; Dobigeon, N.; Parente, M.; Du, Q.; Gader, P.; Chanussot, J. Hyperspectral Unmixing Overview: Geometrical, Statistical, and Sparse Regression-Based Approaches. *IEEE J. Sel. Top. Appl. Earth Obs. Remote Sens.* **2012**, *5*, 354–379.

35. Plaza, A.; Chang, C.I. An improved N-FINDR algorithm in implementation. In Proceedings of the SPIE 5806, Algorithms and Technologies for Multispectral, Hyperspectral, and Ultraspectral Imagery XI, Orlando, FL, USA, 28 March 2005.

36. Bioucas-Dias, J.; Figueiredo, M. Alternating direction algorithms for constrained sparse regression: Application to hyperspectral unmixing. In Proceedings of the 2010 2nd Workshop on Hyperspectral Image Signal Processing: Evolution in Remote Sensing (WHISPERS), Reykjavik, Iceland, 14–16 June 2010.

37. Bishop, C. *Neural Networks for Pattern Recognition*; Oxford University Press: London, UK, 1995.

38. Hornik, K. Approximation capabilities of multilayer feedforward networks. *Neural Netw.* **1991**, *4*, 251–257.

39. Licciardi, G.; Marpu, P.R.; Chanussot, J.; Benediktsson, J.A. Linear Versus Nonlinear PCA for the Classification of Hyperspectral Data Based on the Extended Morphological Profiles. *IEEE Geosci. Remote Sens. Lett.* **2011**, *9*, 447–451.

40. Datt, B.; Vicar, T.R.M.; Niel, T.G.V.; Jupp, D.L.B.; Pearlman, J.S. Preprocessing EO-1 Hyperion Hyperspectral Data to Support the Application of Agricultural Indexes. *IEEE Trans. Geosci. Remote Sens.* **2003**, *41*, 1246–1259.

41. Lima, A.; Vivo, B.D.; Spera, F.J.; Bodnar, R.J.; Milia, A.; Nunziata, C.; Belkin, H.E.; Cannatelli, C. Thermodynamic model for uplift and deflation episodes (bradyseism) associated with magmatic-hydrothermal activity at the Campi Flegrei (Italy). *Earth Sci. Rev.* **2009**, *97*, 44–58.

42. Deino, A.L.; Orsi, G.; de Vita, S.; Piochi, M. The age of the Neapolitan Yellow Tuff caldera-forming eruption (Campi Flegrei caldera—Italy) assessed by 40Ar/39Ar dating method. *J. Volcanol. Geotherm. Res.* **2004**, *133*, 157–170.

43. Orsi, G.; Di Vito, M.A.; Isaia, R. Volcanic hazard assessment at the restless Campi Flegrei caldera. *Bull. Volcanol.* **2004**, *66*, 514–530.

44. Chiodini, G.; Caliro, S.; Cardellini, C.; Granieri, D.; Avino, R.; Baldini, A.; Donnini, M.; Minopoli, C. Long-term variations of the Campi Flegrei, Italy, volcanic system as revealed by the monitoring of hydrothermal activity. *J. Geophys. Res. Solid Earth* **2010**, *115*, doi:10.1029/2008JB006258

geosciences

MDPI

Article

Proximal Monitoring of the 2011–2015 Etna Lava Fountains Using MSG-SEVIRI Data

Stefano Corradini [1,*], Lorenzo Guerrieri [2], Valerio Lombardo [1], Luca Merucci [1], Massimo Musacchio [1], Michele Prestifilippo [3], Simona Scollo [3], Malvina Silvestri [1], Gaetano Spata [3] and Dario Stelitano [1]

[1] Centro Nazionale Terremoti (CNT)-Istituto Nazionale di Geofisica e Vulcanologia (INGV), 00143 Rome, Italy; valerio.lombardo@ingv.it (V.L.); luca.merucci@ingv.it (L.M.); massimo.musacchio@ingv.it (M.M.); malvina.silvestri@ingv.it (M.S.); dario.stelitano@ingv.it (D.S.)

[2] Istituto di Scienze dell'Atmosfera e del Clima (ISAC)-Consiglio Nazionale delle Ricerche (CNR), 40129 Bologna, Italy; lorenzo.guerrieri74@gmail.com

[3] Osservatorio Etneo, Istituto Nazionale di Geofisica e Vulcanologia (INGV), 95123 Catania, Italy; michele.prestifilippo@ingv.it (M.P.); simona.scollo@ingv.it (S.S.); gaetano.spata@ingv.it (G.S.)

* Correspondence: stefano.corradini@ingv.it; Tel.: +39-06-51860621

Received: 21 December 2017; Accepted: 17 April 2018; Published: 21 April 2018

Abstract: From 2011 to 2015, 49 lava fountains occurred at Etna volcano. In this work, the measurements carried out from the Spinning Enhanced Visible and InfraRed Imager (SEVIRI) instrument, on board the Meteosat Second Generation (MSG) geostationary satellite, are processed to realize a proximal monitoring of the eruptive activity for each event. The SEVIRI measurements are managed to provide the time series of start and duration of eruption and fountains, Time Averaged Discharge Rate (TADR) and Volcanic Plume Top Height (VPTH). Due to its temperature responsivity, the eruptions start and duration, fountains start and duration and TADR are realized by exploiting the SEVIRI 3.9 μm channel, while the VPTH is carried out by applying a simplified procedure based on the SEVIRI 10.8 μm brightness temperature computation. For each event, the start, duration and TADR have been compared with ground-based observations. The VPTH time series is compared with the results obtained from a procedures-based on the volcanic cloud center of mass tracking in combination with the Hybrid Single-Particle Lagrangian Integrated Trajectory (HYSPLIT) back-trajectories. The results indicate that SEVIRI is generally able to detect the start of the lava emission few hours before the ground measurements. A good agreement is found for both the start and the duration of the fountains and the VPTH with mean differences of about 1 h, 50 min and 1 km respectively.

Keywords: Etna volcano; 2011–2015 Etna lava fountains; remote sensing; SEVIRI data; eruption start and duration; volcanic plume top height; time averaged discharge rate

1. Introduction

In 2011–2015, the eruptive style of Mt. Etna volcano (Sicily, Italy) showed an intense explosive activity. Strombolian events became more frequent, often associated to magnificent episodes of lava fountains. The most intense phase of these eruptions, commonly indicated as "paroxysmal" episodes, can be very short (from minutes to hours). The low frequency of polar-orbiting satellite observations is often inadequate to detect these paroxysmal episodes. Therefore, the Spinning Enhanced Visible and InfraRed Imager (SEVIRI), the primary instrument aboard Meteosat Second Generation (MSG) geostationary platforms, with its high temporal resolution (15 min for the Earth full disk and 5 min for the rapid scan mode over Europe and Northern Africa) has become an important tool for volcano observation in spite of its coarse spatial resolution [1,2]. SEVIRI has 12 spectral channels from visible

(VIS) to Thermal InfraRed (TIR) with a nadir spatial resolution of 3 km (1 km for the high resolution High Resolution Visible-HRV channel). Measurements from SEVIRI allow the monitoring of the whole evolution of both the proximal volcanic activity and the ash and gas emissions into the atmosphere generated by explosive events, from the near-source plume column to the distal volcanic clouds transported by the winds [1]. Several physical parameters estimated from each SEVIRI image provide a quantitative characterization of the volcanic clouds in terms of ash mass burden, effective radius, aerosol optical depth and SO_2 mass [3–5]. The space-based observations covering an entire eruptive event allow the description of the evolution of these volcanic cloud parameters, and the detection of the horizontal and vertical extent of the atmospheric volume affected by the ash cloud [1]. Moreover, they provide a record of the proximal thermal history of the event [6–8] and a precise timing of the early phase of the eruption [9]. All these space-based parameters enable continuous monitoring of the volcanic activity which can be complemented and validated by the available ground-based observations [1].

In this work a review of the lava fountaining events that occurred at Etna volcano from the beginning of 2011 to the end of 2015 is presented with a focus on the induced proximal activity analyzed by using SEVIRI data. Here both the SEVIRI instruments aboard MSG platforms positioned at 0 and 9.5° E, with a repeat cycle of 15 min (Earth full disk) and 5 min (rapid scan mode) are considered.

The paper is organized as follows: Section 2 outlines and references the Etna 2011–2015 lava fountains, and Section 3 describes the methods used to retrieve the eruptions and fountains beginning and duration, the Time Averaged Discharge Rate (TADR) and the Volcanic Plume Top Height (VPTH) from SEVIRI data. In Section 4 the results of these analyses are reported, while in Section 5 they are validated by exploiting the comparison with ground-based data, or by applying different and independent retrieval methods based on satellite measurements and model simulations. Final conclusions are drawn in Section 6.

2. The 2011–2015 Etna Lava Fountains

Since 2011 Etna was very active with 49 lava fountain events produced from the central craters with the most violent eruptions forming high plumes that overtake the tropopause. Those eruptions were characterized by three well defined phases: in the first one, there was the rising of Strombolian activity and lava flow emission; the second phase was characterized by the formation of lava fountains which produced abundant tephra fallout entirely covering the volcano flanks and finally, in the third and last phase, there was a decreasing of the explosive activity up to the end of the eruption [9,10]. The most frequent events were produced from the New South East Crater (NSEC), a new vent that opened in 2010 at the base of the South East Crater (SEC). During those events eruption columns were well visible from the video-surveillance system [11] and, in some cases, were also retrieved by lidar [9] that showed volcanic ash concentrations higher than the ash concentration thresholds for safe airspace defined by the International Civil Aviation Organization (ICAO) in the 2010 Volcanic Ash Contingency Plan [12]. Tephra fallout from the NSEC events gives total masses between ~10^8 and ~10^9 kg [13] and this variability is function of the ratio between ash and lava amount produced during the eruption [14]. Among the NSEC events, the 23 November 2013 lava fountain had a great impact because the high mass eruption rate associated to strong winds allowed to larger clasts to fall at distances of 5–6 km from the vent, hitting hikers and tourists [13] and affecting the airspace [1]. It is noteworthy that similar violent events were produced from the Voragine Crater (VOR) that produced four events in less than three days from 3 to 5 December 2015. This eruption produced columns rising up to 15 km a.s.l. [15] and copious tephra fallout deposit having a volume of 7.1×10^6 m^3 [16].

3. Proximal Monitoring of Volcanic Eruptions

In this work the proximal monitoring is referred to the analysis of SEVIRI data on the Etna summit craters area. In this section, the procedures developed for the estimation of eruptions and fountains start and duration, TADR and VPTH are described.

3.1. Eruption Start and Duration

Infrared remotely sensed data can be used to evaluate the surface thermal state of active volcanoes [17–20]. Because the spectral radiance emitted by hot spots reaches its maximum in the region of Mid InfraRed (MIR), the early detection of an impending eruption is realized by exploiting the SEVIRI 3.9 μm channel. Despite its relatively coarse spatial resolution (3 × 3 km^2 at sub satellite point) the presence of a high temperature source, even affecting only a small portion of one large pixel, causes a dramatic increase of the emitted MIR radiance [18].

A procedure named MS2RWS (MeteoSat to Rapid Response Web Service), has been developed to exploit the capability to detect the beginning and duration of an eruption. The algorithm is an improvement of the procedure presented in Musacchio et al. [21,22] applied to the SEVIRI 3.9 μm measurements. The procedure starts from the assumption that in a remote sensing image a pixel may assume a limited number of values ranging from 0 up to the saturation. During a continuous daily acquisition, the radiance of a given pixel, in clear sky condition and no eruption, follows a characteristic trend related to the Sun irradiance. By considering the 3.9 μm SEVIRI channel, five years of images have been analyzed for each 15 min SEVIRI acquisition and the maximum radiance values of the pixel centered on Etna craters (red pixel in the inset zoom of Figure 1), and the maximum average radiance in a region of 5 × 5 pixels around it (blue pixels in the inset zoom of Figure 1) have been computed.

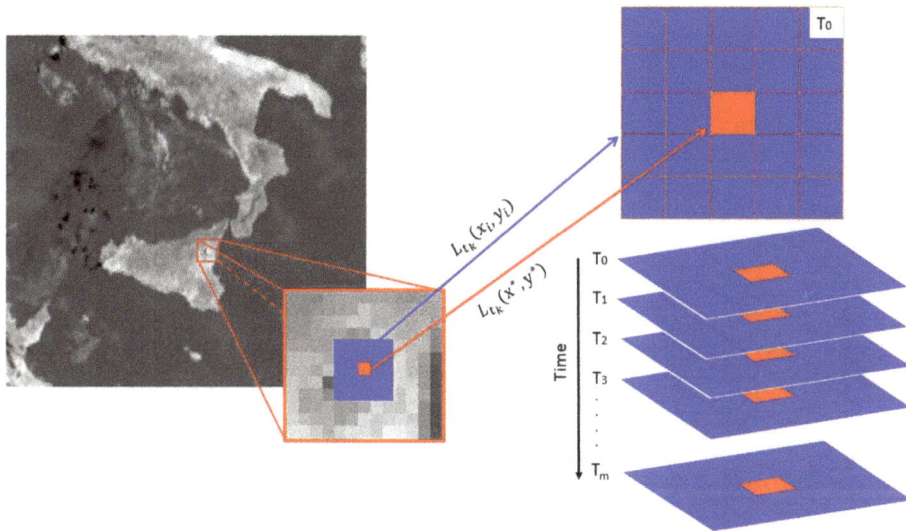

Figure 1. On the left: Spinning Enhanced Visible and InfraRed Imager (SEVIRI) image at 3.9 μm and zoom on Etna area. On the right: details of the pixels centered on Etna craters (red) and the surrounding 5 × 5 pixels (blue) considered for the start and duration eruption computation.

In this space and time domain an "historical" threshold called "Dynamic Threshold" (DT) is defined:

$$DT(t) = Upper\_Limit(t) - Lower\_Limit(t) \tag{1}$$

where Upper_Limit is the maximum radiance values for the pixel centered on Etna craters defined as:

$$Upper\_Limit(t) = \max(L_{t_k}(x^*, y^*))_{t_k \in T} \tag{2}$$

With $L_{t_k}(x^*, y^*)$ is the radiance at 3.9 μm of the pixel centered on Etna craters at time t_k; T = {$t_0, t_1, \ldots, t_m$} with m = 365 days × 5 years.

The Lower_Limit(t) is the maximum of the mean value computed in a region of 5 × 5 pixels around the pixel centered on Etna:

$$\text{Lower\_Limit}(t) = \max \left\{ \frac{[\sum_{i,j=1}^{n} L_{t_k}(x_i, y_j)] - L_{t_k}(x^*, y^*)}{n^2 - 1} \right\}_{t_k \in T} \tag{3}$$

with n = 5.

DT(t) is then compared with the "Difference of Radiances" (DR(t)) defined as DT(t), but, instead of the maximum historical values, the real time radiance values are considered.

Finally, by making the difference DT(t)–DR(t), two solutions are possible:

$$DT(t) \geq DR(t), \text{ no eruption occurs;} \tag{4}$$

$$DT(t) < DR(t), \text{ eruption occur.} \tag{5}$$

Figure 2 shows two examples of no eruption (17 April 2013, upper plot) and eruption (5 January 2012, lower plot) test cases respectively. In these plots, DT(t), the radiance measured at 3.9 μm for the central pixel (L(x^*, y^*)) and DR(t) are represented by the dashed, solid and dotted lines respectively. As the upper plate of Figure 2 shows, the dotted line is always below the dashed line, therefore no eruption was detected. On the contrary, the lower plate of Figure 2 shows an abrupt increase of the 3.9 μm radiance until the saturation value (2.37 W/m²/sr/μm). The beginning of the eruption is identified at 01:10 UTC when DR(t) became greater than DT(t), while the end of the eruption is detected at 20:15 UTC, when DR drops back to values lower than DT(t). The trend of DR(t) and L(x^*, y^*) identifies also minor oscillations due to fluctuations of the volcanic activity, and a deep absorption between 5:50 to 6:55 UTC that indicate the start and the end of the fountaining with the formation of an eruptive plume. In fact, the volcanic plume, absorbing the underlying radiation, produce the decrease of the radiance measured from the satellite.

(a)

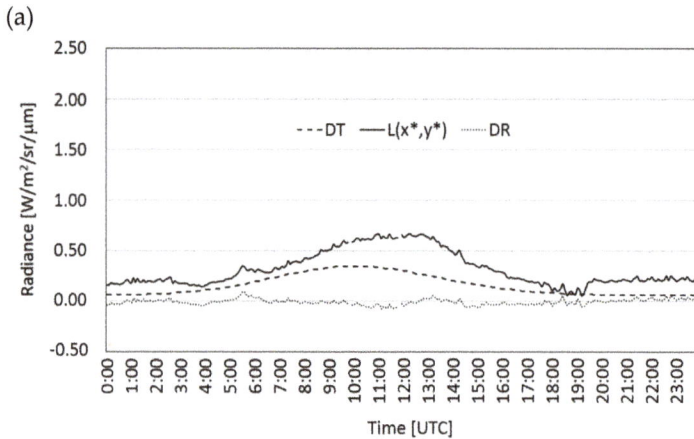

Figure 2. *Cont.*

(b)

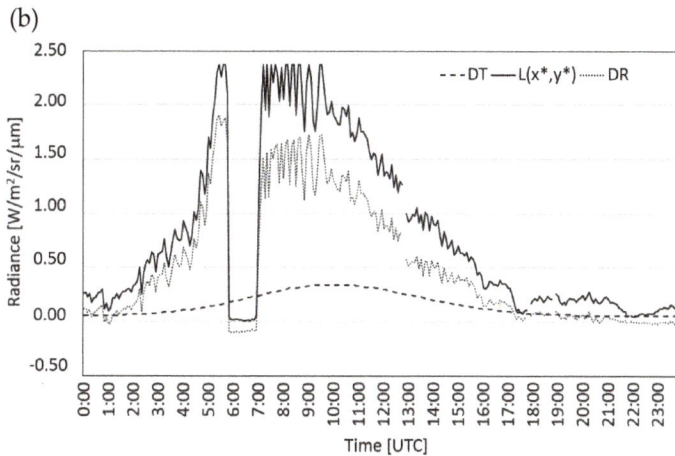

Figure 2. Data for 17 April 2013 (**a**) and 5 January 2012 (**b**) test cases. In these plots, the dashed, solid and dotted lines represent the Dynamic Threshold (DT(t)), the radiance at 3.9 μm of the pixel centered on Etna summit craters, and the Differences of Radiance (DR(t)) respectively.

3.2. Lava Discharge during Etna's Lava Fountains

Following Gouhier et al. [23], Wright et al. [24] and Harris et al. [6], the Time Averaged Discharge Rate (TADR) is estimated during Etna's 2011–2015 eruptive events using SEVIRI 3.9 μm measurements. The data are processed using AVHotRR routine developed by Lombardo [25] to monitor volcanic activities in near-real time. AVHotRR allows for automatic hot-spot detection and heat flux estimate (Q_{tot}). To convert Q_{tot} to TADR using the satellite thermal data, the well-established conversion of Harris et al. [6] is applied. The conversion to TADR reduces to an empirical relation, whereby [24]:

$$TADR = mA/c, \tag{6}$$

in which A is the area of active lava flow derived from the satellite image, and m and c are coefficients set on a case-by-case basis [26]. Following Gouhier et al. [23], A is estimated from radiant pixels containing lava from:

$$A = \frac{L(x^*, y^*) - L(T_a)}{L(T_c) - L(T_a)} A_{pix}, \tag{7}$$

where A_{pix} is the pixel area, $L(T_a)$ and $L(T_c)$ are the 3.9 μm radiances at ambient (T_a) and lava (T_c) temperature respectively. T_a is computed from adjacent lava-free pixels using the TIR channels. Because of the small fraction of the SEVIRI pixel occupied by lava and considering that the radiant peak of energy is centered at MIR wavelengths, there is no anomaly in the TIR. Therefore, T_c is set by considering a suitable range of values [27] that lead to a solution spanning over a wide range of TADR values. Uncertainties in TADR estimates can be reduced using data from higher spatial and spectral resolution sensors such as the Moderate Resolution Imaging Spectroradiometer (MODIS) or the Advanced very-high-resolution radiometer (AVHRR) [28–30], for which the pixel size of about 1 km² is generally sufficient to detect anomalies also in the TIR.

3.3. Volcanic Plume Top Height

The VPTH is determined by using a simplified procedure based on the computation of the brightness temperature at 10.8 μm ($T_{b,10.8}$) of the most opaque pixels of the volcanic plume, and considered as a proxy for the ambient temperature at the same height [31]. This value can be compared with a temperature profile (as close as possible in time and space) to obtain the height where the temperature best matches the plume-top temperature [31,32]. For each SEVIRI image, the VPTH is estimated by computing the minimum $T_{b,10.8}$ value in an area of 9×9 pixels centered over the volcanic vents. This procedure, labelled in the following Dark Pixel (DP), is very simply to apply despite it works reliably only when the cloud behaves as a black-body and the atmospheric profile is representative. The highest uncertainties occur for plume heights near the tropopause where the temperature variation as a function of height is small. In this work, the atmospheric temperature profiles used for the VPTH estimation are those derived from the mesoscale model of the hydrometeorological service of Agenzia Regionale per la Protezione Ambientale (ARPA) Emilia Romagna named ARPA-SIM. An hourly model output from 72-h weather forecast provided every 12 h is considered. The ARPA-SIM grid spans from 12.5° to 18.5° E and from 34.5° to 40.5° N and has 22 isobaric levels. Data are provided as GRIB of 101×101 points stepped by 0.0625°.

Figure 3 shows an example of the VPTH retrieval obtained from the SEVIRI image collected the 23 November 2013 at 10:00 UTC and the corresponding ARPASIM temperature profile. In this case, the brightness temperature of the most opaque pixels is −53 °C and yields to an altitude of 11.1 km. The VPTH uncertainty is obtained by computing the altitudes for $T_{b,10.8}$ +/− 2 K (dashed gray lines), in which 2 K take into account the statistical variability of the most opaque pixels of the cloud. This leads to a final result of 11.1 +/− 0.7 km.

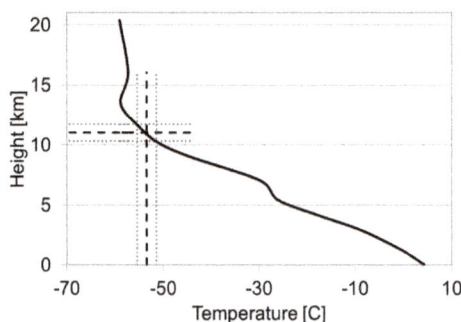

Figure 3. Atmospheric temperature profile derived from the ARPA-SIM model at 10:00 UTC the 23 November 2013. The dashed vertical line represents the $T_{b,10.8}$ of the most opaque pixels obtained from the SEVIRI image collected at 10:00 UTC (−53 °C). The horizontal dashed line represents the estimated VPTH (11.1 km). The vertical and horizontal dotted gray lines represent the uncertainty on $T_{b,10.8}$ and VPTH respectively.

4. Results

Figure 4 shows the time series of DT(t), L(x*,y*) and DR(t) for the 2–5 December 2015 events. In the figure the start and end of eruptions, the 3.9 µm signal saturation, the presence of volcanic plumes (lava fountains) and the meteorological clouds over the vents are emphasized. The SEVIRI images on the top highlight the different plot signatures. In particular, in the first image on the left, the 3.9 µm channel shows the high. The other RGB composite images (R: $T_{b,12}-T_{b,10.8}$; G: $T_{b,10.8}-T_{b,8.6}$; B: $T_{b,10.8}$) emphasize the presence of the volcanic and/or meteorological clouds in the area of interest.

Figure 4 clarifies possibilities and limits of the SEVIRI 3.9 µm channel analysis. For the 2–3 December event, the eruption start, end and volcanic cloud presence are clearly detected, while for the 4–5 December events the situation is much more critic due to the presence of a wide meteorological cloud system in the area. In this latter case, it is not possible to identify the end of the 4 December and the start of the 5 December events: from the 3.9 µm analysis these two events are merged into one.

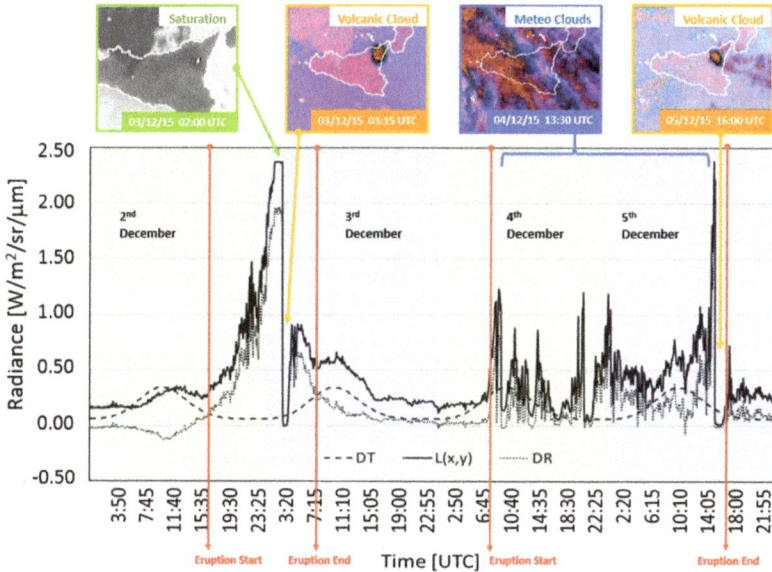

Figure 4. The 3.9 μm radiance from 2 to 5 December 2015. Different characteristics of the eruption are shown: start, end, saturation, presence of volcanic plume and meteorological clouds. The satellite images displayed on the top highlight the different plot signatures.

Table 1 (formatted in agreement with Behncke et al. [10] and De Beni et al. [33]) summarizes the timing and duration of the different phases identified for all the 2011–2015 events with the MS2RWS procedure applied to the 3.9 μm SEVIRI measurements. As the table shows, in some cases the estimations are impossible because of the presence of meteorological clouds over the volcanic area that covers completely the eruption signal (gray rows). As an example, the MS2RWS approach cannot discriminate the start and the end of all the events from 19 to 23 February 2013 and from 4 to 5 December 2015 because of the cloudy scenarios. The table shows also that several start/end fountains cannot be detected. The reason of that can be threefold: the presence of meteorological clouds over the plume that mask completely the ash signal, a plume too thin to sufficiently darken the 3.9 μm signal, and a plume too quickly transported out from the volcano area due to the high wind speed.

The last column of Table 1 shows the VPTHs estimated with the DP technique and the associated errors. It indicates that more than the 60% of the VPTHs are between 8 and 12 km a.s.l. and that the mean uncertainty is about 0.5 km.

Figure 5 shows the unconstrained TADR obtained for all the Etna lava fountains in which T_c has been set in the range 100–600 °C. As the figure shows the big uncertainty on T_c values lead to big uncertainties on the TADR, ranging from 3 to 35 m^3/s. Moreover, TADR estimates suffer from radiance saturation of the SEVIRI 3.9 μm channel and therefore TADR values result underestimated. Figure indicate also the occurrence of many short-term but quite intense fire fountains and a significant event from 6 to 8 December 2015 (bottom plot) not considered in Table 1. The reason is that this event produced lava flow emission without the formation of stable lava fountain. In Section 5.2 the AVHRR higher spatial resolution measurements will be used to reduce the SEVIRI TADR uncertainty.

Table 1. Timing of the different phases and duration recognized for all the 2011–2015 events for both the eruption and lava fountains derived from the MS2RWS procedure applied to the 3.9 μm SEVIRI measurements. The last column indicates the Volcanic Plume Top Height (VPTH) estimated by considering the DP approach. In grey the date with a complete meteorological cloud cover. All start and end times are in UTC.

Reference Date	Episode Number	Start Eruption	Start Fountaining	End Fountaining	End Eruption	Duration Lava Emission	Duration Fountaining	VPTH [Km]
12/01/2011	1	12/01/2011 19:30	12/01/2011 22:00	12/01/2011 22:30	13/01/2011 03:00	7:30	00:30	9.0 +/− 0.4
18/02/2011	2	/	/	/	/	/	/	
10/04/2011	3	09/04/2011 19:00	10/04/2011 11:00	10/04/2011 15:40	11/04/2011 23:50	52:50	04:40	6.7 +/− 0.3
12/05/2011	4	11/05/2011 18:35	/	/	12/05/2011 23:50	29:15	/	5.2 +/− 0.3
09/07/2011	5	09/07/2011 09:25	09/07/2011 14:40	09/07/2011 15:00	10/07/2011 11:25	26:00	00:20	9.6 +/− 0.3
19/07/2011	6	18/07/2011 20:45	/	/	20/07/2011 04:00	31:15	/	/
25/07/2011	7	24/07/2011 20:50	/	/	26/07/2011 02:45	29:55	/	5.2 +/− 0.3
30/07/2011	8	30/07/2011 07.25	/	/	31/07/2011 10:40	27:15	/	/
05/08/2011	9	05/08/2011 19:45	/	/	06/08/2011 10:35	14:50	/	13.1 +/− 1.0
12/08/2011	10	12/08/2011 05:50	/	/	13/08/2011 08:20	26:30	/	8.2 +/− 0.3
20/08/2011	11	20/08/2011 03:40	20/08/2011 07:10	20/08/2011 07:50	21/08/2011 05:35	25:55	00:40	11.2 +/− 0.3
29/08/2011	12	28/08/2011 23:25	/	/	29/08/2011 16:20	16:55	/	9.6 +/− 0.3
08/09/2011	13	08/09/2011 06:15	/	/	09/09/2011 09:15	27:00	/	11.0 +/− 0.3
19/09/2011	14	/	/	/	/	/	/	/
28/09/2011	15	/	/	/	/	/	/	/
08/10/2011	16	08/10/2011 12:25	08/10/2011 13:25	08/10/2011 14:50	09/10/2011 03:40	15:15	01:25	/
23/10/2011	17	23/10/2011 17:40	/	/	24/10/2011 04:30	10:50	/	5.5 +/− 0.3
15/11/2011	18	15/11/2011 10:15	/	/	15/11/2011 20:45	10:30	/	9.9 +/− 0.4
05/01/2012	19	05/01/2012 01:10	05/01/2012 05:50	05/01/2012 06:55	05/01/2012 20:15	19:05	01:05	16.2 +/− 1.7
09/02/2012	20	08/02/2012 19:00	09/02/2012 03:15	09/02/2012 06:45	09/02/2012 06:15	11:15	03:30	8.8 +/− 0.6
04/03/2012	21	/	/	/	/	/	/	/
18/03/2012	22	18/03/2012 04:50	/	/	19/03/2012 04:15	23:25	/	11.0 +/− 0.4
01/04/2012	23	31/03/2012 22:40	/	/	02/04/2012 01:35	26:55	/	10.8 +/− 0.4
12/04/2012	24	12/04/2012 11:45	/	/	12/04/2012 23:30	11:45	/	/
24/04/2012	25	23/04/2012 03:30	/	/	24/04/2012 12:50	33:20	/	10.7 +/− 0.2
19/02/2013	26	19/02/2013 02:30	/	/	/	/	/	7.8 +/− 0.3
20/02/2013	27	/	/	/	/	/	/	/
20/02/2013	28	/	/	/	/	/	/	/
21/02/2013	29	/	21/02/2013 04:00	21/02/2013 06:00	/	/	02:00	/
23/02/2013	30	/	23/02/2013 19:00	23/02/2013 20:15	23/02/2013 18:45	/	01:15	6.7 +/− 0.3
28/02/2013	31	28/02/2013 08:45	/	/	28/02/2013 23:30	13:45	/	9.3 +/− 0.9
05/03/2013	32	05/03/2013 20:00	05/03/2013 23:00	05/03/2013 23:20	07/03/2013 03:05	31:05	00:20	/

Table 1. *Cont.*

Reference Date	Episode Number	Start Eruption	Start Fountaining	End Fountaining	End Eruption	Duration Lava Emission	Duration Fountaining	VPTH [Km]
16/03/2013	33	16/03/2013 11:20	/	/	17/03/2013 04:45	17:25	/	/
03/04/2013	34	03/04/2013 10:40	/	/	04/04/2013 04:40	17:59	/	5.8 +/− 0.3
12/04/2013	35	11/04/2013 11:15	/	/	13/04/2013 04:00	40:45	/	7.3 +/− 0.3
18/04/2013	36	18/04/2013 07:25	18/04/2013 11:35	18/04/2013 12:05	19/04/2013 04:30	21:05	00:30	6.6 +/− 0.3
20/04/2013	37	19/04/2013 23:00	/	/	21/04/2013 03:05	28:05	/	11.7 +/− 0.4
27/04/2013	38	27/04/2013 13:35	/	/	28/04/2013 07:30	17:55	/	5.2 +/− 0.4
26/10/2013	39	25/10/2013 19:20	/	/	27/10/2013 21:50	50:30	/	8.1 +/− 0.3
11/11/2013	40	11/11/2013 08:00	11/11/2013 01:05	11/11/2013 0 7:30	12/11/2013 04:40	20:40	06:25	
17/11/2013	41	16/11/2013 15:15	/	/	18/11/2013 03:20	36:05	/	10.5 +/− 0.8
23/11/2013	42	23/11/2013 05:15	/	/	24/11/2013 01:00	19:45	/	11.1 +/− 0.7
28/11/2013	43	28/11/2013 16:50	/	/	29/11/2013 10:30	17:40	/	/
02/12/2013	44	02/12/2013 17:00	02/12/2013 19:20	02/12/2013 20:20	03/12/2013 17:05	24:05	01:00	/
28/12/2014	45	28/12/2014 18:15	28/12/2014 20:50	28/12/2014 21:50	29/12/2014 16:20	22:05	01:00	/
03/12/2015	46	02/02/2015 16:40	03/12/2015 02:55	03/12/2015 03:40	03/12/2015 09:55	17:20	00:45	12.5 +/− 0.3
04/12/2015	47	04/12/2015 07:55	04/12/2015 09:20	04/12/2015 11:10	/	/	01:50	17.6 +/− 1.1
04/12/2015	48	/	04/12/2015 20:55	04/12/2015 21:10	/	/	00:15	16.1 +/− 0.8
05/12/2015	49	/	05/12/2015 15:10	05/12/2015 16:45	08/02/2015 18:05	/	01:35	14.1 +/− 2.0

It is important to note that the duration of the 3 December 2015 eruption measured from SEVIRI data (bottom plot of Figure 5) is longer than the duration derived from ground observations [16]. This can be explained by the presence of volcanic products that are still hot after the end of the eruption. Sensors measure the thermal emission from fallout deposits and effusive materials even if the activity has been over for days. As a result, the application of Equation (6) can yield to false TADR.

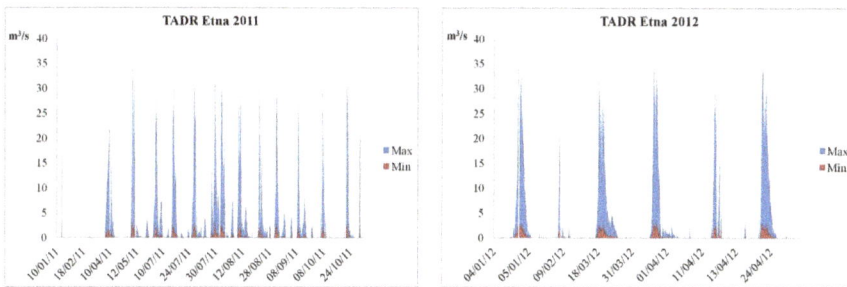

Figure 5. *Cont.*

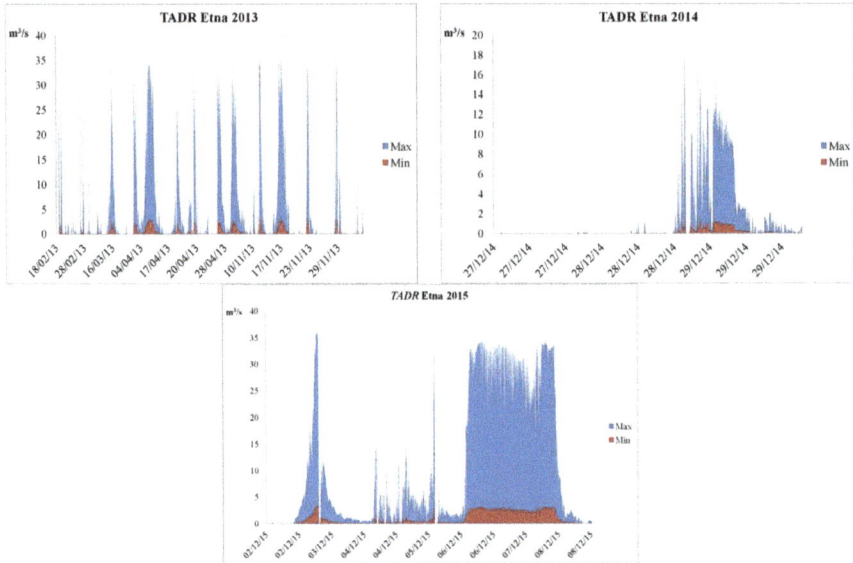

Figure 5. Time Averaged Discharge Rate (TADR) time series derived from SEVIRI 3.9 μm images during 2011–2015 Etna's lava fountains. Blue and red values are maximum and minimum derived TADR respectively.

5. Validation

The start of the eruption and the start and duration of the lava fountains are compared with ground-based measurements, while the VPTH with estimations obtained merging procedures based on tracking the volcanic cloud center of mass and on HYSPLIT backwards trajectories.

5.1. Eruption and Fountains Start and Duration

Being the high temperature source mainly caused by the presence of lava emission, the Eruption Start retrieved from SEVIRI (ES) has been compared with the "Start Lava emission" (SL) results published by Behncke et al. [10] and De Beni et al. [33], that were obtained from the analysis of the VIS-TIR ground based cameras placed at Etna and volcanological observations hereafter named ground-based observations.

The blue and red bars in Figure 6 represent the time differences (SL-ES) and (ES-SL) respectively. Following the latter definition, the blue bars indicate that, for a single event, SEVIRI is able to detect the lava emission before the VIS-TIR cameras, while the contrary is true for the red bars. As the figure shows, for the most of the eruption, the SEVIRI alert is given, on average, about 3 h before the ground-based alert. Possible reasons of this early SEVIRI alert could be the presence of the magma in the conduit or an increase of the strombolian activity that usually precede the lava emission.

Figure 7 indicates a good agreement for the start of the lava fountains, while greater differences are found for the duration time (Figure 8). The fountaining duration retrieved from SEVIRI can be lower than the duration obtained from the cameras because the plume could be too transparent to cause the drop of the SEVIRI 3.9 μm signal. The opposite (SEVIRI duration greater than ground-based observations) can be due to the presence of meteorological clouds not correctly detected.

Figures 7 and 8 show time start and duration of the lava fountains estimated from SEVIRI and from ground-based observations.

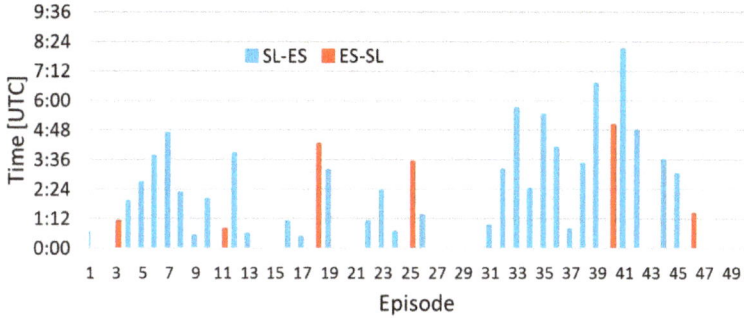

Figure 6. Time differences between the eruption start retrieved from SEVIRI (ES) data and the lava emission start time as derived from ground-based observations (SL). The blue and red bars represent the time differences (SL-ES) and (ES-SL) respectively.

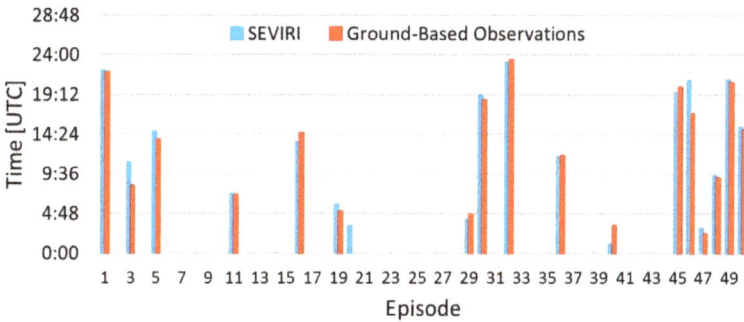

Figure 7. Fountaining starts retrieved from SEVIRI (red bars) and from ground-based observations (blue bars).

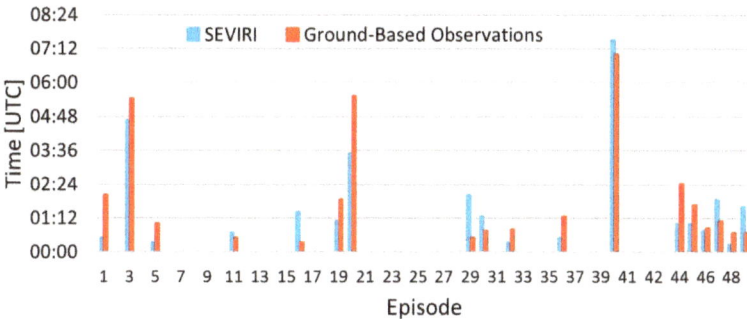

Figure 8. Lava fountains durations retrieved from SEVIRI (red bars) and from ground-based visible (VIS) to Thermal InfraRed (TIR) (VIS/TIR) observations (blue bars).

5.2. TADR Improvement Using AVHRR Data

The TADR accuracy can be improved by using remote sensing systems with higher spatial resolution compared to SEVIRI. Here the AVHRR data have been considered to constrain the SEVIRI time series. Figure 9 shows the minimum (black bars) and maximum (yellow bars) TADR derived from 5 images collected from 6 to 8 December 2015 by AVHRR. The green line shows the maximum SEVIRI derived TADR, rescaled using the maximum AVHRR derived TADR. SEVIRI values are corrected with each new AVHRR acquisition (green stars) as in the real-time case. Minimum values are not shown in the graph for clarity. Figure 9 shows that, even if only five cloud-free AVHRR images are available in this time frame, the TADR derived from AVHRR allows to reduce uncertainties in SEVIRI estimates by 40%. Constrained SEVIRI time series show a mean TADR of 12 m^3/s that is in good agreement with the effusion rates of 10–15 m^3/s measured by Corsaro et al. [16] from field data.

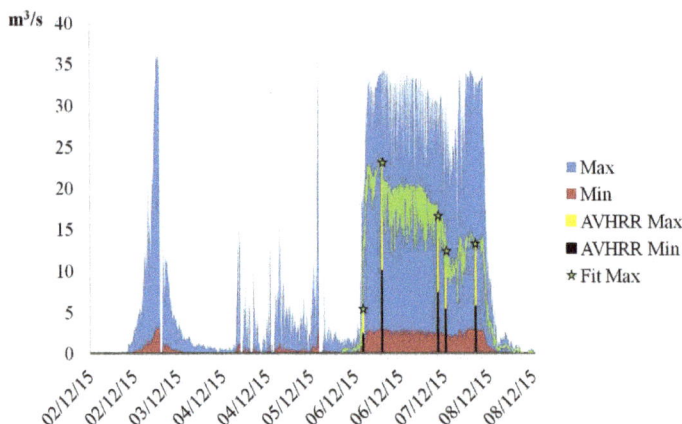

Figure 9. TADR estimated from 5 advanced very-high-resolution radiometer (AVHRR) images collected from 6 to 8 December 2015 (yellow and black values are maximum and minimum TADR respectively). The green line shows the maximum SEVIRI derived TADR constrained using the maximum AVHRR derived TADR.

5.3. VPTH by Using the Tracking of the Ash Cloud Center of Mass and HYSPLIT Backtrajectories

The high data frequency of the SEVIRI images (every 15 or 5 min) can be exploited to retrieve wind speed and direction of the volcanic clouds for each event [34]. These terms were derived by applying the following steps:

- retrieval of the ash abundance map from a given SEVIRI image [34];
- identification of the ash cloud centre of mass;
- computation of the ash centre of mass distance from the top of the volcano and the angle relative to the North.

The previous three steps are repeated for some subsequent SEVIRI images (at least 2–3 h from the start of the eruption). Using a linear fit, the speed (from distance and image time acquisition) and direction (from angle) of the volcanic cloud were obtained.

Basic assumption of this method is that the estimated peak speed is assumed to be the whole plume speed, which is the true wind speed at cloud altitude. Then, by comparing the wind speed and direction with the wind speed and direction of an atmospheric profile collected in the same time and position, the volcanic cloud altitude can be derived. For this purpose the National Centers for Environmental Prediction (NCEP)/National Center for Atmospheric Research (NCAR) reanalysis profiles [35] (resolution 2.5° × 2.5°) centered near the Etna volcano (37.5° N, 15.0° E) and collected close in time with the eruption start (time resolution 6 h) have been considered.

Due to the characteristics of the atmospheric wind speed and direction profiles, more than a single intersection with the wind speed and direction computed from the volcanic ash center of mass, can be found. For this reason, a procedure [36] based on HYSPLIT backtrajectories has been also considered [37–39]. By plotting several backtrajectories at different altitudes starting from a volcanic cloud detected several hours after the eruption, the correct volcanic cloud top height has been identified as the one corresponding to the trajectory that intersects Etna at the time of the eruption start.

Figure 10 shows an example of the volcanic cloud height estimation obtained by combining of the described methods (labelled in the following CM-HYSPLIT). The not univocal results obtained from wind speed (from 4.3 to 5.7 km and 16 km) and direction (4.3 and 16.5 km) are constrained by using the HYSPLIT backward trajectories obtaining for the plume height a final result of 4.3 km.

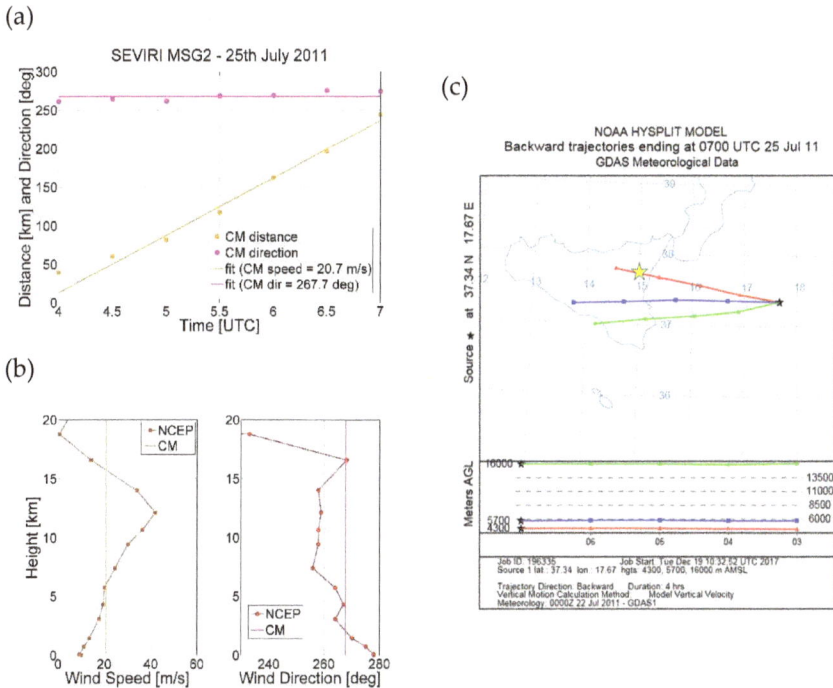

Figure 10. Example of the CM and HYSPLIT procedures combination for the 25 July 2011 event. (**a**): distance (yellow) and direction (magenta) of the cloud center of mass referring to the Etna position. (**b**): comparison between the volcanic cloud speed (yellow vertical line) and direction (magenta vertical line) with the wind speed (orange profile) and direction (red profile) extracted from NCEP. (**c**): HYSPLIT backtrajectories for different altitudes (the yellow star indicates the Etna position).

Figure 11 shows the comparison between the results obtained from the DP and the CM-HYSPLIT procedures. The figure shows a general good agreement, except for few dates (episode numbers: 9 and 19, respectively 5 August 2011 and 5 January 2012) and for the events of December 2015. These significant discrepancies can be due to an inertial overshoot of the plume and a consequent settling at the level of neutral buoyancy. Another reason can be that the volcanic cloud reaches the tropopause where the temperature profile is almost constant. This leads to big uncertainties on the VPTH estimation obtained with the DP procedure. For the December 2015 events, also the estimation made by a ground-based radar system collocated at Catania airport have been plotted [15]. As the figure shows, these measurements lie in between the DP and CM-HYSPLIT retrievals.

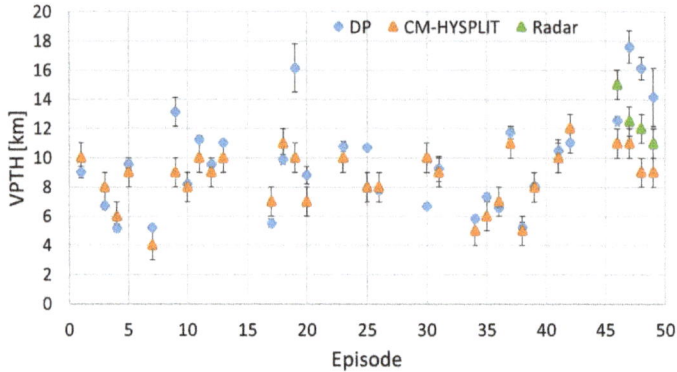

Figure 11. Comparison between the VPTH estimated from DP and CM-HYSPLIT procedures. Also, the ground based radar VPTH estimation, for the December 2015 events, have been considered.

6. Conclusions

In this work the MSG-SEVIRI space-based measurements have been used for a proximal monitoring of the Etna lava fountain events occurred from 2011 to 2015.

Results show that the SEVIRI 3.9 μm radiance measurements can be exploited to estimate eruption start and duration and the volcanic cloud presence. In particular, the ash plume presence (recognized by the drop of the 3.9 μm signal) can be used to estimate the start and end of the lava fountains. The procedure considered is based on the comparison between the SEVIRI measurements with an historical threshold computed from the analysis of 5 years of satellite data.

The VPTHs, estimated by a simplified procedure based on the brightness temperature at 10.8 μm, indicates that more than the 60% of the VPTH's are between 8 and 12 km with a height mean error of about 0.5 km.

The TADR results show the big uncertainties due to the unknown lava temperature and confirm the short-term but quite intense events occurred from 2011 to 2015. SEVIRI TADR estimates have been improved using the higher resolution AVHRR data, considering five cloud-free images collected during the 6–8 December 2015 event. The AVHRR TADR retrieved reduces the uncertainties in SEVIRI estimates by 40% with a mean value of 12 m^3/s which is in good agreement with the effusion rates measured by Corsaro et al. [16].

As expected, the main limitation for all the procedures is due to the presence of the meteorological clouds over the volcanic area that can partially or completely mask the eruption signal.

The start and duration of eruption and fountaining activities have been compared with ground-based observations. The results indicate that, for the most of the eruption, SEVIRI is able to detect the start of the lava emission about 3 h before the ground measurements, while there is a good agreement for both the start and duration of the lava fountains with a mean difference of about 1 h and 50 min respectively.

VPTH has been compared with the results obtained with a procedure based on the combination of an algorithm based on the volcanic cloud center of mass tracking and the HYSPLIT back-trajectories. The results indicate a general good agreement with a mean difference of about 1 km. For the 2015 December events, the differences are greater because the volcanic plume reached the tropopause where the temperature profile is almost constant.

Acknowledgments: The work was part of the Conv. Allegato B2 INGV-DPC 2017 funded by the Italian Civil Protection. The authors gratefully acknowledge the NOAA Air Resources Laboratory (ARL) for the provision of the HYSPLIT transport and dispersion model and/or READY website (http://www.ready.noaa.gov) used in this publication.

Author Contributions: Stefano Corradini conceived and coordinated the work; Lorenzo Guerrieri analyzed the SEVIRI data for the computation of the volcanic cloud top height using CM-HYSPLIT procedures; Valerio Lombardo processed the SEVIRI 3.9 μm data for the TADR estimations; Luca Merucci and Dario Stelitano took part in the discussion of the presented work and prepared the SEVIRI data needed for the different processing; Massimo Musacchio and Malvina Silvestri processed the SEVIRI 3.9 μm data for the estimation of start/end of the eruptions and lava fountains; Michele Prestifilippo, Simona Scollo and Gaetano Spata processed the SEVIRI 10.8 μm data for the volcanic plume top height estimation using DP procedure.

Conflicts of Interest: The authors declare no conflict of interest.

References

1. Corradini, S.; Montopoli, M.; Guerrieri, L.; Ricci, M.; Scollo, S.; Merucci, L.; Marzano, F.S.; Pugnaghi, S.; Prestifilippo, M.; Ventress, L.J.; et al. Multi-Sensor Approach for Volcanic Ash Cloud Retrieval and Eruption Characterization: The 23 November 2013 Etna Lava Fountain. *Remote Sens.* **2016**, *8*. [CrossRef]
2. Merucci, L.; Zakšek, K.; Carboni, E.; Corradini, S. Stereoscopic estimation of volcanic cloud-top height from two geostationary satellites. *Remote Sens.* **2016**, *8*, 206. [CrossRef]
3. Prata, A.J.; Kerkmann, J. Simultaneous retrieval of volcanic ash and SO₂ using MSG-SEVIRI measurements. *Geophys. Res. Lett.* **2007**, *34*, L05813. [CrossRef]
4. Corradini, S.; Merucci, L.; Prata, A.J. Retrieval of SO₂ from Thermal Infrared Satellite Measurements: Correction Procedures for the Effects of Volcanic Ash. *Atmos. Meas. Tech.* **2009**, *2*, 177–191. [CrossRef]
5. Merucci, L.; Burton, M.; Corradini, S.; Salerno, G. Reconstruction of SO₂ flux emission chronology from space-based measurements. *J. Volcanol. Geotherm. Res.* **2011**. [CrossRef]
6. Harris, A.J.L.; Blake, S.; Rothery, D.A.; Stevens, N.F. A chronology of the 1991 to 1993 Etna eruption using AVHRR data: Implications for real time thermal volcano monitoring. *J. Geophys. Res.* **1997**, *102*, 7985–8003. [CrossRef]
7. Wooster, M.J.; Rothery, D.A. Time series analysis of effusive volcanic activity using the ERS along track scanning radiometer: The 1995 eruption of Fernandina volcano, Galapagos Island. *Remote Sens. Environ.* **1997**, *69*, 109–117. [CrossRef]
8. Ganci, G.; Harris, A.J.L.; Del Negro, C.; Guehenneux, Y.; Cappello, A.; Labazuy, P.; Calvari, S.; Gouhier, M. A year of lava fountaining at Etna: Volumes from SEVIRI. *Geophys. Res. Lett.* **2012**, *39*, L06305. [CrossRef]
9. Scollo, S.; Boselli, A.; Coltelli, M.; Leto, G.; Pisani, G.; Prestifilippo, M.; Spinelli, N.; Wang, X. Volcanic ash concentration during the 12 August 2011 Etna eruption. *Geophys. Res. Lett.* **2015**, *42*. [CrossRef]
10. Behncke, B.; Branca, S.; Corsaro, R.A.; De Beni, E.; Miraglia, L.; Proietti, C. The 2011–2012 summit activity of Mount Etna: Birth, growth and products of the new SE crater. *J. Volcanol. Geotherm. Res.* **2014**, *270*, 10–21. [CrossRef]
11. Scollo, S.; Prestifilippo, M.; Pecora, E.; Corradini, S.; Merucci, L.; Spata, G.; Coltelli, M. Eruption column height estimation of the 2011–2013 Etna lava fountains. *Ann. Geophys.* **2014**, *57*. [CrossRef]
12. International Civil Aviation Organization (ICAO). *Volcanic Ash Contingency Plan—Eur and Nat Regions*; EUR Doc 019–NAT Doc 006, Part II; International Civil Aviation Authority: Montreal, QC, Canada, 2010.
13. Andronico, D.; Scollo, S.; Cristaldi, A. Unexpected hazards from tephra fallouts at Mt Etna: The 23 November 2013 lava fountain. *J. Volcanol. Geotherm. Res.* **2015**, *304*, 118–125. [CrossRef]
14. Bonaccorso, S.; Calvari, S.; Linde, A.; Sacks, S. Eruptive processes leading to the most explosive lava fountain at Etna volcano: The 23 November 2013 episode. *Geophys. Res. Lett.* **2014**, *41*, 4912–4919. [CrossRef]
15. Vulpiani, G.; Ripepe, M.; Valade, S. Mass discharge rate retrieval combining weather radar and thermal camera observations. *J. Geophys. Res. Solid Earth* **2016**, *121*, 5679–5695. [CrossRef]
16. Corsaro, R.A.; Andronico, D.; Behncke, B.; Branca, S.; Caltabiano, T.; Ciancitto, F.; Cristaldi, A.; De Beni, E.; La Spina, A.; Lodato, L.; et al. Monitoring the December 2015 summit eruptions of Mt. Etna (Italy): Implications on eruptive dynamics. *J. Volcanol. Geotherm. Res.* **2017**. [CrossRef]
17. Matson, M.; Dozier, J. Identification of subresolution high temperature sources using a thermal IR sensor. *Photogramm. Eng. Remote Sens.* **1981**, *47*, 1311–1318.
18. Dozier, J. A method for satellite identification of surface temperature fields of subpixel resolution. *Remote Sens. Environ.* **1981**, *11*, 221–229. [CrossRef]
19. Rothery, D.A.; Francis, P.; Wood, W. Volcano monitoring using short wavelength infrared data from satellite. *J. Geophys. Res.* **1988**, *93*, 7993–8008. [CrossRef]

20. Sobrino, J.A.; Romaguera, M. Land surface temperature retrieval from MSG1-SEVIRI data. *Remote Sens. Environ.* **2004**, *92*, 247–254. [CrossRef]
21. Musacchio, M.; Silvestri, M.; Buongiorno, M.F. Use of radiance value from MSG SEVIRI and MTSAT data: Application for the monitoring on volcanic area. In Proceedings of the 2011 EUMETSAT Meteorological Satellite Conference, Oslo, Norway, 5–9 September 2011.
22. Musacchio, M.; Silvestri, M.; Buongiorno, M.F. RT Monitoring of active volcanoes: MT Etna. In Proceedings of the Sea Space International Remote Sensing Conference, Seaspace Corporation, San Diego, CA, USA, 21–24 October 2012.
23. Gouhier, M.; Harris, A.J.L.; Calvari, S.; Labazuy, P.; Guéhenneux, Y.; Donnadieu, F.; Valade, S. Erratum to: Lava discharge during Etna's January 2011 fire fountain tracked using MSG-SEVIRI. *Bull. Volcanol.* **2012**, *74*, 1261. [CrossRef]
24. Wright, R.; Blake, S.; Harris, A.J.L.; Rothery, D. A Simple explanation for the space-based calculation of lava eruptions rates. *Earth Planet. Sci. Lett.* **2001**, *192*, 223–233. [CrossRef]
25. Lombardo, V. AVHotRR: Near-real time routine for volcano monitoring using IR satellite data. *Geol. Soc. Lond.* **2015**, *426*. [CrossRef]
26. Harris, A.J.L.; Baloga, S.M. Lava discharge rates from satellite measured heat flux. *Geophys. Res. Lett.* **2009**, *36*, L19302. [CrossRef]
27. Harris, A.J.L.; Favalli, M.; Steffke, A.; Fornaciai, A.; Boschi, E. A relation between lava discharge rate, thermal insulation, and flow area set using Lidar data. *Geophys. Res. Lett.* **2010**, *37*, L20308. [CrossRef]
28. Ganci, G.; Vicari, A.; Fortuna, L.; Del Negro, C. The HOTSAT volcano monitoring system based on a combined use of SEVIRI and MODIS multispectral data. *Ann. Geophys.* **2011**, *54*. [CrossRef]
29. Hirn, B.; Ferrucci, F.; Di Bartola, C. Near-tactical eruption rate monitoring of Pu'u O'o (Hawaii) 2000–2005 by synergetic merge of payloads ASTER and MODIS. In Proceedings of the IEEE International Geoscience and Remote Sensing Symposium, IGARSS 2007, Barcelona, Spain, 23–28 July 2007; pp. 3744–3747. [CrossRef]
30. Govaerts, Y.; Arriaga, A.; Schmetz, J. Operational vicarious calibration of the MSG/SEVIRI solar channels. *Adv. Space Res.* **2001**, *28*, 21–30. [CrossRef]
31. Prata, A.J.; Grant, I.F. Retrieval of microphysical and morphological properties of volcanic ash plumes from satellite data: Application to Mt Ruapehu, New Zealand. *Q. J. R. Meteorol. Soc.* **2001**, *127*, 2153–2179. [CrossRef]
32. Corradini, S.; Spinetti, C.; Carboni, E.; Tirelli, C.; Buongiorno, M.F.; Pugnaghi, S.; Gangale, G. Mt. Etna tropospheric ash retrieval and sensitivity analysis using Moderate Resolution Imaging Spectroradiometer measurements. *J. Appl. Remote Sens.* **2008**, *2*, 023550. [CrossRef]
33. De Beni, E.; Behncke, B.; Branca, S.; Nicolosi, I.; Carluccio, R.; D'Ajello Caracciolo, F.; Chiappini, M. The continuing story of Etna's New Southeast Crater (2012–2014): Evolution and volume calculations based on field surveys and aerophotogrammetry. *J. Volcanol. Geotherm. Res.* **2015**, *303*, 175–186. [CrossRef]
34. Guerrieri, L.; Merucci, L.; Corradini, S.; Pugnaghi, S. Evolution of the 2011 Mt. Etna ash and SO$_2$ lava fountain episodes using SEVIRI data and VPR retrieval approach. *J. Volcanol. Geotherm. Res.* **2015**, *291*, 63–71. [CrossRef]
35. The NCEP/NCAR Reanalysis Project at the NOAA/ESRL Physical Sciences Division. Available online: http://www.esrl.noaa.gov/psd/data/reanalysis/reanalysis.shtml (accessed on 1 December 2017).
36. Pardini, F.; Burton, M.; de' Michieli Vitturi, M.; Corradini, S.; Salerno, G.; Merucci, L.; Di Grazia, G. Retrieval and intercomparison of volcanic SO$_2$ injection height and eruption time from satellite maps and ground-based observations. *J. Volcanol. Geotherm. Res.* **2016**. [CrossRef]
37. Draxler, R.R. *HYSPLIT4 User's Guide*; NOAA Tech. Memo. ERL ARL-230; NOAA Air Resources Laboratory: Silver Spring, MD, USA, 1999.
38. Stein, A.F.; Draxler, R.R.; Rolph, G.D.; Stunder, B.J.B.; Cohen, M.D.; Ngan, F. NOAA's HYSPLIT atmospheric transport and dispersion modeling system. *Bull. Am. Meteorol. Soc.* **2015**, *96*, 2059–2077. [CrossRef]
39. Rolph, G.; Stein, A.; Stunder, B. Real-time Environmental Applications and Display sYstem: READY. *Environ. Model. Softw.* **2017**, *95*, 210–228. [CrossRef]

geosciences

MDPI

Review

Ground-Based Remote Sensing and Imaging of Volcanic Gases and Quantitative Determination of Multi-Species Emission Fluxes

Ulrich Platt [1,2,*], Nicole Bobrowski [1,2] and Andre Butz [3,4]

[1]   Institute of Environmental Physics, Heidelberg University, D-69120 Heidelberg, Germany; Nicole.Bobrowski@iup.uni-heidelberg.de
[2]   Max Planck Institute for Chemistry, D-55128 Mainz, Germany
[3]   Meteorologisches Institut, Ludwig-Maximilians-Universität München, D-80 539 Munich, Germany; Andre.Butz@dlr.de
[4]   Institut für Physik der Atmosphäre, DLR Deutsches Zentrum für Luft-und Raumfahrt e. V., D-82 234 Oberpfaffenhofen, Germany
*   Correspondence: ulrich.platt@iup.uni-heidelberg.de

Received: 15 November 2017; Accepted: 17 January 2018; Published: 26 January 2018

Abstract: The physical and chemical structure and the spatial evolution of volcanic plumes are of great interest since they influence the Earth's atmospheric composition and the climate. Equally important is the monitoring of the abundance and emission patterns of volcanic gases, which gives insight into processes in the Earth's interior that are difficult to access otherwise. Here, we review spectroscopic approaches (from ultra-violet to thermal infra-red) to determine multi-species emissions and to quantify gas fluxes. Particular attention is given to the emerging field of plume imaging and quantitative image interpretation. Here UV SO_2 cameras paved the way but several other promising techniques are under study and development. We also give a brief summary of a series of initial applications of fast imaging techniques for volcanological research.

Keywords: volcanology; gases; remote sensing

1. Introduction

The physical and chemical structure and the spatial evolution of volcanic plumes is of great interest for a number of reasons beyond scientific curiosity:

(A) Volcanic gas emissions influence the atmosphere and therefore also the climate and other Earth system parameters in a number of ways and on different temporal and spatial scales (e.g., [1–3]). Investigations of plume chemistry and plume dispersal will help constrain these influences (see e.g., [4,5]).

(B) The composition and emission rate of volcanic gases are linked to processes occurring in the Earth's interior, therefore measuring volcanic gases provides insights into these otherwise largely inaccessible processes. For instance, already Noguchi and Kamiya [6] showed that eruptions of Mt. Asama (Japan) could be forecast with some degree of accuracy by measuring the variable partitioning of sulphur species, chlorine species, and carbon dioxide in the emissions from the active crater. Malinconico [7] showed first that also the amount of gas, in particular the amount of emitted SO_2 varies when the volcanic activity changes. Several authors (e.g., [8,9]) used the measured SO_2 emission also to calculate the amount of magma involved in the simultaneously observed volcanic activity.

These remote sensing techniques have a number of decisive advantages over in-situ observations: The largest being that under most conditions the total amount of gas in the plume can be determined, rather than the gas concentration at the plume edge (usually probed by ground-based in-situ instruments). Since measurements can be made from distances of typically a few kilometres remote sensing is also much safer than in-situ sampling. Moreover, the technology allows easy automation and thus continuous measurements in real time are readily possible, even during periods of explosive activity. In fact continuous measurements of SO_2 and BrO fluxes are realised by the global Network of Observation of Volcanic and Atmospheric Change (NOVAC) [10–12].

Recent advances in technology, in particular for spectroscopic techniques, allow remote analysis of many species in volcanic plumes. Specifically, the two-dimensional or even three-dimensional distribution of gases within volcanic plumes and their temporal evolution can now be determined in real time. In other words, recent technology allows "imaging" of trace gas distributions and their motion in plumes. Compared to earlier techniques, which were only capable of measuring the total column density of a particular gas along a single line of sight within the plume (e.g., [13–15]), new imaging techniques (see e.g., [5,16]) give much better insight into transport and mixing processes, as well as into chemical transformations within plumes. Moreover, much more accurate, quantitative determinations of trace gas fluxes are now possible in most cases. If the imaging is performed at adequate time resolution, the plume speed can be derived directly from image series, either by correlation techniques (e.g., [13,17–19]) or by more advanced image processing techniques (e.g., [20–22]).

The most abundant volcanic gases, i.e., H_2O, CO_2, SO_2, HCl, HF, H_2S and many other trace species like BrO, ClO, OClO, NO_2 CO, COS, SiF_4, can be measured either by UV/vis spectroscopy (i.e., H_2O, SO_2, BrO, ClO, OClO, NO_2 are readily measured by DOAS) or IR spectroscopy (e.g., H_2O, CO_2, SO_2, HCl, HF, H_2S, CO, COS, SiF_4). Note, however, that the two species, which are usually most abundant in volcanic plumes (i.e., H_2O and CO_2) also occur in high concentrations in the atmosphere surrounding the plume. Therefore it is frequently difficult to determine these gases by passive spectroscopy (see Section 2.3.3), since the contrast between the background atmospheric signal and background atmosphere plus plume is very small (sometimes in the 10^{-3} range or below).

In the following we give an overview of the principles of optical remote sensing and of plume imaging techniques. These techniques can be applied from the ground, from (manned or unmanned) aircraft (e.g., [23]), or from satellite platforms. Although satellite imaging constitutes an interesting and important branch of these techniques (e.g., [24–26]) we restrict ourselves to the first type of application.

At present ground-based remote sensing techniques are in various stages of development, ranging from methods that are already routinely being applied to study volcanic degassing, to those which are so new that only first field test have been done. In addition we discuss applications to flux measurements.

2. Remote Sensing of Volcanic Gases

Remote sensing of volcanic plume composition relies on measuring the absorption, scattering or emission of radiation by the gases (and particulates) in the plume. At this point it is convenient to make a distinction between active (using an artificial light source) and passive (using natural light sources, e.g., the sun or thermal emission) remote sensing. In general, the radiative transfer equation (see e.g., [27]) underlying all these techniques would need to be solved for a remote sensing observer collecting radiation from a certain direction. The remote sensing observer could be located on a satellite in space, on an aircraft or another airborne platform, or on ground. The complexity of the remote sensing problem strongly depends on the chosen vantage point and viewing geometry, the spectral ranges covered, and the abundances of particulates and the target gases in the plume and in the background atmosphere. Figure 1 shows some basic ground based plume sensing schemes.

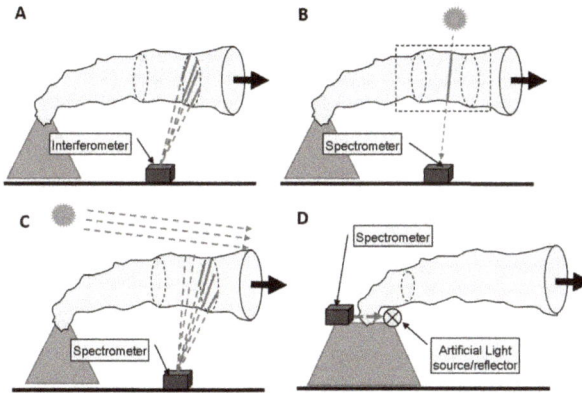

Figure 1. Overview of the different plume sensing schemes: (**A**) Thermal emission is recorded by passive spectroscopy in the thermal infra-red; (**B**) Direct light absorption spectroscopy in the UV, visible, or near IR; (**C**) Scattered sunlight spectroscopy in the UV or visible spectral ranges; (**D**) Active absorption spectroscopy in the UV, visible, or near IR.

In the following we give an overview of the principles of optical remote sensing and of plume imaging techniques. We limit ourselves to showcasing ground-based techniques that are regularly used for volcanic plume remote sensing. We recapitulate the basic aspects of absorption (Section 2.1) and thermal emission (Section 2.2) spectroscopic techniques and highlight some applications (Section 2.3). Imaging techniques are discussed in Section 3, below.

2.1. Absorption Spectroscopy

The classic absorption experiment, illustrated in Figure 2, puts a light source at a distance L from the observer such that the absorbing medium is in-between. Neglecting light-scattering and thermal emission by the medium itself, Beer-Lambert-Bouguer's law describes the change of the radiance spectrum $I(\lambda, x)$ due to absorption along the light path x,

$$\frac{dI(\lambda, x)}{dx} = -k_a(\lambda, x)\, I(\lambda, x) \tag{1}$$

with wavelength λ and absorption coefficient $k_a(\lambda, x)$. For a gaseous medium, the latter is given by

$$k_a(\lambda, x) = \sum_i \sigma_i(\lambda,\, p(x),\, T(x),\, ...)\, n_i(x) \tag{2}$$

where $n_i(x)$ is the number density of absorbing gas species i (of which there might be multiple), and $\sigma_i(\lambda,\, p(x),\, T(x),\, ...)$ is the respective absorption cross section that generally depends on the path via ambient pressure p, temperature T, or other variables like water vapour that causes spectroscopic effects such as foreign line broadening.

Assuming that the light source emits a background spectrum $I(\lambda, 0) = I_0(\lambda)$ at location $x = 0$, integration of Equation (1) yields the radiance spectrum at the location $x = L$ of the remote sensing observer,

$$I(\lambda,\, L) = I_0(\lambda) \exp(-\tau(\lambda, 0, L)) \tag{3}$$

where we define absorption optical density $\tau(\lambda, x_1, x_2)$ as

$$\tau(\lambda,\, x_1, x_2) = \int_{x_1}^{x_2} k_a(\lambda, x)\, dx = \int_{x_1}^{x_2} \sum_i \sigma_i(\lambda,\, p(x),\, T(x),\, ...)\, n_i(x)\, dx \tag{4}$$

If the absorption cross sections can be assumed independent of ambient conditions along the path ($p(x)$, $T(x)$, ... $= const.$), the absorption optical density is approximately given by

$$\tau(\lambda, 0, L) \approx \sum_i \sigma_i(\lambda) \int_0^L n_i(x)\, dx = \sum_i \sigma_i(\lambda)\, S_i \qquad (5)$$

where S_i denotes the slant column density of trace gas species i defined as the concentration integrated along the light path. For atmospheric remote sensing, this assumption typically holds for absorption cross sections in the UV and visible spectral range and thus, the Differential Optical Absorption Spectroscopy (DOAS) technique described in Section 2.3.1 adopts Equation (5). In the infra-red, the assumption is not generally applicable and absorption spectroscopic techniques need to start out from Equation (4) (Section 2.3.2), unless the absorbing gases can be safely assumed to only exist in a single plume layer with homogeneous pressure, temperature etc.

Assuming that Equation (5) holds and that there is only a single absorbing gas in the volcanic plume, the respective gas slant column density S can, in principle, be determined by ratioing a background spectrum $I_0(\lambda)$ by the plume spectrum $I(\lambda)$ according to,

$$S = \frac{\tau(\lambda)}{\sigma(\lambda)} = \frac{1}{\sigma(\lambda)} \ln \frac{I_0(\lambda)}{I(\lambda)} \qquad (6)$$

from which the plume average absorber concentration $\bar{n}$ is straightforward to calculate through

$$\bar{n} = \frac{S}{L_p} \qquad (7)$$

if the species only exists inside the plume (and not in the background atmosphere) and if the geometric path length L_p through the plume can be estimated e.g., using the geometric plume extent. In practice, various complications render this approach too simplistic and real-world solutions need to be found (Section 2.3).

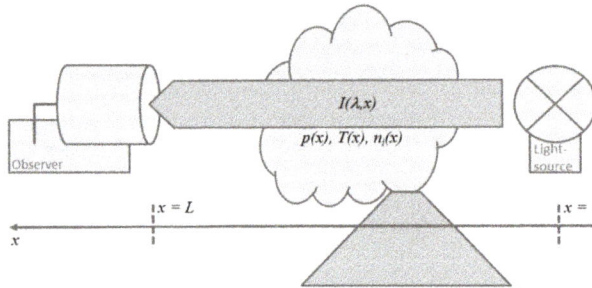

Figure 2. Schematic illustration of the classic absorption experiment with a light source at a distance L from the observer and the absorbing volcano plume in-between.

2.2. Thermal Emission Spectroscopy

Thermal emission spectroscopy makes use of the infra-red emission of the atmospheric gases themselves i.e., it collects radiation emitted along the lines-of-sight through the atmosphere in the infra-red spectral range without targeting a specific light source as illustrated by Figure 3. Typically, Schwarzschild's approximation to the radiative transfer equation is used to describe the emitted radiance spectra,

$$\frac{dI(\lambda, x)}{dx} = -k_a(\lambda, x)\, I(\lambda, x) + k_a(\lambda, x)\, B(\lambda, T(x)) \qquad (8)$$

where $k_a(\lambda, x)$ is the absorption coefficient defined in Equation (2) and $B(\lambda, T(x))$ is Planck's thermal emission spectrum

$$B(\lambda, T(x)) = \frac{2\,h\,\frac{c^2}{\lambda^5}}{e^{\frac{h\,c}{\lambda k T(x)}} - 1} \tag{9}$$

(c = speed of light, h = Planck constant, k = Boltzmann constant, T = temperature). Solving Schwarzschild's equation for an observer at location $x = L$ pointing toward a cold background at location $= 0$ ($I(\lambda, 0) = 0$), with the volcanic plume in-between, yields the observed radiance

$$I(\lambda, L) = \int_0^L k_a(\lambda, x)\,B(\lambda, T(x))\exp(-\tau(\lambda, x, L))\,dx \tag{10}$$

with $\tau(\lambda, x, L)$ the absorption optical density defined in Equation (5). Equation (10) essentially tells that thermal radiation accumulates along the path toward the observer taking into account that radiation emitted at location x gets reabsorbed along the rest of the way to the observer

Dividing the light path into N homogeneous layers k at constant temperature T_k with layer boundaries x_k and x_{k+1}, Equation (10) becomes

$$I(\lambda, L) = \sum_{k=1}^N B(\lambda, T_k) \int_{x_k}^{x_{k+1}} k_a(\lambda, x)\exp(-\tau(\lambda, x, L))\,dx \tag{11}$$

which after calculating the integral reduces to

$$I(\lambda, L) = \sum_{k=1}^N B(\lambda, T_k)(1 - \exp(-\tau(\lambda, x_k, x_{k+1})))\exp(-\tau(\lambda, x_{k+1}, x_{N+1}))) \tag{12}$$

Introducing the transmittance $t_k = \exp(-\lambda(\lambda, x_k, x_{k+1})$ of layer k for short-hand notation, Equation (11) reads

$$I(\lambda, L) = \sum_{k=1}^N B(\lambda, T_k)\,(1 - t_k) \prod_{\substack{l = k+1 \\ k < N}}^N t_l \tag{13}$$

A common view on Equation (13) for volcanic applications is the one shown in Figure 3. The emission along the line-of-sight is assumed to consist of three contributions, the atmospheric background radiance $I_{behind}(\lambda)$ entering the plume from behind, thermal emission by the plume layer (2), and thermal emission by the layer (3) in front of the plume which hosts the observer. Then, Equation (13) simplifies to

$$I(\lambda, L) = I_{behind}(\lambda)t_2 t_3 + B(\lambda, T_2)\,(1 - t_2)t_3 + B(\lambda, T_3)\,(1 - t_3) \tag{14}$$

Equation (14) carries the target gas concentrations in the plume transmittance t_2, and thus can be used to setup plume remote sensing experiments (Section 2.3.3).

For further conceptual insight, we assume that a spectral range can be found where the target gas is the only absorber and that the target gas only exists in the volcanic plume. Then, all layer transmittances except t_2 become unity and the background radiance $I_{behind}(\lambda)$ vanishes, reducing Equation (14) to

$$I(\lambda, L) = B(\lambda, T_2)\,(1 - t_2) = B(\lambda, T_2)\,(1 - \exp(-\tau(\lambda, x_2, x_3))) \tag{15}$$

Following the discussion of Equation (5), for a homogeneous plume layer $(p(x),\ T(x),\ \ldots = const)$, the optical density can be approximated by $\tau(\lambda) = \sigma(\lambda) \times S$. Then, Equation (15) readily yields the slant column density S of the target gas via

$$S = \frac{\tau(\lambda)}{\sigma(\lambda)} = -\frac{1}{\sigma(\lambda)} \ln\left(1 - \frac{I(\lambda, L)}{B(\lambda, T_2)}\right) \tag{16}$$

from which the plume average concentration $\bar{n}$ calculates through Equation (7). Thus, under the above approximations, measuring a thermal emission spectrum $I(\lambda, L)$ and measuring (or knowing from external sources) the plume layer temperature T_2 would allow for estimating the gas concentrations. In practice, infra-red spectral windows are typically packed with overlapping absorption of various background and plume gases and thus, Equation (16) is too simplistic.

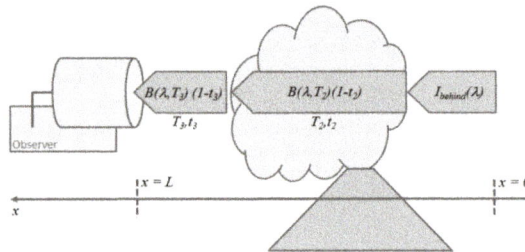

Figure 3. Schematic illustration of thermal emission sounding for a simplified 3-layer atmosphere consisting of a layer behind the volcanic plume, the plume-layer itself, and the layer between the plume and the observer.

2.3. Applications to Remote Sensing of Volcanic Plumes

There is a great variety of ways implement the above theoretical considerations for the purpose of volcanic research and surveillance. The currently available techniques can be broadly grouped into the following categories (see also Figure 4):

- Active or passive spectroscopy (i.e., will there be an artificial light source or a natural one)
- Arrangement of light path (source-detector, topographic reflector, artificial reflector, backscattered (artificial) light, scattered (sun) light)
- Path integrated column measurement or range resolved detection
- Dispersive or non-dispersive detection
  - For dispersive detection: Type of wavelength analysis (grating spectrometer,
  - Fourier transform interferometer, Fabry Pérot interferometer, tunable light source, ...)
  - For non-dispersive detection: Filter, narrow band emitting light source (e.g., laser, LED)
- One dimensional (single column) or two-dimensional (imaging) measurement (see Section 4)

While there is a large number of possible combinations of the above approaches only several of the techniques have actually been used for remote sensing in volcanological environments and therefore became popular, we describe these in the following:

1. Active UV/vis absorption spectroscopy (e.g., DOAS) in Section 2.3.1
2. Passive (i.e., scattered sunlight) UV/vis absorption spectroscopy in Section 2.3.1
3. IR absorption spectroscopy in Section 2.3.2
4. Thermal emission spectroscopy in Section 2.3.3
5. LIght Detection And Ranging (LIDAR) in Section 2.3.4

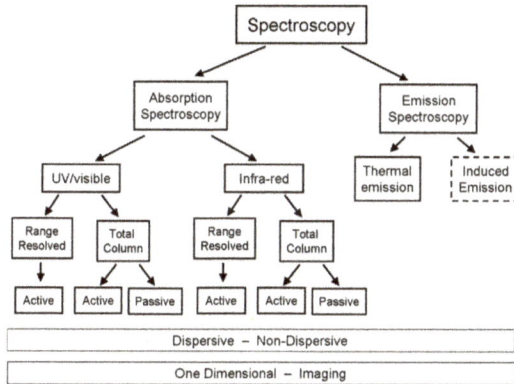

Figure 4. Family tree of spectroscopic remote sensing schemes.

2.3.1. Absorption Spectroscopy in the UV/Visible

Optical absorption spectroscopy (see e.g., [27]) is a technique following the principles outlined in Section 2.1. Absorption cross sections can be assumed to be approximately independent of pressure and temperature in the UV/visible spectral range, and thus, Equation (3) can be combined with Equation (5). However, there is the challenge how to disentangle the minute target gas absorption from overlapping interfering absorption, from spectrally broad extinction of radiation due to light-scattering, and from solar Fraunhofer lines, if the sun is the light source (which is a particularly popular case). Facing these challenges the COSPEC (COrrelation SPECtrometer, e.g., [28]) techniques pioneered UV/vis spectroscopy for volcanic applications from the 1970s. As another approach to address the above challenge, the DOAS (Differential Optical Absorption Spectroscopy) techniques (e.g., [27]) measure absorption spectra in a contiguous range of wavelengths for which the molecular absorption cross sections show several differential structures that are unique fingerprints for the absorbing gases. This implies that absorption cross sections must show strong variations with wavelength to make a gas measurable by DOAS. Today, modern instruments are mostly based on the DOAS technique (e.g., [27]), which our discussions focus on in the following. DOAS can be performed with miniaturized spectrometers and, it is well suited to detect even smallest gas amounts.

Building on the overall idea of using differential absorption structures, DOAS techniques are applicable to direct light-source (sun, moon, or artificial sources, see below) arrangements as well as to applications using scattered (solar) light. In the former case, e.g., when directly pointing into sufficiently bright light sources such as the sun, the path x is the well-defined geometric path between the source and the observer. In the latter case, e.g., when pointing into the sky away from the sun, the path x is the average geometric path that photons travel between the light source and the observer along various individual trajectories such as suggested by Figure 5. This interpretation makes the path x, its length L and in consequence the slant column density S_i in Equation (3) depend on wavelength since the average scattering light path depends on atmospheric light-scattering properties such as the particle scattering cross sections and phase functions as well as on the molecular absorption itself, which are all wavelength dependent. However, the narrow spectral ranges and the frequently small absorption optical densities ($\tau(\lambda) = 10^{-4} \dots 10^{-1}$) encountered in DOAS applications, typically (but not always, see below) allow for dropping wavelength dependencies in x, L, and S_i.

Although solar radiation has considerable intensity in the near IR (up to 2 to 3 µm) typically scattered sunlight absorption spectroscopy is only used in the UV/vis spectral region since it usually relies on Rayleigh scattering, where the scattering efficiency falls off strongly with wavelength (approximately proportional to λ^{-4}). There is usually also a contribution of Mie (aerosol) scattering, however the intensity of this component also falls off with increasing wavelength (typically

Geosciences **2018**, 8, 44

proportional to $\lambda^{-1.3}$). A great advantage of using scattered solar light is its universal applicability with observation from only one point, which greatly simplifies the logistics. Dispersive or non-dispersive detectors can be used, up to now the former are employed for one dimensional work and the latter for imaging applications.

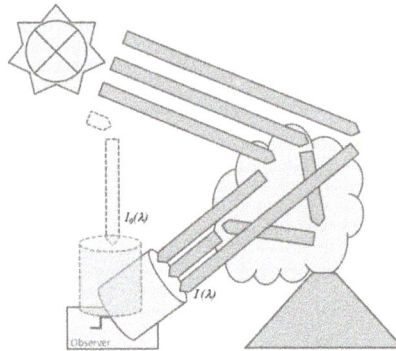

Figure 5. Schematic illustration of scattered skylight DOAS measurements with the sun as a light source.

Thus, conceptually, DOAS is fit to use Equations (6) and (7) to infer gas concentrations in volcanic plumes. As illustrated by Figure 5, the plume spectra $I(\lambda)$ in Equation (6) are the measurements for which the lines-of-sight cross the volcanic plume. The background spectra $I_0(\lambda)$ are obtained by adjusting the observing geometry such that lines-of-sight avoid the volcanic plume or, if such an adjustment is not possible, by resorting to external sources such as solar spectral atlases [29]. In practice, this simple approach faces various complications which need to be overcome.

First and foremost, the absorption optical densities of the target species are typically small ($\tau(\lambda) = 10^{-4}$ 10^{-1}) in the UV/visible spectral range and the target absorption usually interferes with strong Fraunhofer lines in the solar spectrum and with absorption by other gases. As explained above, DOAS relies on fingerprinting the target absorption by measuring contiguous spectral ranges covering a sufficient number of differential absorption structures of the target gas. For accessing the minute target absorption with high accuracy, however, it is further crucial to exclude systematic errors originating from uncertainties in the Fraunhofer spectrum or in the absorption cross sections causing spurious effects on the subjacent gas signature. Therefore, DOAS prefers using background spectra (and, if possible, absorption cross sections) that are measured by the actual remote sensing instrument. Then, the aforementioned error sources efficiently cancel, making minute absorption signals accessible.

Strictly, however, Equation (6), assumes the spectra $I_0(\lambda)$ and $I(\lambda)$ to be available at an 'infinite' spectral resolution i.e., at a spectral resolution much better than the width of the atmospheric spectral features. This is generally not the case when using remote sensing measurements conducted by a field spectrometer. The instrument spectral response function (ISRF) smoothes the atmospheric spectra according to

$$I^*(\lambda) = \int F(\lambda')I(\lambda - \lambda')d\lambda' = F(\lambda) \otimes I(\lambda) \qquad (17)$$

where the symbol $\otimes$ indicates the convolution operation and $F(\lambda)$ is the ISRF that characterizes the observing instrument. The ISRF can be measured in the laboratory or calculated based on the instrument's optical and detector properties. Laser based approaches typically have spectrally narrow ISRF and thus, Equation (17) can indeed be dropped. For DOAS applications, replacing I, I_0, and σ in Equation (6) by the measured (or ISRF convolved) quantities I^*, I_0^* and σ^* holds approximately in optically thin conditions. The approximation is applicable for most UV/visible absorbing gases and work-around's can be found for moderately thick absorption optical densities such that DOAS is widely applicable to trace gas remote sensing in this spectral range.

164

In practice, DOAS is applied to the retrieval of UV/visible absorbing gases such as H_2O, SO_2, BrO, OClO, NO_2 and O_3 using scattered sunlight as a light source or by directly pointing toward the sun or artificial light sources. In the latter case a broad-band (relative to the absorption band or line) artificial light source, typically in a search light arrangement, supplies a light beam with intensity $I_0(\lambda)$, which traverses the plume. At the other side of the plume the light is collected and the column density is calculated form the absorption spectrum. Frequently a variant of this set-up is used, where light source and detector are at one end of the light path, while at the other end there is only a reflector (typically a retro-reflector). This arrangement has two advantages: First, the light beam crosses the plume twice, thus doubling the sensitivity and second (and usually more important), all parts of the instrument, which require power are at one end of the light path, while only a passive reflector is at far end. Prerequisite for this approach is the availability of suitable sites on either side of the plume (e.g., two points at the crater rim) for mounting light source and detector or reflector (see e.g., [30,31]). In addition, the technique largely avoids the problems of the large background column density of e.g., H_2O. In principle dispersive or non-dispersive detectors could be used, however, up to now in all applications only the former type is employed.

If the gas under investigation only exists inside the plume and not in the background atmosphere and if scattering effects are not too severe, estimating n from S can be as simple as Equation (7) i.e., dividing the slant column by the length of the line-of-sight transect through the plume. In practice, for measurements using scattered sunlight estimating the intra-plume concentrations from the slant columns S might be complicated due to several effects: (1) the intra-plume average light path has contributions from multiple scattering inside the plume due to large abundances of plume condensate or aerosols, (2) the overall average light path has contributions from photons that do not transect the plume at all and thus, they do not see any target gas absorption (commonly termed light dilution), (3) the target gas has non-negligible abundances in the background atmosphere and thus, background removal is required, or (4) the absorption optical density is not thin and the average scattering light path x cannot be assumed independent of wavelength and the convolution operation does not commute with the exponential in Beer-Lambert-Bouguer's law.

Following the early introduction of COSPEC (COrrelation SPECtrometer, e.g., [28]), today's instruments are based on the DOAS technique (e.g., [27]), which can be performed with miniaturized spectrometers. There are also variants with simplified spectral evaluation e.g., Flyspec (e.g., [32]). In principle reflection of sunlight on topographic objects or by clouds could be used, but this has not been attempted yet.

2.3.2. Absorption Spectroscopy in the Infra-Red

In the infra-red (from about 1 to 20 μm wavelength), the absorption cross sections usually depend strongly on pressure and temperature and the absorption optical densities are frequently large, this is particularly true for the thermal IR. Thus, the DOAS approach is not readily applicable. So, generally, Equation (3) in combination with Equations (4) and (17) must be used to forward model the measured absorption spectra based on an initial assumption of gas concentration profiles, plume geometry, and ambient meteorological conditions. An inverse estimation technique iteratively adjusts the target gas concentrations to statistically best match the measurements until some convergence criterion is met.

Absorption spectroscopy in the infra-red mostly relies on direct-light approaches i.e., techniques that have a detector which only accepts radiation from a narrow solid angle centred directly on the light source—be it the sun, a hot telluric source (e.g., lava) or an artificial light source. Thus, Figure 2 is directly applicable. The incoming radiance $I_0(\lambda)$ in Equation (3) would be the solar Fraunhofer spectrum for direct sun observations, it would be Planck's thermal emission spectrum $B(\lambda, T)$, Equation (9), for a hot opaque source with temperature T such as lava, or it would be a silicon carbide lamp (Globar), incandescent lamp, LED (light emitting diode), or laser emission spectrum for artificial light source applications.

As described in Section 2.3.1 direct-light techniques have the great advantage that the light path is well-defined. As long as the light source is sufficiently bright to exclude contributions from light scattering and thermal emission along the atmospheric path, the light path computes easily from the positions of the source and the observer and, if necessary, from the atmosphere's refractive properties. The major disadvantage of direct-light techniques (as in the case of active UV/vis spectroscopy) is that one has little flexibility in choosing the observing geometry. The line-of-sight is strictly defined by the position of the light source and the receiving system, which have to be arranged such that the volcanic plume is in-between.

IR spectroscopy is implemented in four popular variants:

(a) With broad-band, thermal light sources, e.g., globars in combination with dispersive detection systems, typically Fourier-Transform interferometers of the Michelson type ([33–35]). At volcanoes hot lava can be used as source of radiation (e.g., [36], see below), which—according to our definition—would be classified as passive absorption spectroscopy.

(b) Tunable Diode laser spectroscopy (TDLS), which (e.g., [37–39]) is a variety of dispersive spectroscopy where a narrow-band emitting light source (i.e., a semiconductor laser) is rapidly wavelength modulated in order to sweep across an absorption line of the gas to be measured. The original, rather unreliable lead salt diodes are now replaced by much more stable (though still expensive) quantum cascade laser diodes [40,41]. Since the light source is wavelength modulated there is no need for dispersive detection.

All variants of the technique share the prerequisites (suitable sites) and advantages (low background) with the active UV/vis spectroscopy. Also the options for the arrangement of light paths (single path, retro reflector) are similar, in addition 'topographic targets' i.e., back-reflection of the transmitted radiation at terrain surfaces (see below) are in use.

Future developments for active IR spectroscopy could use Dual frequency comb spectroscopy (e.g., [42]), which—although it works differently—has some similarities to a combination of TDLS plus FT spectrometer.

(c) Passive IR spectroscopy using the sun (or the moon) as a direct light source is commonly referred to as the solar (or lunar) occultation technique which has been used for remote sensing of volcanic gases such as SO_2, HF, HCl, and SiF_4, e.g., [43–45]. The technique simplifies substantially if the target gas only exists (at relevant amounts) in the volcanic plume and not in the background atmosphere which is, however, not the case for the major volcanic plume constituents CO_2 and H_2O. Due to the rapid downwind dispersion of the volcanic plume, the volcanic enhancements (on top of the large background) become small and thus, increasingly difficult to measure the farther downwind the plume is sampled. Only recently, Butz et al. [46], demonstrated safe-distance remote sensing of volcanic CO_2 in Mt. Etna's plume during passive degassing conditions. They operated a sun-viewing, portable Fourier Transform Spectrometer (FTIR) on a truck in stop-and-go patterns underneath Mt. Etna's plume such that the lines-of-sight to the sun sampled the plume in 5–10 km distance from the crater. Co-measuring O_2 columns helped calibrating spurious variations in the targeted CO_2 columns which were merely due changes in observer position. Sequentially measuring intra-plume and extra-plume spectra and using co-measured HF, HCl and SO_2 as intra-plume tracers helped removing the atmospheric CO_2 background. These current generation instruments were able to discriminate the volcanic CO_2 signal out of a 300-1000 times larger atmospheric background path.

(d) Hot volcanic material such as lava or volcanic rocks have been used heavily in open-path spectroscopic techniques, e.g., [33,35,36,43,47–51] Naughton et al. 1969, Mori et al. 1993, Notsu et al. 1993, Mori et al. 1995, Francis et al. 1998, Mori and Notsu, 1997, Burton et al. 2000, Gerlach et al. 2002, Allard et al. 2005] targeting volcanic SO_2, HF, HCl, SiF_4, CO, CO_2, COS. Generally, the technique requires that the hot material or the lamp locates behind the plume and that it can be sighted by the observer. For many volcanoes, this requirement implies deploying instrumentation in the proximity of the crater or at the crater rim which spoils the general remote sensing advantage of avoiding hazards and hostile environments to operators and instrumentation. Recently, laser-based techniques for CO_2 have been developed using topographic reflection targets, e.g., [52,53], promising greater deployment flexibility but still requiring plume sampling close to the source where CO_2 enhancements are large compared to the atmospheric background.

2.3.3. Thermal Emission Spectroscopic Techniques

Typically, Equation (14) forms the basis for volcanic plume remote sensing by thermal emission spectroscopy, e.g., [54,55]. Using a forward modelling approach similar to the one used for direct-light absorption spectroscopic techniques, one starts out with an initial guess for the ambient and plume temperatures as well as for the transmittances. The transmittances contain the layer-wise optical densities and thus, the information on the layer-wise gas composition. Iteratively, the targeted plume gas abundances are adjusted to yield a best match of measured and forward modelled spectra.

Thermal emission spectroscopy allows for the detection of other species than UV/vis absorption spectroscopy (see Section 2.3.1) and for observation independent of sunlight, i.e., also at night time.

Disadvantages include complicated set-up (FT-IR), usually cryogenic cooling of the detector is required. However recently also non-dispersive detection with detectors at room temperature was demonstrated [56,57]. The classic approach is the differential one sketched in Figure 6, using the difference between two measurements, one pointing into the plume and one pointing next to the plume into the background sky, e.g., [54,58]. For these two measurements, the recorded measurements approximately differ by temperature T_2 and transmittance t_2 in Equation (14) either referring to intra-plume or background conditions. Thus, the difference spectra isolate the information on the targeted plume composition which is straightforward to retrieve as outlined in Goff et al. [54], Stremme et al. [55], and Krueger et al. (2013) [59], used a thermal emission FTIR to sequentially scan the SO_2 and SiF_4 plume (and the adjacent background sky) of Popocatepetl volcano from 12 km distance providing a series of two-dimensional images of the volcanic gas columns. Then, they used the series of images to simultaneously estimate the plume average wind speed and the constituent outflux from the crater. Current developments toward imaging FTIRs (e.g., [57,60]) are promising for use in volcano plume monitoring since they allow for rapidly imaging the thermal emission spectrum emerging from two-dimensional scenes without the need for scanning a telescope.

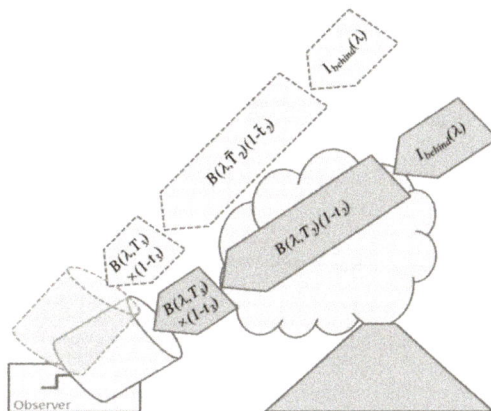

Figure 6. Schematic illustration of thermal emission techniques exploiting the difference between lines-of-sight pointing through the volcanic plume and those pointing into the background sky.

2.3.4. Range Resolved (LIDAR) Techniques

A further, well known approach to active remote sensing trace gases is the LIght Detection and Ranging (LIDAR) technique, e.g., [61,62]. The usual set-up is the mono-static LIDAR where transmitter and receiving system are at the same location. The (laser) light source emits short (<few 100 ns) pulses into the atmosphere. By analyzing the time profile of the intensity back-scattered from the atmosphere the spatial distribution of scattering and extinction along the direction of the emitted (and received) radiation can be deduced. With single-wavelength lasers aerosol backscatter and extinction are measured.

In order to determine trace gas distributions a more complicated set-up is required. These devices have become known as Differential Absorption Lidar (DIAL) using at least two wavelengths, one "on absorption" of the trace gas and one "off absorption" [13,14,63,64]. The approach is similar to the SO_2-camera (see Section 3.2.1). While DIAL systems have been bulky, relatively complicated and power consuming recent developments including micro-pulse diode lasers (e.g., [65]) promise smaller, lower power, and more portable systems.

Alternatively to the customary mono-static approach a bi-static LIDAR scheme has been developed (e.g., [4,66,67]). Here, radiation source and radiation detector are separated by a small distance (on the order of 10 m for a few km range) and the range resolution is accomplished by changing the elevation angle of either the radiation source or of the radiation detector. Advantages are the possibility to use continuous light sources (like LEDs) and simpler detection.

In principle two-dimensional or even three-dimensional push-broom imaging of volcanic plumes is possible with LIDAR instruments and to some extent has been demonstrated at volcanic plumes (e.g., [13,14,63]).

3. Imaging of Plumes

Many of the developments for volcanic plume remote sensing have started with collecting a series of individual plume (column density) measurements in order to sample one-dimensional plume cross sections. Modern technology allows to 'image' plumes, i.e., to provide two-dimensional arrays of optical densities (or 'images') of a scene containing volcanic gases. This approach offers many advantages over the simpler one-dimensional plume scanning:

- Complex situations (multiple plumes, change of wind direction, etc.) can be recognised and analyzed (e.g., [68])
- Advanced techniques of image analysis (e.g., segmentation and optical flow analysis) can be applied, thus enabling more precise flux determination (e.g., [22,69,70]), see also Section 4, below.
- Redundant measurements can be made by e.g., making trace gas flux determinations at several planes along the plume propagation direction allowing e.g., internal consistency checks (see e.g., [22])
- Redundant flux measurements can be used to determine the exact plume propagation direction (see [22])
- Last not least: the human visual system has powerful analysis capacities which can be used once images are available ('seeing is believing')

For these (and more) reasons a series of imaging systems for volcanic plumes have been developed and applied (see e.g., [5,16]). In this context the requirements for plume imaging instruments include:

- Capability to specifically detect the desired gas (or parameter)
- To provide sufficient sensitivity (e.g., for SO_2 measurements a detection limit for SO_2-column densities of the order of 10^{17} molecules/cm^2 or ca. 40 ppmm is required for volcanic emissions observations)
- To provide sufficient spatial resolution to allow discrimination of the relevant features within the plume(s)

- Sufficient time resolution (typically of the order of seconds) of the measurement is further required to be able to resolve the motion of the plume and variations in the volcanic source strength

Some of these requirements are in conflict with each other. For example enhanced specificity may be achieved by recording of the intensity at several wavelengths, since this typically can only be done sequentially the time resolution will be reduced. Likewise, lower detection limits (i.e., higher sensitivity) are achieved by collecting a larger number of photons per pixel, which comes at the expense of time resolution (or, alternatively, spatial resolution). In particular, the number of gases which can be detected by imaging instruments is limited in comparison to one dimensional instruments.

It should be noted that there is a connection between spatial and temporal resolution in that a given spatial resolution requires sufficient temporal resolution to be useful. The motion of the plume $\Delta x = v \cdot \Delta t$ within the acquisition time Δt of an image (v = wind speed) should be at least of the order of the size of one picture element in the plane of the plume. In other words, high spatial resolution makes sense only if the technique also offers sufficient temporal resolution (for a given wind speed). Note that the opposite is not necessarily true: A temporal resolution exceeding the above limit would still result in useful (though partly oversampled) data.

As described in Section 2 several remote sensing techniques for the imaging of plumes have been developed during recent years. As described in Sections 2.3.2 and 2.3.3 the various techniques can be grouped into active and passive as well as in dispersive and non-dispersive approaches. For dispersive approaches several scanning schemes are possible as detailed in Section 3.1, below.

The characteristics of dispersive techniques is the determination of a spectrum (i.e., a series of intensities as function of wavelength) for each pixel of the image. The trace gas column density for each pixel is then derived from the measured spectrum using the algorithms described in Section 2, above. Non-dispersive approaches, on the other hand, derive the column density from the ratio of just a few (typically two) intensity measurements integrated over suitable (narrow) wavelength intervals. Frequently, a reference intensity ratio is recorded with the instrument pointed away from the plume. The trace gas column densities are then derived from the ratio of the sample and reference intensity ratios.

Dispersive approaches include variations of Differential Optical Absorption Spectroscopy (DOAS, see Section 2.3.1, above, and e.g., [27]) and Fourier-Transform Infra-red (FTIR) spectroscopy (e.g., [47]) adapted to allow two-dimensional measurements.

Dispersive approaches are typically more complex and since usually scanning is required slower than non-dispersive techniques. However, they have the advantage of being less vulnerable to spectroscopic interferences and to be able to also measure minor constituents like BrO due to higher spectral resolution enabling higher specificity for gas detection. Furthermore, analysis using traceable, well quantified absorption cross-sections increases the rigour of the technique. Many gas absorption features are naturally narrow (of the order of 1–2 nm), particularly for small molecules, and this limits the effectiveness (i.e., the sensitivity as well as the capability to specifically detect a particular species) of non-dispersive techniques.

Note that any combination of dispersive/non-dispersive techniques is in principle possible and several combinations have been realized, such as the combined use of a dispersive UV spectrometer alongside a non-dispersive UV imaging system (see e.g., [71]). In fact, an important aim of present research is the development of techniques combining the speed of non-dispersive techniques with the rigor of dispersive techniques. Overall, despite impressive accomplishments, techniques for quantitative imaging of volcanic plumes are still in an early stage of development and much progress can be expected in the near future.

3.1. Categories of Plume Imaging

Two dimensional images of trace gas column amounts are derived from three-dimensional data: two spatial dimensions and one wavelength (or interferogram) dimension are recorded. At present three different imaging techniques are known (see Figure 7 and Table 1):

(1) Pixel at a time scanning ("whiskbroom" imaging, see Figure 7A): In this approach all pixels of an image scene are scanned sequentially according to a particular scheme (e.g., line by line as in early TV cameras), for each pixel a spectrum is determined. In this approach all (of e.g., 10^5 pixels) have to be scanned individually, such that it is potentially a rather slow approach. Michelson interferometers have been combined with whisk-broom scanners to obtain 2-D images (e.g., [34,55,72]).

(2) Column (or row) at a time scanning ("push-broom" imaging, see Figure 7B): Here all pixels of an image column are scanned simultaneously while the image columns are scanned sequentially. Because each column of the image is recorded at once, only of the order of several hundred columns have to be measured sequentially. Since only one scan-dimension is required the scanning mechanisms can become simpler though it is not necessarily faster. In fact, as will be explained below the amount of radiation collected by the entrance optics has to be split between all pixels of a column, thus, when the time to acquire an image is determined by the available number of photons the technique will not generally be faster than whiskbroom imaging. Push broom scanners have been realized with DOAS instruments as e.g., described by Lohberger et al. [73], Bobrowski et al. [74], Louban et al. [75] and Lee et al. [76], see Section 3.3, below. Michelson interferometers have also been combined with push-broom scanners (i.e., moving platform) to obtain 2-D images (e.g., [77]). A recent development based on a special type of interferometer (Sagnac interferometer) is the Thermal Hyperspectral Imager, which produces a spatial interferogram across the field of view, which is scanned across the image. After Fourier transformation a high resolution spectrum for each image pixel is obtained [57] (Gabrieli et al. 2016), which can be analysed for spectral signatures of volcanic gases. Gabrieli et al. [57,78] used such a device to produce images of the SO_2 distribution derived from spectra around 8.6 μm at a spectral resolution of about 0.25 μm. Explorative measurements were made at Kilauea Halema'uma'u crater (Hawaii) with a scan duration of 1 s.

(3) Frame at a time scanning ("full frame" imaging, see Figure 7C): Here the entire frame is recorded at once (or in a sequence of steps in time). While in principle there could be large arrays of spectrometers determining the spectrum of each pixel (and in the future arrays of integrated micro-spectrometers could become viable), in practice the spectral information is usually determined by collecting sequential images with different wavelength selective elements (e.g., suitable filters, see Figure 8) in front of the camera sensors.

Current research also aims to employ an array of detectors for two-dimensional Michelson interferometers, where each pixel can effectively be thought of having its own interferometer. An example is GLORIA (e.g., [60]), which uses a 256 × 256 element Mercury Cadmium Telluride focal plane array (FPA) cooled to 60 K. The spectral coverage of the instrument is from 7.1 μm to 12.8 μm.

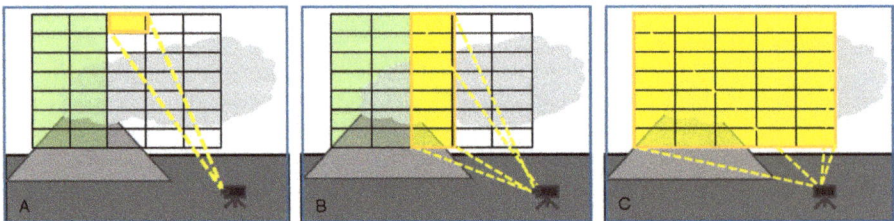

Figure 7. Image scanning schemes (**A**) Pixel at a time ("whiskbroom imaging"); (**B**) Column at a time ("push-broom imaging"); (**C**) Frame at a time.

Table 1. Characteristics of the three basic imaging techniques.

Imaging Principle	Detector Type	Examples/Comments
Whisk-broom	Spectrometer or Michelson Interferometer	Experimental instruments [55]
	Filter	not used
	Fabry-Pérot Interferometer	Theoretical studies, [79,80]
	Gas Correlation	Presently not used
Push-broom	Spectrometer	I-DOAS, [74], Imaging Sagnac-Interferometer [57] FTIR
	Filter	Presently not used [A]
	Fabry-Pérot Interferometer	Presently not used [A]
	Gas Correlation	Presently not used [A]
Full-Frame	Wavelength sensitive pixels	Future Technology (e.g., [81])
	Filter	UV SO$_2$-camera e.g., [19,82] and references in the text
	Fabry-Pérot Interferometer	Theoretical studies in the UV [79], in use in the IR [60]
	Gas correlation	Presently not used [A]

[A] For volcanic plume imaging.

3.2. Non-Dispersive Plume Imaging

Non-dispersive imaging usually uses one or several two-dimensional image sensor(s), being sensitive in the desired spectral range in combination with a device offering (limited) spectral selectivity. This compromise favours spatial resolution and imaging speed over (usually) sensitivity and trace gas selectivity.

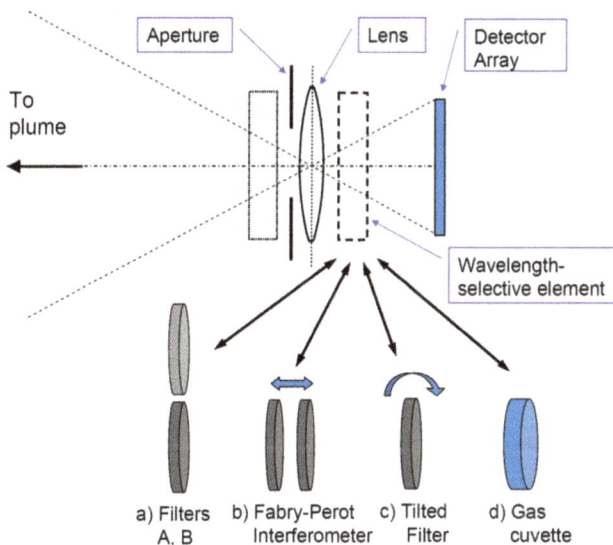

Figure 8. Trace gas camera schemes: The basic set-up consists of a (UV) lens, aperture, wavelength selective element (WSE) and suitable 2-D detector. The WSE can either be mounted in front of the lens (dotted box) or between lens and detector (dashed box), the relative merits of the two approaches are discussed by Kern et al. [82]. The different approaches are distinguished by their WSE, which can be one of the following: (**a**) two narrow band filters alternatingly being brought into the beam; (**b**) a Fabry-Pérot interferometer with adjustable transmission wavelength; (**c**) a narrow-band interference filter, which can be rotated; (**d**) a cell (cuvette) containing the gas to be measured, periodically introduced into the light beam.

3.2.1. Quantifying Column Densities Using Classic (UV) SO$_2$ Cameras

The theory developed in Section 2.1 is best illustrated by briefly inspecting a "classic" SO$_2$ Camera. Such an instrument consists of an UV-sensitive camera viewing e.g., the area around an actively degassing volcanic vent through a UV band-pass filter that selectively transmits radiation at a wavelength λ_A which is strongly absorbed by SO$_2$ (about $\lambda_A \approx 310$ nm). Thus the sections of the image receiving radiation which has passed through the volcanic plume (assumed to contain high amounts of SO$_2$) receive a lower radiation intensity $I(\lambda_A)$ than the background sky (i.e., sections of the image where the radiation did not pass the plume), which ideally has the intensity $I_0(\lambda_A)$. The attenuation depends on the SO$_2$ column density and is quantitatively described by Equations (3) to (5) in Section 2.1. While this simple approach works in principle, in reality a higher sensitivity is reached by using two filters (usually referred to as "Channel A" and "Channel B") transmitting radiation at different wavelengths as described e.g., by Kern et al. [82,83], Platt et al. [5], Smekens et al. [84], McGonigle et al. [16].

In order to improve the absolute accuracy of SO$_2$-cameras it has become customary to add a 1-D (DOAS) spectrometer pointing at a point within the field of view of the camera (typically the centre). This approach has been pioneered by Lübcke et al. [71].

3.2.2. Non-dispersive IR Imaging of Plumes

Plume imaging by thermal emission is a passive technique relying on the thermal emission from the plume (at ambient temperature) itself, see Section 2.3.3. Unlike the case for the SO$_2$ camera described above, the equivalent infra-red system has received far less attention, although several approaches are being studied:

(1) IR-cameras with two or more filters similar to the SO$_2$ camera principle (e.g., Prata and Bernardo 2009) have been developed for SO$_2$ retrieval and ash detection ([56,85]). A four filter IR camera was actually used by Lopez et al. [86] to simultaneously determine plume the SO$_2$ SCD, temperature, and ash content of the plume of Stromboli (Italy), Karymsky (Russia), and Láscar (Chile) volcanoes.

(2) Gas-correlation spectroscopy has also been applied in the IR for real time imaging of ammonia and ethylene (e.g., [87,88]).

In the following we give a brief description of an IR-camera system already in use for volcanic imaging. The absorption spectrum in the thermal infra-red region (7–13 µm) exhibits many features attributable to gases present in volcanic emissions, including SO$_2$, CO$_2$, H$_2$O, HCl among others. Of these gases SO$_2$ is the easiest to detect because of its generally low background concentration and broad absorption peaks at 7.3 µm and 8.6 µm. Atmospheric volcanic ash also absorbs and scatters infra-red radiation in this region of the electromagnetic spectrum and its characteristic features have been exploited by using infra-red satellite measurements for many years (e.g., [89]). There are some important differences when sensing gas and ash clouds from the ground compared to measurements from space. The most important of these is the thermal contrast between target (gas or ash cloud) and background, for example the warm surface below the plume in the case of satellite sensing, and the cold clear sky behind the plume in the case of ground-based sensing. A typical infra-red camera system consists of a cooled or uncooled detector array, focusing optics, infra-red interference filters and data recording electronics. For instance, Prata and Bernardo [56] describe the development of an uncooled multispectral imaging camera system based on a 320 by 240 bolometric detector and F/1 Germanium (Ge) optics. The system incorporates four narrowband interference filters with selectable central wavelengths and ~1 µm bandwidths. The usual configuration is to use 8.6 µm for SO$_2$; 10, 11 and or 12 µm for ash detection and quantification; and a broadband channel (8–12 µm) for plume and background scene temperature measurements.

3.3. Dispersive Imaging

Dispersive imaging instruments have been realized for the UV/visible as well as for the IR spectral ranges.

The Imaging DOAS (I-DOAS) technique brings advantages from different techniques together: a good spatial resolution, like the SO_2 camera, and high spectral information similar to the one used in conventional DOAS [73]. I-DOAS measurements result in a three dimensional data set, including two-dimensions of spatial information (i.e., each picture element of an image (pixel) corresponds to a defined solid angle of space), and a third dimension containing highly resolved spectral information within each pixel. For this reason imaging spectroscopic instruments require both imaging and dispersive optical components. Today I-DOAS has been applied using various platforms: satellites (e.g., OMI [90]), airplane (e.g., [23,91]) and ground based (e.g., [73–76]). An imaging satellite instrument operating in the thermal IR was proposed by Wright et al. [77].

The major advantage of I-DOAS is the capability of the technique to measure several trace gases simultaneously, enabling plume dispersal and chemical transformations within the plume to be studied. Also, I-DOAS is less dependent on meteorological conditions compared to SO_2 cameras. Furthermore, no calibration is needed and radiative transfer corrections can be calculated [83,92]. Drawbacks in comparison to the classic SO_2 camera might be the higher complexity of hardware, as well as the fact that it usually takes much longer (minutes rather than seconds) to acquire a full image of the plume, also the computation time for spectral evaluation is longer.

Both the push-broom and whiskbroom applications can be employed for the I-DOAS technique. The Whiskbroom approach is usually implemented by adding 2-D scanning entrance optics to a DOAS instrument consisting of either two motors turning two mirrors (or prisms) or one motor and moving the entire instrument in one direction. The push-broom approach (see Figures 7B and 9) uses a 2 dimensional CCD while the second spatial direction can be implemented by a scanning mirror or by moving the instrument (e.g., [73–76]).

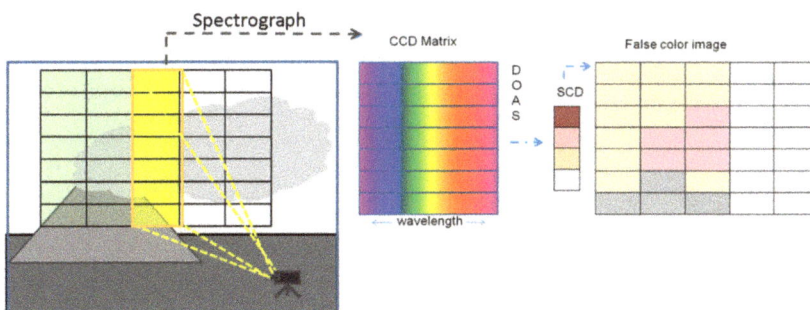

Figure 9. Schematics of an Imaging-DOAS (I-DOAS) whisk-broom instrument and sketch of the evaluation procedure: The I-DOAS instrument simultaneously creates spectra for each pixel in a column of the image (**left**), these are recorded by a 2-dimensional CCD (**centre**). Upon DOAS evaluation trace-gas column density values for each pixel in the image column are derived. After completion of a scan all image columns are combined into a two-dimensional image (**right**), which is typically presented as "false colour image".

3.4. Combining both Approaches

Obviously, it would be very desirable to combine the good specificity of dispersive techniques (see Section 3.3) with the speed (and ideally simplicity) of the non-dispersive approaches (see Section 3.2). In fact, there exist several technical solutions which are essentially non-dispersive and thus fast and simple while making use of the details of the spectral features of the gas to be measured. Two relatively popular (although not in the area of volcanic gas measurements) solutions are Fabry-Pérot Interferometers and gas correlation sensors. Both show much superior discrimination power than simple filter cameras while being rather simple and fast devices. In particular both approaches allow high-speed (seconds) two-dimensional imaging of trace gas distributions.

Fabry-Pérot Interferometers can be seen as filters with a periodic transmission pattern, i.e., transmission maxima are regularly spaced at given wavelength intervals $\Delta\lambda$, while the width of each transmission maximum is $\delta\lambda$ (with $\delta\lambda < \Delta\lambda$). By manufacturing a Fabry-Pérot interferometer the quantities $\Delta\lambda$ and $\delta\lambda$ can be tailored within rather wide limits, in particular the can be chosen such as to match the periodic structure of the absorption cross section of many trace gases. Thus, such a device (see Figure 8, case b) is most sensitive for a gas with a given separation between the maxima of the absorption cross section. This is described in detail by Kuhn et al. [79]. A demonstration of an actual device built on this principle is reported by Kuhn et al. [80], the same authors describe an extension of the principle for the detection of BrO (and other gases) in volcanic plumes.

A similar general idea is behind the gas correlation sensor (see e.g., [93]), which consists of a cuvette containing the gas to be measured (e.g., SO_2) mounted within a camera (see Figure 8, case d). The cell can be moved in and out the optics and the intensity ratio of two images recorded with the cell in the camera and the cell removed contains the trace gas signal. The principle relies on the fact that, while the intensity is always reduced when the cell is moved in, the intensity reduction is smaller when the gas is present in the observed scene. This is due to the fact that the intensity $I(\lambda)$ reaching the camera is already attenuated at certain wavelengths (i.e., where the gas has its absorption maxima) and thus cannot be attenuated much more by the gas in the cuvette. Although the technique is essentially limited to gases that are sufficiently stable to be sealed into the absorption cuvette, which is part of the instrument, it is technically simple, and lends itself to imaging applications. Imaging by the gas correlation method applied in the IR are reported e.g., by Sandsten et al. [88].

4. Volcanic Gas Flux Determination

Usually the gas flux from a volcanic source is more important than the gas concentration (or column density through the plume), since the former is an indicator of the activity while the latter quantities are influenced by (varying) dilution and other processes which are not related to volcanic activity.

4.1. The Principle of Volcanic Gas Flux Determination

As described above, conventional one-dimensional (1-D) spectroscopic measurements typically measure the trace gas column density of the species of interest along a single line of sight. Integration of the gas concentration along this line occurs intrinsically (see Equation (6) in Section 2.1), and the measured quantity is the number of molecules along the line of sight. In order to derive an emission rate, the number of molecules Q in a cross-section of the plume perpendicular to its propagation direction is needed. Conventional 1D instruments measure this quantity by traversing under the plume or scanning the viewing direction across the plume and integrating the derived column densities (e.g., [15]).

The procedure is sketched in Figure 10 it consists of measuring a series of column densities S_i through the plume (ideally an infinite number of column densities) and integrating over the derived columns within the width w of the plume (Equation (18)).

$$J = v_P \cdot \int_W S(y) \cdot dy = Q \cdot v_P \qquad (18)$$

From the quantity Q the emission rate J can be readily determined by multiplication with the plume speed v_P.

While determining the position y across the plume is no problem when traversing under the plume (e.g., with a spectrometer mounted on a car) for instruments scanning the plume from a fixed point (e.g., all NOVAC stations work like this, see [10]) only the observation angle is known. In this case an estimate of the distance L_P to the plume is required and the perpendicular position y can be derived from geometrical considerations (in the simplest case as $\Delta y = L_P \cdot \Delta \gamma$, where $\Delta \gamma$ is the change in scan angle).

The plume speed (i.e., wind speed v_P at the position of the plume) and the wind direction can be determined in a number of ways:

(1) From local measurements
(2) From large scale wind fields, which are available from regional or global data bases (e.g., ECMWF or MERRA).
(3) From measured correlation data within the plume itself (see e.g., [18,64,94,95])

Approaches (1) and (2) suffer from various shortcomings: Local wind speed measurements at the crater rim are usually not available and wind speed measurements further away are frequently disturbed by orographic influences. Wind speeds from large scale wind fields usually are on a very coarse grid (e.g., 1° by 1°), while they are typically less affected by local orography these data may just not be representative for the site of the volcano. In contrast to that approach (3) is based on the fact that the gas (e.g., SO_2) emission strength from a volcano usually varies considerably (typically several 10%) with time on time scales of seconds to minutes as sketched in Figure 11. Thus, the column densities $S_1(t)$ and $S_2(t)$ are measured at two points within the plume which are located at different distances d_1 and d_2 downwind from the volcanic source (i.e., usually the crater). From the two time series the time lag Δt_P for maximum correlation between the two time series (which typically may encompass a time period of several minutes) is calculated. From Δt_P and the difference $d_2 - d_1$ in downwind distance of the two measurement points the plume speed is given by:

$$v_P = \frac{d_2 - d_1}{\Delta t_P} \qquad (19)$$

For the above reasons approach (3) is the preferred method of determining the wind speed in the plume within the NOVAC network.

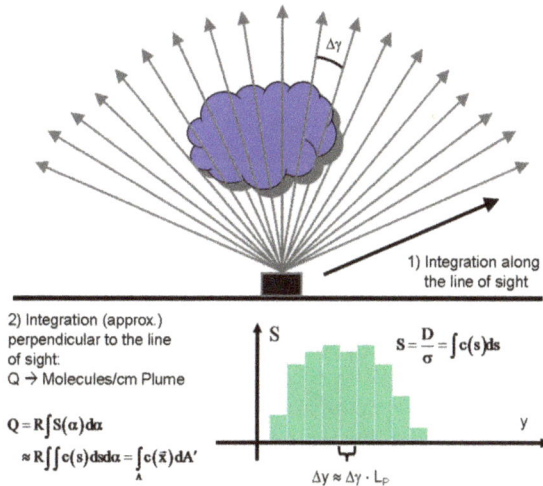

Figure 10. Schematics of flux determination from a plume scan.

Imaging techniques, however, have the advantage that they already record two spatial dimensions. Therefore, scanning is only required if the imaging technique itself is based on scanning technology (e.g., Imaging DOAS, see Section 3). Trace gas cameras (e.g., SO_2 cameras described above and in Section 3) based on band-pass filters do not require scanning. Instead, the number of SO_2 molecules Q in the plume cross-section can be derived by spatially integrating the recorded column images across the plume width w, perpendicular to the plume direction (see Figure 10). In case the plume cross-section is not taken perpendicular to its propagation direction a trigonometric correction has to be applied as described by several authors (e.g., [10,82,88]) and is discussed in detail in Section 4.2 below.

In this case the plume speed can be determined directly from the camera data through analysis of sequentially recorded images (e.g., [68,69,84,96]), providing a significant advantage over non-imaging techniques, which typically employ independent measurements of the wind speed at the appropriate plume height. Special software packages were developed e.g., by Valade et al. [97] and a direct validation of thus derived SO_2 emissions were reported by Smekens et al. [84].

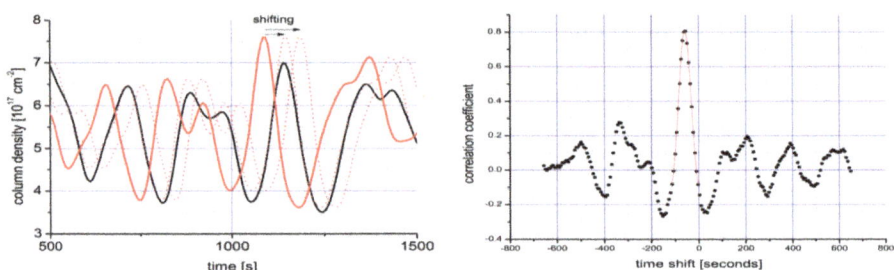

Figure 11. Sketch of the determination the wind speed v_P inside a plume. Left panel: Two time series (continuous red and black lines) are recorded at two different positions (difference $d_2 - d_1 \approx 900$ m) downwind the plume of Mt. Etna. The dotted red lines indicate the shifting time series 2 versus series 1. In the graph, both series are low-pass filtered, to make the relevant structures clearly visible. Right panel: Correlation coefficient between the two time series as a function of time lag. The highest correlation is found for a time shift of $\Delta t = 59 \pm 6.5$ s resulting in a wind speed of 15.2 ± 1.6 m/s. Adapted from [98].

4.2. A More Detailed View—Determination of the Wind Direction

The simple approach sketched in Section 4.1—while applicable directly in certain cases—must be extended to compensate for particular conditions of the measurement:

(1) Light dilution may affect the accuracy of the column density retrieval
(2) Multiple scattering inside the plume can also affect the column density retrieval
(3) The effect of the plume propagation direction being non-perpendicular to the viewing direction must be corrected

The first two effects are discussed in several publications, e.g., [30,82,83,92,99,100] with the bottom line being that corrections are possible on the basis of a detailed analysis of the recorded trace gas (e.g., SO_2) spectra. Correction of the third effect is addressed in the following. The problem lies in the correct determination of the plume propagation direction. While this is usually not a problem when the trace gas column density is determined by a zenith looking instrument during traverses, since in these cases the position of the source (e.g., the crater) and the position of the measurement are usually well known. Consequently the angle α between the plume propagation direction (i.e., the wind direction) and the direction of the traverse is also known with high precision.

However, if measurements are made from a fixed point, either by a scanning system or a two dimensional trace gas camera (e.g., a SO_2 Camera), then the angle α between the plume propagation direction and the scanning direction or the image plane of the camera are not a priori known.

As sketched in Figure 12 there are several effects if the plume propagation direction is not perpendicular to the scanning direction or not in the image plane, respectively.

Following the discussion in Klein et al. [22], we consider data derived by a plume imaging system. For simplicity we initially also assume a cylindrical plume (i.e., a plume where the cross section does not change with distance from the origin and with time). Later we will generalize for an arbitrary plume cross section and its evolution. As can be seen in Figure 12 top view (lower left panel) in the centre of the image plane the determined flux will be independent of the "tilt" angle α between image plane and plume propagation direction. This is due to the fact that two effects cancel (as noted e.g., by Mori and Burton [96]):

(1) The length of the light path through the plume increases as $1/\cos(\alpha)$.
(2) The determined apparent wind speed is reduced by the factor $\cos(\alpha)$ since the determined $d_2 - d_1$ appears shorter by this factor (see bottom right panel of Figure 12) while the determined time lag Δt_P stays the same.

However, this only holds for one point (e.g., the centre) within the field of view. Across the field of view there are two effects:

(1) Still the light path through the plume is larger towards the edges of the image, while the determined velocity v_P stays the same. Thus the flux appears to increase somewhat towards the edges of the image compared to the centre.
(2) A further effect of at "tilt" is due to geometry in that the closer part of the plume appears larger than the part which is further away from the camera (see top right panel of Figure 12) thus the integral (Equation (18)) will extend over a larger extent and thus be larger.

When image series ("movies") of the plume are available these effects can be fortunately not only corrected, but rather the data can be used to determine the "tilt" angle α of the plume. This is done by (1) determining the flux in each image column of the image (where the plume is covered) and (2) making use of the usually well justified assumption that the amount of trace gas (e.g., SO_2) within the plume is conserved over a time period of a few minutes. Therefore the flux must be the same across the entire field of view (once the effect of the centre to edge enhancement (see above) is corrected for). If the thus corrected flux is not constant under the initial assumption of $\alpha = 0$ then α is varied until the flux is constant across the field of view (see Figure 13).

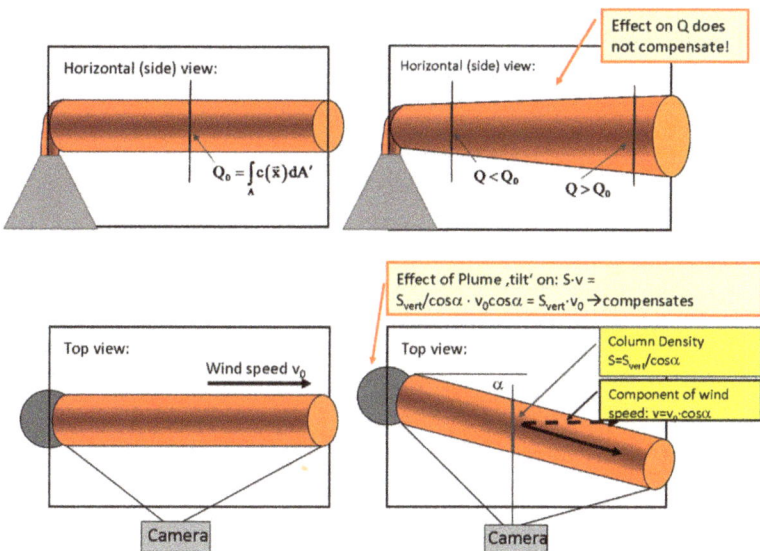

Figure 12. Schematics of flux determination from a plume scan. See Section 4.2 for details.

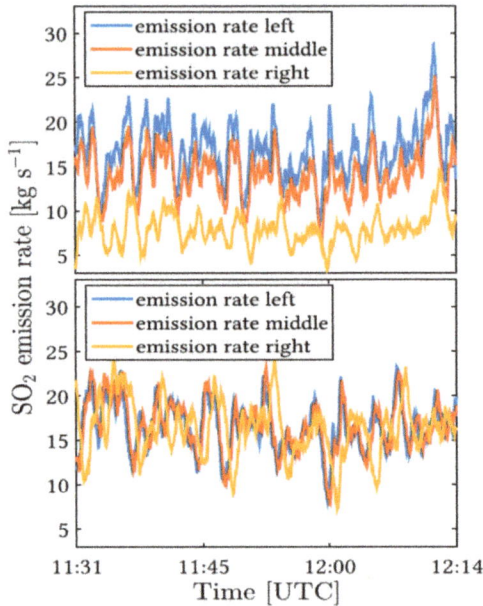

Figure 13. Sample application of the plume "tilt" correction on a set taken at Mount Etna on the 9th of July 2014. Top panel: Deviations of the apparent SO2 fluxes in three different vertical cross sections through the plume. These apparent deviations are caused by wrongly assuming an inclination ("tilt") of the plume with respect to the image plane of $\alpha = 0$. Using the conservative flux condition an inclination angle of $\alpha = 38°$ was determined. Figure from [22].

5. Sample Applications

Technical advances in remote sensing and volcanic gas imaging have led to high temporal and spatial resolution data, which now allow the investigation of volcanic source behaviour taking place below the Earth surface by combining the surface observation of gas release with geophysical data and fluid dynamic models for melt and atmosphere including numerical models and laboratory data. Apart from the volcanological application also the volcanic plume chemistry and therefore the impact on the atmosphere can be investigated in more detail for instance by studying the heterogeneity of volcanic plumes and also by visualizing their mixing with the atmosphere. However, to our knowledge this possible application has not been realised so far. In the following section we will illustrate the new possibilities for volcanological investigations using some recent example studies.

5.1. UV Spectroscopy

Measurements of gases emitted from volcanoes, in particular the amount of SO_2, have traditionally taken place using UV-light applications, starting with COSPEC [28] in the 70s and after 2002 measurements were more widely undertaken by applying DOAS technique. However, both approaches were rarely used for two-dimensional imaging and SO_2 fluxes were determined with a time resolution of >5 min. An exception are the wide field of view UV spectrometer applications [101,102]. Highly improved time resolution and reduced error for plume velocities are possible since the onset of the SO_2 camera application [82,96]. Using an SO_2 camera leads to the determination of SO_2 emissions with time resolution of the order of a second and therefore it opens the possibility to correlate gas emission measurements directly with geophysical signals (tremor, very long period events etc.) and thus adding information to test current models of physical processes below the Earth surface.

A recent example is the work of Moussallam et al. [103], describing different degassing patterns of lava lakes in relation with their underlying conduit flow processes (See Figure 14). The authors point out that there is no similarity between gas emissions of three out of four volcanoes (Erebus, Ambrym, Kilauea, Villarica) containing lava lakes, which were all studied with a comparable time resolution. The reason proposed are differences in magma viscosity and conduit geometry, which cause the different character of emission flux behaviour. No regular periodicity of gas fluxes as well as composition could be determined in data of Villarica volcano, Chile. Those surficial gas measurements, which were combined with thermal infra-red data, point to flow instabilities inside the conduit and a turbulent mixing between the ascending gas rich and the descending gas poor magma. Part of the observed high frequency variability of the emission strength of Villarica volcano is suggested to originate from atmospheric transport processes. Instead, periodic oscillations of flux and composition were found at Erebus and Ambrym (e.g., [101,104]) indicating, combined with geophysical data, a bidirectional magma flow, whereas at Kilauea [105], oscillation of degassing amounts has been interpreted as shallow degassing patterns due to gas percolation.

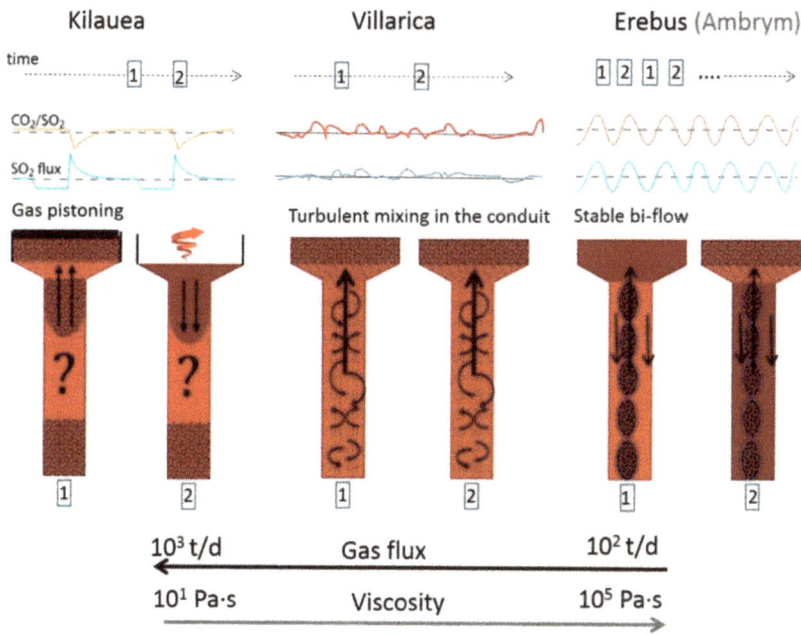

Figure 14. Three suggested degassing mechanisms (Kilauea, Villarica, and Erebus) with different conduit dynamics caused by the difference in magma input and viscosity. After [103], Figure 7.

However, not only has lava lake degassing been investigated so far, high time resolution SO_2 camera measurements were also done to investigate puffing and strombolian explosions at various volcanoes, showing similarly large differences in degassing characteristics of different volcanoes [19,106,107], came to the conclusion that explosions are only second order degassing processes compared with quiescent degassing. Those studies were undertaken at Stromboli and Assama volcano, respectively, showing that less than 16% of the gas amount is emitted by explosions. In contrast studies at Fuego volcano by Waite et al. [108] suggest that 95% of the total degassing amount is contributed by explosions.

Table 2 is an expanded overview of the one of Burton et al. [109]. Summarizing comparison studies between SO_2 camera-derived SO_2 fluxes and some geophysical parameters.

Table 2. Comparison studies between SO_2 camera-derived SO_2 fluxes and geophysical parameters.

Volcano	Seismic VLP	Seismic Tremor	Acoustic	Thermal	Reference
Pacaya, Guatemala	−	−	−	−	[110]
Asama, Japan	+	−	−	−	[107]
Fuego, Guatemala	−	+ (time shifted)	−	−	[68]
Stromboli, Italy	+	−	−	−	[106]
Etna, Italy	−	+	−	−	[107]
Karymsky, Kamchatka	−	−	0	0	[111]
Fuego, Guatemala	+	−	−	−	[108]
Etna, Italy	−	+	−	−	[112]
Stromboli, Italy	−	−	−	−	[113]
Stromboli, Italy	+	−	−	−	[109]
Hawaii, USA	−	+	−	−	[105]
Etna, Italy	−	0	0	−	[114]
Stromboli, Italy	−	−	+	0	[115]
Stromboli, Italy	+	−	0	0	[116]

+: Correlation is reported; −: no correlation has been found; 0: the parameter were independently reported and used together for a volcanological interpretation no investigation of the correlation between the parameter has been presented.

5.2. IR Spectroscopy

Although IR techniques have complex analytical requirements and still need relatively bulky and cost intensive instruments, IR techniques are highly desirable to investigate volcanic gas compositions, because nearly all major volcanic gas compounds show characteristic absorption bands in the infra-red.

Unfortunately, the two main volcanic gases, water vapour and CO_2, although they have strong absorption features in the near IR are quite challenging to detect, mainly due to their high atmospheric background and its spatial variation. Nevertheless, the value of such information has been already demonstrated by using light path arrangements which avoid large atmospheric distances, i.e., light paths close to the source and using either artificial lamps or hot ground surfaces (e.g., lava lakes). For instance, at Erebus (e.g., [117]), Etna (e.g., [36,118,119]), Ambrym ([104]), and Stromboli (e.g., [120,121]), high frequency variations of major plume composition could be successfully determined over time frames of hours to months. Further, progress has been done during the last few years (e.g., [46,53]) and even from space volcanic CO_2 emissions of at Yasur (on Tanna Island, Vanuatu), Aoba and Ambrym volcanoes (Vanuatu) have been detected by OCO-2 [122]. Although these last examples are far from routinely usable measurements they show new technical developments for future research.

An interesting example for the advanced understanding of physical processes inside the feeding system of a volcano is the work by La Spina et al. [121]. Only with spatial and high time resolved investigations by FT-IR spectroscopy the authors could show that the previously proposed explanation of CO_2 flushing, a commonly proposed process, is unlikely in the case of Stromboli. The authors could come to this conclusion by observing the differences in the composition of the three investigated craters at Stromboli (CC, NEC, SWC) and the high H_2O/CO_2 ratios. They therefore suggested an alternative model (see Figure 15) to explain the enhanced CO_2/SO_2 ratios weeks before major explosions and the decreasing CO_2/SO_2 ratios during the hours just before this activity (without the need of CO_2 flushing) and also give an alternative explanation how a crystal pure magma may be transformed into a crystal rich one. This later fact could be explained by H_2O exsolution and crystallization driven by depressurization when magma rises to the surface. La Spina et al. [121] assume that the increase of the CO_2/SO_2 ratio is caused by more magma input into the shallow system. This should lead to a drop of permeability in the deep system which then causes the decrease of CO_2/SO_2 ratios as the mixture of deep degassing and shallow degassing is shifted to a larger percentage of shallow degassing. According to the calculation of the authors this seems far more realistic than the incredible large amounts of CO_2 needed to produce significant isobaric dehydration to reproduce the measured

H_2O/CO_2 values. Further, La Spina et al. [121] utilize the observed spatial and temporal heterogeneity to produce a new model describing the geometrical structure of the shallow plumbing system of Stromboli (see Figure 15).

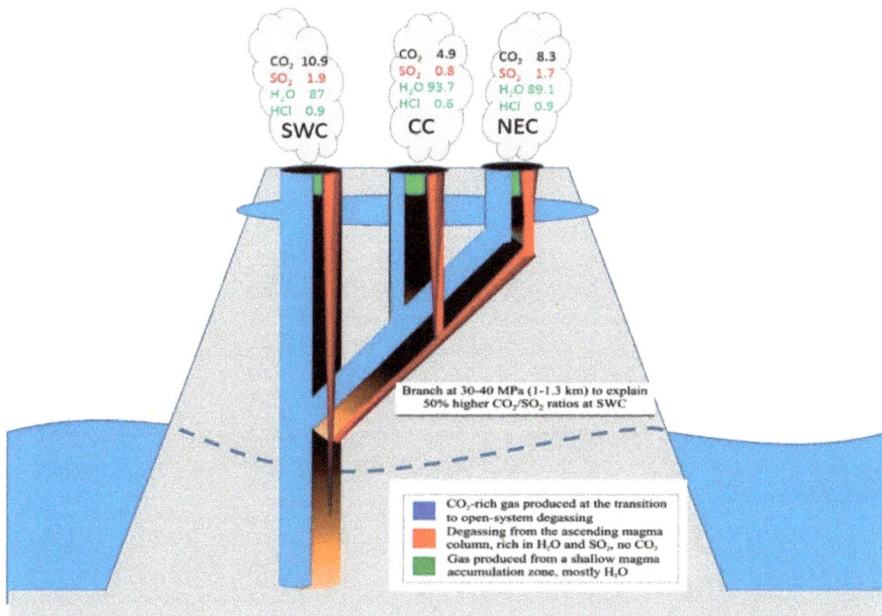

Figure 15. Illustrating a schematic model of the steady-state degassing system on Stromboli (described in the text). Bottom of figure: Point of transition to open system degassing, the gas exsolution is colour coded as explained in the box. The SWC crater (top left of figure) carries the largest contribution of gas produced in closed system degassing, and is therefore richest in CO_2, while the NEC crater has the weakest connection to the closed-system gas source, and is therefore relatively poor in CO_2. The branching depth of 30–40 MPa is the minimum depth for which such an CO_2/SO_2 fractionation could occur between the SWC and other craters. The final gas observed from each crater is therefore the superposition of different relative amounts of the three gas sources: CO_2-rich gas from depth, H_2O-SO_2 rich gas released from the ascending magma in the conduit and H_2O-rich gas released from the shallow accumulation zone. Adapted from Figure 6 of [121].

In contrast to the observation of [121] at Stromboli, in another study of Oppenheimer et al. [123] did not find a significant variation of the CO_2/SO_2 ratio at Yasur volcano between explosive and passive degassing activity. However, large variations (more than a magnitude) are observed in the HCl/SO_2 ratio with a ratio of about two during quiescent degassing and about 30 during explosions. These results are explained with gas accumulation below the HCl exsolution.

In addition, less abundant gases (e.g., SiF_4) where recently successfully studied using FTIR. After the first measurements of SiF_4 in a volcanic plume by Francis et al. [124], two dimensional SiF_4 distributions have been presented by [45] using thermal emission spectroscopy. Taquet et al. [125] also used thermal emission based spectroscopy to study the SiF_4/SO_2 ratio over a period of 6 month during dome growth at Popocatepetl. The authors found high variability of this ratio in coincidence with seismic data. SiF_4 is a secondary gas formed due to the interaction of HF with rocks or ash. Nevertheless, there is a high potential to study certain types of activity (e.g., dome growth) and eventually using this approach for monitoring in the future. In the article by Taquet et al. [125] the variation of the SiF$_4$, SO$_2$ and the ratio SiF$_4$/SO$_2$ helped to decipher the different processes related to the dome permeability.

6. Summary and Outlook

There is a clear development from occasional volcanic gas concentration measurements (typically by "wet" chemical techniques or gas chromatography) to continuous observations as pioneered by the COSPEC technique (e.g., [28]) and brought to a large scale by the NOVAC network (e.g., [10]) and other local networks (e.g., [109,126]). In addition gas concentration (or column density) measurements have been extended to flux measurements (see e.g., NOVAC). Another dimension is marked by the extension of SO$_2$ measurements to continuous multi-species observations. In this context also extremely affordable spectrometers based on mass produced smart phone imagers [127] can play a role.

From here a next quantitative jump (or paradigm shift, see [16]) is marked by gas imaging techniques like the SO$_2$ camera (see e.g., [109]), which allow—compared to scanning approaches—a much more detailed analysis of volcanic emissions and plume behaviour. Besides more precise flux measurements for instance short term periodicity in emission flux [16] or the angle between plume direction and image plane can be derived. Moreover, the spatial inhomogeneity of volcanic sources can be investigated.

High Frequency imaging of volcanic plumes was first realized with UV cameras and instruments based on this technique can be produced rather cost effectively and—as was recently shown by Wilkes et al. [128] have the potential to become very cheap instruments which may be ubiquitous in volcanic research. However, the approach has two drawbacks.

(1) The relatively crude spectroscopic approach, basically just ratioing the intensities in two wavelength intervals leads to interferences by aerosol, (stratospheric) ozone, and possibly other species. Moreover, radiation transport issues may also influence the accuracy of the technique (see e.g., [71,82,129]).

(2) UV cameras rely on sunlight and can thus only operate during daylight hours.

There are techniques under development (or being adapted for the purpose of volcanic observations) which promise to overcome these weaknesses. These include gas correlation spectroscopy (see Section 3.4), titled filter imaging, and Fabry Pérot imaging (see [79,80]), which both make better use of the spectral finger print of the target gas than the classic UV camera, thus reducing interferences and potentially improving the accuracy of the measurement (see also [5]).

Infra-red thermal emission spectroscopy is independent of daylight and may be the technology of the future in either scanning or imaging applications. At present cost and logistic requirements (cooling of the IR detector, use of interferometers with moving parts) are still high. Here technological development aimed at lowering complexity and cost of the instruments would be very welcome.

At present, imaging Michelson interferometers exist and are used for certain purposes, but the instruments are still extremely expensive. Alternatively, the UV-camera approach can be extended to the thermal IR region (see e.g., [56]), while this approach suffers from similar problems as the UV camera (see above) the rapidly falling prices of thermal IR cameras may make this approach feasible for volcanological applications.

With all spectroscopic techniques radiation transport issues are important and may be limiting the achievable accuracy (see Section 2). Here, better radiation transport models and new approaches like using polarisation of the radiation and ratioing techniques based on the absorption due to "known" absorbers like O$_2$ or O$_4$ (oxygen dimers, see e.g., [130,131]) will be helpful.

Overall, it becomes clear that spectroscopic plume analysis is just at the beginning of its development and it constitutes a new and rapidly evolving technique to analyze magmatic degassing, which is (coupled with viscosity and crystallisation) the main driver of volcanic processes.

Acknowledgments: The authors like to thank two anonymous reviewers for very helpful comments and corrections.

Conflicts of Interest: The authors declare no conflict of interest.

References

1. Robock, A. Volcanic Eruptions and Climate. *Rev. Geophys.* **2000**, *38*, 191–219. [CrossRef]
2. Von Glasow, R.; Bobrowski, N.; Kern, C. The effects of volcanic eruptions on atmospheric chemistry. *Chem. Geol.* **2009**, *263*, 131–142. [CrossRef]
3. Kutterolf, S.; Hansteen, T.H.; Appel, K.; Freundt, A.; Krüger, K.; Pérez, W.; Wehrmann, H. Combined bromine and chlorine release from large explosive volcanic eruptions: A threat to stratospheric ozone? *Geology* **2013**, *41*, 707–710. [CrossRef]
4. Platt, U.; Bobrowski, N. Quantification of volcanic reactive halogen emissions. In *Volcanism and Global Change*; Schmidt, A., Fristad, K., Elkins-Tanton, L., Eds.; Cambridge University Press: Cambridge, UK, 2015, ISBN 9781107058378.
5. Platt, U.; Lübcke, P.; Kuhn, J.; Bobrowski, N.; Prata, F.; Burton, M.R.; Kern, C. Quantitative Imaging of Volcanic Plumes—Results, Future Needs, and Future Trends. *J. Volcanol. Geotherm. Res.* **2015**, *300*, 7–21. [CrossRef]
6. Noguchi, K.; Kamiya, H. Prediction of volcanic eruption by measuring the chemical composition and amounts of gases. *Bull. Volcanol.* **1963**, *26*, 367–378. [CrossRef]
7. Malinconico, L.L., Jr. On the Variation of SO$_2$ emission from volcanoes. *J. Volcanol. Geotherm. Res.* **1987**, *33*, 231–237. [CrossRef]
8. Sutton, A.J.; Elias, T.; Gerlach, T.M.; Stokes, J.B. Implications for eruptive proecesses as indicated by sulfur dioxide emissions from Kilauea Volcano, Hawai'i, 1979–1997. *J. Volcanol. Geotherm. Res.* **2001**, *108*, 283–302. [CrossRef]
9. Burton, M.R.; Allard, P.; Mure, F.; Oppenheimer, C. FTIR remote sensing of fractional magma degassing at Mt. Etna, Sicily. *Geol. Soc.* **2003**, *213*, 281–293. [CrossRef]
10. Galle, B.; Johansson, M.; Rivera, C.; Zhang, Y.; Kihlman, M.; Kern, C.; Lehmann, T.; Platt, U.; Arellano, S.; Hidalgo, S. Network for Observation of Volcanic and Atmospheric Change (NOVAC)—A global network for volcanic gas monitoring: Network layout and instrument description. *J. Geophys. Res.* **2010**, *115*, D05304. [CrossRef]
11. Lübcke, P.; Bobrowski, N.; Arellano, S.; Galle, B.; Garzon, G.; Vogel, L.; Platt, U. BrO/SO$_2$ ratios from the NOVAC Network. *Solid Earth* **2014**, *5*, 409–424. [CrossRef]
12. Dinger, F.; Bobrowski, N.; Warnach, S.; Bredemeyer, S.; Hidalgo, S.; Arellano, S.; Galle, B.; Platt, U.; Wagner, T. Periodicity in the BrO/SO$_2$ molar ratios in volcanic gas plumes and its correlation with the Earth tides, Part 1: Observation during the Cotopaxi eruption 2015. *Solid Earth Discuss.* **2017**. [CrossRef]
13. Weibring, P.; Andersson, M.; Edner, H.; Svanberg, S. Remote Monitoring of Industrial Emissions by Combination of Lidar and Plume Velocity Measurements. *Appl. Phys. B* **1998**, *66*, 383–388. [CrossRef]
14. Weibring, P.; Swartling, J.; Edner, H.; Svanberg, S.; Caltabiano, T.; Condarelli, D.; Cecchi, G.; Pantani, L. Optical Monitoring of Volcanic Sulphur Dioxide Emissions—Comparison between four Different Remote Sensing Techniques. *Opt. Lasers Eng.* **2002**, *37*, 267–284. [CrossRef]
15. Galle, B.; Oppenheimer, C.; Geyer, A.; McGonigle, A.J.; Edmonds, M.; Horrocks, L. A miniaturised ultraviolet spectrometer for remote sensing of SO$_2$ fluxes: A new tool for volcano surveillance. *J. Volcanol. Geotherm. Res.* **2003**, *119*, 241–254. [CrossRef]
16. McGonigle, A.J.S.; Pering, T.D.; Wilkes, T.C.; Tamburello, G.; D'Aleo, R.; Bitetto, M.; Aiuppa, A.; Willmott, J.R. Ultraviolet Imaging of Volcanic Plumes: A New Paradigm in Volcanology. *Geosciences* **2017**, *7*, 68. [CrossRef]
17. McGonigle, A.J.S.; Hilton, D.R.; Fischer, T.P.; Oppenheimer, C. Plume velocity determination for volcanic SO$_2$ flux measurements. *Geophys. Res. Lett.* **2005**, *32*, L11302. [CrossRef]
18. McGonigle, A.J.S.; Inguaggiato, S.; Aiuppa, A.; Hayes, A.R.; Oppenheimer, C. Accurate measurement of volcanic SO$_2$ flux: Determination of plume transport speed and integrated SO$_2$ concentration with a single device. *Geochem. Geophys. Geosyst.* **2005**, *6*, Q02003. [CrossRef]
19. Mori, T.; Burton, M. Quantification of the gas mass emitted during single explosions on Stromboli with the SO$_2$ imaging camera. *J. Volcanol. Geotherm. Res.* **2009**, *188*, 395–400. [CrossRef]
20. Kern, C.; Sutton, J.; Elias, T.; Lee, L.; Kamibayashi, K.; Antolik, L.; Werner, C. An automated SO$_2$ camera system for continuous, real-time monitoring of gas emissions from Klauea Volcano's summit Overlook Crater. *J. Volcanol. Geotherm. Res.* **2015**, *300*, 81–94. [CrossRef]

21. Peters, N.; Hoffmann, A.; Barnie, T.; Herzog, M.; Oppenheimer, C. Use of motion estimation algorithms for improved flux measurements using SO_2 cameras. *J. Volcanol. Geotherm. Res.* **2015**, *300*, 58–69. [CrossRef]

22. Klein, A.; Lübcke, P.; Bobrowski, N.; Kuhn, J.; Platt, U. Plume Propagation Direction Determination with SO_2 Cameras. *Atmos. Meas. Tech.* **2017**, *10*, 979–987. [CrossRef]

23. General, S.; Pöhler, D.; Sihler, H.; Bobrowski, N.; Frieß, U.; Zielcke, J.; Horbanski, M.; Shepson, P.; Stirm, B.; Simpson, W.; et al. The Heidelberg Airborne Imaging DOAS Instrument (HAIDI) A Novel Imaging DOAS Device for 2-D and 3-D Imaging of Trace Gases. *J. Atmos. Meas. Tech.* **2014**, *7*, 3459–3485. [CrossRef]

24. Krueger, A.J. Sighting of El Chichon sulfur dioxide clouds with the Nimbus 7 Total Ozone Mapping Spectrometer. *Science* **1983**, *220*, 1377–1378. [CrossRef] [PubMed]

25. Hörmann, C.; Sihler, H.; Bobrowski, N.; Beirle, S.; Penning de Vries, M.; Platt, U.; Wagner, T. Systematic investigation of bromine monoxide in volcanic plumes from space by using the GOME-2 instrument. *Atmos. Chem. Phys.* **2013**, *13*, 4749–4781. [CrossRef]

26. Carn, S.A.; Clarisse, L.; Prata, A.J. Multi-decadal satellite measurements of global volcanic degassing. *J. Volcanol. Geotherm. Res.* **2016**, *311*, 99–134. [CrossRef]

27. Platt, U.; Stutz, J. *Differential Optical Absorption Spectroscopy, Principles and Applications*; Springer: Heidelberg, Germany, 2008; p. 597, ISBN 978-3-540-21193-8.

28. Moffat, A.J.; Millán, M.M. The application of optical correlation techniques to the remote sensing of SO_2 plumes using sky light. *Atmos. Environ.* **1971**, *5*, 677–690. [CrossRef]

29. Lübcke, P.; Lampel, J.; Arellano, S.; Bobrowski, N.; Dinger, F.; Galle, B.; Garzón, G.; Hidalgo, S.; Ortiz, Z.C.; Vogel, L. Retrieval of absolute SO_2 column amounts from scattered-light spectra—Implications for the evaluation of data from automated DOAS Networks. *Atmos. Meas. Tech.* **2016**, *9*, 5677–5698. [CrossRef]

30. Kern, C.; Sihler, H.; Vogel, L.; Rivera, C.; Herrera, M.; Platt, U. Halogen oxide measurements at Masaya volcano, Nicaragua using Active Long Path Differential Optical Absorption Spectroscopy. *Bull. Volcanol.* **2009**, *71*, 659–670. [CrossRef]

31. Vita, F.; Kern, C.; Inguaggiato, S. Development of a portable active long-path differential optical absorption spectroscopy system for volcanic gas measurements. *J. Sens. Sens. Syst.* **2014**, *3*, 355–367. [CrossRef]

32. Horton, K.A.; Williams-Jones, G.; Garbeil, H.; Elias, T.; Sutton, A.J.; Mouginis-Mark, P.; Porter, J.N.; Clegg, S. Real-time measurement of volcanic SO_2 emissions: Validation of a new UV correlation spectrometer (FLYSPEC). *Bull. Volcanol.* **2005**, *68*, 323–327. [CrossRef]

33. Mori, T.; Notsu, K. Remote CO, COS, CO_2, SO_2, HCl detection and temperature estimation of volcanic gas. *Geophys. Res. Lett.* **1997**, *24*, 2047–2050. [CrossRef]

34. Oppenheimer, C.; Francis, P.; Burton, M.; Maciejewski, A.; Boardman, L. Remote measurement of volcanic gases by Fourier transform infrared spectroscopy. *Appl. Phys. B* **1998**, *67*, 505–515. [CrossRef]

35. Burton, M.R.; Oppenheimer, C.; Horrock, L.A.; Francis, P.W. Remote sensing of CO_2 and H_2O emission rates from Masaya volcano, Nicaragua. *Geology* **2000**, *28*, 915–918. [CrossRef]

36. Allard, P.; Burton, M.; Muré, F. Spectroscopic evidence for a lava fountain driven by previously accumulated magmatic gas. *Nature* **2005**, *433*, 407–410. [CrossRef] [PubMed]

37. Carapezza, M.L.; Barberi, F.; Ranaldi, M.; Ricci, T.; Tarchini, L.; Barrancos, J.; Fischer, L.C.; Perez, N.; Weber, K.; Gattuso, A.; et al. Diffuse CO_2 soil degassing and CO_2 and H_2S concentrations in air and related hazards at Vulcano Island (Aeolian arc, Italy). *J. Volcanol. Geotherm. Res.* **2011**, *207*, 130–144. [CrossRef]

38. Pedone, M.; Aiuppa, A.; Giudice, G.; Grassa, F.; Cardellini, C.; Chiodini, G.; Valenza, M. Volcanic CO_2 flux measurement at Campi Flegrei by tunable diode laser absorption spectroscopy. *Bull. Volcanol.* **2015**, *76*, 812. [CrossRef]

39. Chiarugi, A.; Viciani, S.; D'Amato, F.; Burton, M. Diode laser-based gas analyser for the simultaneous measurement of CO_2 and HF in volcanic plumes. *Atmos. Meas. Tech.* **2018**, *11*, 329–339. [CrossRef]

40. Weidmann, D.; Wysocki, G.; Oppenheimer, C.; Tittel, F.K. Development of a compact quantum cascade laser spectrometer for field measurements of CO_2 isotopes. *Appl. Phys. B Lasers Opt.* **2005**, *80*, 255–260. [CrossRef]

41. Richter, D.; Erdelyi, M.; Curl, R.F.; Tittel, F.K.; Oppenheimer, C.; Duffell, H.J.; Burton, M. Field measurements of volcanic gases using tunable diode laser based mid-infrared and Fourier transform infrared spectrometers. *Opt. Lasers Eng.* **2002**, *37*, 171–186. [CrossRef]

42. Waxman, E.M.; Cossel, K.C.; Truong, G.-W.; Giorgetta, F.R.; Swann, W.C.; Coburn, S.; Wright, R.J.; Rieker, G.B.; Coddington, I.; Newbury, N.R. Intercomparison of open-path trace gas measurements with two dual-frequency-comb spectrometers. *Atmos. Meas. Tech.* **2017**, *10*, 3295–3311. [CrossRef] [PubMed]

43. Francis, P.; Burton, M.R.; Oppenheimer, C. Remote measurements of volcanic gas compositions by solar occultation spectroscopy. *Nature* **1998**, *396*, 567–570. [CrossRef]

44. Burton, M.R.; Oppenheimer, C.; Horrocks, L.A.; Francis, P.W. Diurnal changes in volcanic plume chemistry observed by lunar and solar occultation spectroscopy. *Geophys. Res. Lett.* **2001**, *28*, 843–846. [CrossRef]

45. Duffell, H.; Oppenheimer, C.; Burton, M. Volcanic gas emission rates measured by solar occultation spectroscopy. *Geophys. Res. Lett.* **2001**, *28*, 3131–3134. [CrossRef]

46. Butz, A.; Dinger, A.S.; Bobrowski, N.; Kostinek, J.; Fieber, L.; Fischerkeller, C.; Giuffrida, G.B.; Hase, F.; Klappenbach, F.; Kuhn, J.; et al. Remote sensing of volcanic CO_2, HF, HCl, SO_2, and BrO in the downwind plume of Mt. Etna. *Atmos. Meas. Tech.* **2017**, *10*, 1–14. [CrossRef]

47. Naughton, J.J.; Derby, J.V.; Glover, R.B. Infrared measurements on volcanic gas and fume: Kilauea eruption, 1968. *J. Geophys. Res.* **1969**, *74*, 3273–3277. [CrossRef]

48. Mori, T.; Notsu, K.; Tohjima, Y.; Wakita, H. Remote detection of HCl and SO_2 in volcanic gas from Unzen volcano, Japan. *Geophys. Res. Lett.* **1993**, *20*, 1355–1358. [CrossRef]

49. Notsu, K.; Mori, T.; Igarishi, G.; Tohjima, Y.; Wakita, H. Infrared spectral radiometer: A new tool for remote measurement of SO_2 of volcanic gas. *Geochem. J.* **1993**, *27*, 361–366. [CrossRef]

50. Mori, T.K.; Notsu, Y.; Tohjima, H.; Wakita, P.M.; Nuccio, M.; Italiano, F. Remote detection of fumarolic gas chemistry at Vulcano, Italy, using an FT-IR spectral radiometer. *Earth Planet. Sci. Lett.* **1995**, *134*, 219–224. [CrossRef]

51. Gerlach, T.M.; McGee, K.A.; Elias, T.; Sutton, A.J.; Doukas, M.P. Carbon dioxide emission rate of Kilauea Volcano: Implications for primary magma and the summit reservoir. *J. Geophys. Res.* **2002**, *107*, 2189. [CrossRef]

52. Aiuppa, A.; Fiorani, L.; Santoro, S.; Parracino, S.; Nuvoli, M.; Chiodini, G.; Minopoli, C.; Tamburello, G. New groundbased lidar enables volcanic CO_2 flux measurements. *Sci. Rep.* **2015**, *5*, 13614. [CrossRef] [PubMed]

53. Queisser, M.; Burton, M.; Allan, G.; Chiarugi, A. Portable laser spectrometer for airborne and ground-based remote sensing of geological CO_2 emissions. *Opt. Lett.* **2017**, *42*, 2782–2785. [CrossRef] [PubMed]

54. Goff, F.; Love, S.P.; Warren, R.G.; Counce, D.; Obenholzner, J.; Siebe, C.; Schmidt, S.C. Passive infrared remote sensing evidence for large, intermittent CO_2 emissions at Popocatépetl volcano, Mexico. *Chem. Geol.* **2001**, *177*, 133–156. [CrossRef]

55. Stremme, W.; Krueger, A.; Harig, R.; Grutter, M. Volcanic SO_2 and SiF_4 visualization using 2-D thermal emission spectroscopy-Part 1: Slant-columns and their ratios. *Atmos. Meas. Tech.* **2012**, *5*, 275–288. [CrossRef]

56. Prata, A.; Bernardo, C. Retrieval of sulphur dioxide froma ground-based thermal infrared imaging camera. *Atmos. Meas. Tech.* **2014**, *7*, 2807–2828. [CrossRef]

57. Gabrieli, A.; Wright, R.; Lucey, P.G.; Porter, J.N.; Garbeil, H.; Pilger, E.; Wood, M. Characterization and initial field test of an 8–14 μm thermal infrared hyperspectral imager for measuring SO_2 in volcanic plumes. *Bull. Volcanol.* **2016**, *78*, 73. [CrossRef]

58. Love, S.P.; Goff, F.; Counce, D.; Siebe, C.; Delgado, H. Passive infrared spectroscopy of the eruption plume at Popocatepetl volcano, Mexico. *Nature* **1998**, *396*, 563–567. [CrossRef]

59. Krueger, A.; Stremme, W.; Harig, R.; Grutter, M. Volcanic SO_2 and SiF_4 visualization using 2-D thermal emission spectroscopy—Part 2: Wind propagation and emission rates. *Atmos. Meas. Tech.* **2013**, *6*, 47–61. [CrossRef]

60. Friedl-Vallon, F.; Gulde, T.; Hase, F.; Kleinert, A.; Kulessa, T.; Maucher, G.; Neubert, T.; Olschewski, F.; Piesch, C.; Preusse, P.; et al. Instrument concept of the imaging Fourier transform spectrometer GLORIA. *Atmos. Meas. Tech.* **2014**, *7*, 3565–3577. [CrossRef]

61. Hinkley, E.D. (Ed.) *Laser Monitoring of the Atmosphere*; Topics in Applied Physics; Springer: Berlin/Heidelberg, Germany, 1976; Volume 14.

62. Svanberg, S. *Atomic and Molecular Spectroscopy*, 2nd ed.; Springer Series on Atoms and Plasmas; Springer: Berlin/Heidelberg, Germany, 1992.

63. Edner, H.; Ragnarson, P.; Svanberg, S.; Wallinder, E.; Ferrara, R.; Cioni, R.; Raco, B.; Taddeucci, G. Total Fluxes of Sulphur Dioxide from the Italian Volcanoes Etna, Stromboli and Vulcano Measured by Differential Absorption Lidar and Passive Differential Optical Absorption Spectroscopy. *J. Geophys. Res.* **1994**, *99*, 18827–18838. [CrossRef]

64. Weibring, P.; Edner, H.; Svanberg, S.; Cecchi, G.; Pantani, L.; Ferrara, R.; Caltabiano, T. Monitoring of Volcanic Sulphur Dioxide Emissions using Differential Absorption Lidar (DIAL), Differential Optical Absorption Spectroscopy (DOAS) and Correlation Spectroscopy (COSPEC). *Appl. Phys. B* **1998**, *67*, 419–426. [CrossRef]

65. Weckwerth, T.M.; Weber, K.J.; Turner, D.D.; Spuler, S.M. Validation of a Water Vapor Micropulse Differential Absorption Lidar (DIAL). *J. Atmos. Ocean Technol.* **2016**, *33*, 2353–2372. [CrossRef]

66. Barnes, J.E.; Bronner, S.; Beck, R.; Parikh, N.C. Boundary layer scattering measurements with a charge-coupled device camera lidar. *Appl. Opt.* **2003**, *42*, 2647–2652. [CrossRef] [PubMed]

67. Flock, S. LED-Lidar—Theoretische und Experimentelle Machbarkeitsstudie zur Realisierung Eines Bistatischen, LED-Basierten Lidarsystems. Master's Thesis, University of Heidelberg, Heidelberg, Germany, 2012. (In German)

68. Nadeau, P.A.; Palma, J.L.; Waite, G.P. Linking Volcanic Tremor, Degassing, and Eruption Dynamics via SO_2 Imaging. *Geophys. Res. Lett.* **2011**, *38*, 1. [CrossRef]

69. Kern, C.; Lübcke, P.; Bobrowski, N.; Campion, R.; Mori, T.; Smekens, J.-F.; Stebel, K.; Tamburello, G.; Burton, M.; Platt, U.; et al. Intercomparison of SO_2 camera systems for imaging volcanic gas plumes. *J. Volcanol. Geotherm. Res.* **2015**, *300*, 22–36. [CrossRef]

70. Gliß, J.; Stebel, K.; Kylling, A.; Sudbø, A. Optical flow gas velocity analysis in plumes using UV cameras—Implications for SO_2-emission-rate retrievals investigated at Mt. Etna, Italy, and Guallatiri, Chile. *Atmos. Meas. Tech. Discuss.* **2017**. [CrossRef]

71. Lübcke, P.; Bobrowski, N.; Illing, S.; Kern, C.; Vogel, L.; Platt, U. On the absolute calibration of SO_2 Cameras. *Atmos. Meas. Tech.* **2013**, *6*, 677–696. [CrossRef]

72. Rusch, P.; Harig, R. 3-D Reconstruction of Gas Clouds by Scanning Imaging IR Spectroscopy and Tomography. *IEEE Sens. J.* **2010**, *10*, 599–603. [CrossRef]

73. Lohberger, F.; Hönninger, G.; Platt, U. Ground Based Imaging Differential Optical Absorption Spectroscopy of Atmospheric Gases. *Appl. Opt.* **2004**, *43*, 4711–4717. [CrossRef] [PubMed]

74. Bobrowski, N.; Hönninger, G.; Lohberger, F.; Platt, U. IDOAS: A new monitoring technique to study the 2D distribution of volcanic gas emissions. *J. Volcanol. Geotherm. Res.* **2006**, *150*, 329–338. [CrossRef]

75. Louban, I.; Bobrowski, N.; Rouwet, D.; Inguaggiato, S.; Platt, U. Imaging DOAS for Volcanological Applications. *Bull. Volcanol.* **2009**, *71*, 753–765. [CrossRef]

76. Lee, H.-L.; Kim, J.-H.; Ryu, J.; Kwon, S.; Noh, Y.; Gu, M. 2-dimensional Mapping of Sulfur Dioxide and Bromine Oxide at the Sakurajima Volcano with a Ground Based Scanning Imaging Spectrograph System. *J. Opt. Soc. Korea* **2010**, *14*, 204–208. [CrossRef]

77. Wright, R.; Lucey, P.; Crites, S.; Horton, K.; Wood, M.; Garbeil, H. BBM/EM design of the thermal hyperspectral imager: An instrument for remote sensing of Earth's surface, atmosphere and ocean, from a microsatellite platform. *Acta Astronaut.* **2013**, *87*, 182–192. [CrossRef]

78. Gabrieli, A.; Porter, J.N.; Wright, R.; Lucey, P.G. Validating the accuracy of SO_2 gas retrievals in the thermal infrared (8–14 μm). *Bull. Volcanol.* **2017**, *79*, 80. [CrossRef]

79. Kuhn, J.; Bobrowski, N.; Lübcke, P.; Vogel, L.; Platt, U. A Fabry-Pérot Interferometer Based Camera for the two-dimensional Mapping of SO_2-Distributions. *J. Atmos. Meas. Tech.* **2014**, *7*, 3705–3715. [CrossRef]

80. Kuhn, J.; Platt, U.; Bobrowski, N.; Lübcke, P.; Wagner, T. Fabry-Perot interferometer based imaging of atmospheric trace gases. In Proceedings of the 10th EARSeL SIG Imaging Spectroscopy Workshop, Zurich, Switzerland, 19–21 April 2017.

81. Eisenhauer, F.; Raab, W. Visible/infrared imaging spectroscopy and energy-resolving detectors. *Annu. Rev. Astron. Astrophys.* **2015**, *53*, 155–197. [CrossRef]

82. Bluth, G.J.S.; Shannon, J.M.; Watson, I.M.; Prata, A.J.; Realmuto, V.J. Development of an ultra-violet digital camera for volcanic SO_2 imaging. *J. Volcanol. Geotherm. Res.* **2007**, *161*, 47–56. [CrossRef]

83. Kern, C.; Deutschmann, T.; Werner, C.; Sutton, A.J.; Elias, T.; Kelly, P.J. Improving the accuracy of SO_2 column densities and emission rates obtained from upward-looking UV-spectroscopic measurements of volcanic plumes by taking realistic radiative transfer into account. *J. Geophys. Res.* **2012**, *117*, D20302. [CrossRef]

84. Smekens, J.-F.; Burton, M.R.; Clarke, A.B. Validation of the SO_2 Camera for High Temporal and Spatial Resolution Monitoring of SO_2 Emissions. *J. Volcanol. Geotherm. Res.* **2015**, *300*, 37–47. [CrossRef]

85. Prata, A.; Bernardo, C. Retrieval of volcanic ash particle size, mass and optical depth from a ground-based thermal infrared camera. *J. Volcanol. Geotherm. Res.* **2009**, *186*, 91–107. [CrossRef]

86. Lopez, T.; Thomas, H.E.; Prata, A.J.; Amigo, A.; Fee, D.; Moriano, D. Volcanic plume characteristics determined using an infrared imaging camera. *J. Volcanol. Geotherm. Res.* **2015**, *300*, 148–166. [CrossRef]

87. Sandsten, J.; Edner, H.; Svanberg, S. Gas imaging by infrared gas-correlation spectrometry. *Opt. Lett.* **1996**, *21*, 1945–1947. [CrossRef] [PubMed]

88. Sandsten, J.; Edner, H.; Svanberg, S. Gas visualization of industrial hydrocarbon emissions. *Opt. Express* **2004**, *12*, 1443–1451. [CrossRef] [PubMed]

89. Prata, A. Infrared radiative transfer calculations for volcanic ash clouds. *Geophys. Res. Lett.* **1989**, *16*, 1293–1296. [CrossRef]
90. Levelt, P.F.; Hilsenrath, E.; Leppelmeier, G.W.; van Den Oord, G.H.J.; Bhartia, P.K.; Tamminen, J.; De Haan, J.F.; Veefkind, P. Science Objectives of the Ozone Monitoring Instrument. *Geosci. Remote Sens.* **2006**, *44*, 1199–1208. [CrossRef]
91. General, S.; Bobrowski, N.; Pöhler, D.; Weber, K.; Fischer, C.; Platt, U. Airborne I-DOAS measurements at Mt. Etna BrO and OClO evolution in the plume. *J. Volcanol. Geotherm. Res.* **2015**, *300*, 175–186. [CrossRef]
92. Kern, C.; Deutschmann, T.; Vogel, L.; Wöhrbach, M.; Wagner, T.; Platt, U. Radiative transfer corrections for accurate spectroscopic measurements of volcanic gas emissions. *Bull. Volcanol.* **2010**, *72*, 233–247. [CrossRef]
93. Ward, T.V.; Zwick, H.H. Gas cell correlation spectrometer: GASPEC. *Appl. Opt.* **1975**, *14*, 2896–2904. [CrossRef] [PubMed]
94. Bobrowski, N. Volcanic Gas Studies by Multi Axis Differential Optical Absorption Spectroscopy. Diploma Thesis, University of Heidelberg, Heidelberg, Germany, 2002.
95. Johansson, M.; Galle, B.; Zhang, Y.; Rivera, C.; Chen, D.; Wyser, K. The dual-beam mini-DOAS technique—Measurements of volcanic gas emission, plume height and plume speed with a single instrument. *Bull. Volcanol.* **2009**, *71*, 747–751. [CrossRef]
96. Mori, T.; Burton, M. The SO_2 camera: A simple, fast and cheap method for ground-based imaging of SO_2 in volcanic plumes. *Geophys. Res. Lett.* **2006**, *33*. [CrossRef]
97. Valade, S.A.; Harris, A.J.L.; Cerminara, M. Plume Ascent Tracker: Interactive Matlab Software for Analysis of Ascending Plume. *Comput. Geosci. C* **2014**, *66*, 132–144. [CrossRef]
98. Fickel, M. Measurement of Trace Gas Fluxes from Point Sources with Multi-Axis Differential Optical Absorption Spectroscopy. Diploma Thesis, University of Heidelberg, Heidelberg, Germany, 2008.
99. Millán, M.M. Remote sensing of Air Pollutants. A Study of some Atmospheric Scattering Effects. *Atmos. Environ.* **1980**, *14*, 1241–1253. [CrossRef]
100. Kern, C. Spectroscopic Measurements of Volcanic Gas Emissions in the Ultra-Violet Wavelength Region. Ph.D. Thesis, Institute of Environmental Physics, The Faculty of Physics and Astronomy, University of Heidelberg, Heidelberg, Germany, 2009. [CrossRef]
101. Boichu, M.; Oppenheimer, C.; Tsanev, V.; Kyle, P.R. High temporal resolution SO_2 flux measurements at Erebus volcano, Antarctica. *J. Volcanol. Geotherm. Res.* **2010**, *190*, 325–336. [CrossRef]
102. McGonigle, A.J.S.; Aiuppa, A.; Ripepe, M.; Kantzas, E.P.; Tamburello, G. Spectroscopic capture of 1 Hz volcanic SO_2 fluxes and integration with volcano geophysical data. *Geophys. Res. Lett.* **2009**, *36*, L21309. [CrossRef]
103. Moussallam, Y.; Philipson, B.; Curtis, A.; Barnie, T.; Moussallam, M.; Peters, N.; Schipper, C.I.; Aiuppa, A.; Giudice, G.; Amigo, A.; et al. Sustaining persistent lava lakes: Observations from high-resolution gas measurements at Villarrica volcano, Chile. *Earth Planet. Sci. Lett.* **2016**, *454*, 237–247. [CrossRef]
104. Allard, P.; Burton, M.; Sawyer, G.; Bani, P. Degassing dynamics of basaltic lava lake at a top-ranking volatile emitter: Ambrym volcano, Vanuatu arc. *Earth Planet. Sci. Lett.* **2016**, *448*, 69–80. [CrossRef]
105. Nadeau, P.A.; Werner, C.A.; Waite, G.P.; Carn, S.A.; Brewer, I.D.; Elias, T.; Sutton, A.J.; Kern, C.; Patrick, M.R. Using SO_2 camera imagery and seismicity to examine degassing and gas accumulation at Kīlauea Volcano, May 2010. *J. Volcanol. Geotherm. Res.* **2015**, *300*, 70–80. [CrossRef]
106. Tamburello, G.; Aiuppa, A.; Kantzas, E.P.; McGonigle, A.J.S.; Ripepe, M. Passive vs. active degassing modes at an open-vent volcano (Stromboli, Italy). *Earth Planet. Sci. Lett.* **2012**, *359*, 106–116. [CrossRef]
107. Tamburello, G.; Aiuppa, A.; McGonigle, A.J.S.; Allard, P.; Cannata, A.; Giudice, G.; Kantzas, E.P.; Pering, T.D. Periodic volcanic degassing behavior: The Mount Etna example. *Geophys. Res. Lett.* **2013**, *40*, 4818–4822. [CrossRef]
108. Waite, G.P.; Nadeau, P.A.; Lyons, J.J. Variability in eruption style and associated very long period events at Fuego volcano, Guatemala. *J. Geophys. Res. Solid Earth* **2013**, *118*, 1526–1533. [CrossRef]
109. Burton, M.R.; Prata, F.; Platt, U. Volcanological applications of SO_2 cameras. *J. Volcanol. Geotherm. Res.* **2015**, *300*, 2–6. [CrossRef]
110. Dalton, M.P.; Waite, G.P.; Watson, I.M.; Nadeau, P.A. Multiparameter quantification of gas release during weak Strombolian eruptions at Pacaya Volcano, Guatemala. *Geophys. Res. Lett.* **2010**, *37*. [CrossRef]
111. Lopez, T.; Wilson, D.F.; Prata, F.; Dehn, J. Characterization and interpretation of volcanic activity at Karymsky Volcano, Kamchatka, Russia, using observations of infrasound, volcanic emissions, and thermal imagery. *Geochem. Geophys. Geosyst.* **2013**, *14*. [CrossRef]

112. Pering, T.D.; Tamburello, G.; McGonigle, A.J.S.; Aiuppa, A.; Cannata, A.; Giudice, G.; Patanè, D. High time resolution fluctuations in volcanic carbon dioxide degassing from Mount Etna. *J. Volcanol. Geotherm. Res.* **2014**, *270*, 115–121. [CrossRef]

113. Barnie, T.; Bombrun, M.; Burton, M.R.; Harris, A.J.L.; and Sawyer, G. Quantification of gas and solid emissions during Strombolian explosions using simultaneous sulphur dioxide and infrared camera observations. *J. Volcanol. Geothermal Res.* **2015**, 167–174. [CrossRef]

114. D'Aleo, R.; Bitetto, M.; Delle Donne, D.; Tamburello, G.; Battaglia, A.; Coltelli, M.; Aiuppa, A. Spatially resolved SO_2 flux emissions from Mt Etna. *Geophys. Res. Lett.* **2016**, *43*, 7511–7519. [CrossRef] [PubMed]

115. Delle Donne, D.; Ripepe, M.; Lacanna, G.; Tamburello, G.; Bitetto, M.; Aiuppa, A. Gas mass derived by infrasound and UV cameras: Implications for mass flow rate. *J. Volcanol. Geotherm. Res.* **2016**, *325*, 169–178. [CrossRef]

116. Delle Donne, D.; Tamburello, G.; Aiuppa, A.; Bitetto, M.; Lacanna, G.; D'Aleo, R.; Ripepe, M. Exploring the explosive-effusive transition using permanent ultra-violet cameras. *J. Geophys. Res. Solid Earth* **2017**, *122*, 4377–4394. [CrossRef]

117. Oppenheimer, C.; Kyle, P.R. Probing the magma plumbing of Erebus volcano, Antarctica, by open-path FTIR spectroscopy of gas emissions. *J. Volcanol. Geotherm. Res.* **2008**, *177*, 743–754. [CrossRef]

118. La Spina, A.; Burton, M.; Salerno, G.G. Unravelling the processes controlling gas emissions from the central and northeast craters of Mt. Etna. *J. Volcanol. Geotherm. Res.* **2010**, *198*, 368–376. [CrossRef]

119. La Spina, A.; Burton, M.; Allard, P.; Alparone, S.; Muré, F. Open-path FTIR spectroscopy of magma degassing processes during eight lava fountains on Mount Etna. *Earth Planet. Sci. Lett.* **2015**, *413*, 123–134. [CrossRef]

120. Burton, M.; Allard, P.; Muré, F.; La Spina, A. Magmatic gas composition reveals the source depth of slug-driven Strombolian explosive activity. *Science* **2007**, *317*, 227–230. [CrossRef] [PubMed]

121. La Spina, A.; Burton, M.R.; Harig, R.; Mure, F.; Rusch, P.; Jordan, M.; Caltabiano, T. New insights into volcanic processes at Stromboli from Cerberus, a remote-controlled open-path FTIR scanner system. *J. Volcanol. Geotherm. Res.* **2013**, *249*, 66–76. [CrossRef]

122. Schwandner, F.M.; Gunson, M.R.; Miller, C.E.; Carn, S.A.; Eldering, A.; Krings, T.; Verhulst, K.R.; Schimel, D.S.; Nguyen, H.M.; Crisp, D.; et al. Spaceborne detection of localized carbon dioxide sources. *Science* **2017**, *358*, eaam5782. [CrossRef] [PubMed]

123. Oppenheimer, C.; Bani, P.; Calkins, J.; Burton, M.; Sawyer, G. Rapid FTIR sensing of volcanic gases released by Strombolian explosions at Yasur volcano, Vanuatu. *Appl. Phys. B* **2006**, *85*, 453–460. [CrossRef]

124. Francis, P.; Chaffin, C.; Maciejewski, A.; Oppenheimer, C. Remote determination of SiF4 in volcanic plumes: A new tool for volcano monitoring. *Geophys. Res. Lett.* **1996**, *23*, 249–252. [CrossRef]

125. Taquet, N.; Meza Hernández, I.; Stremme, W.; Bezanilla, A.; Grutter, M.; Campion, R.; Palm, M.; Boulesteix, T. Continuous measurements of SiF_4 and SO_2 by thermal emission spectroscopy: Insight from a 6-month survey at the Popocatépetl volcano. *J. Volcanol. Geotherm. Res.* **2017**, *341*, 255–268. [CrossRef]

126. Edmonds, M.; Herd, R.A.; Galle, B.; Oppenheimer, C. Automated, high time-resolution measurements of SO_2 flux at Soufrière Hills Volcano, Montserrat. *Bull. Volcanol.* **2003**, *65*, 578–586. [CrossRef]

127. Wilkes, T.C.; McGonigle, A.J.R.; Willmott, T.D.; Pering, T.D.; Cook, J.M. Low-cost 3D printed 1 nm resolution smartphone sensor-based spectrometer: Instrument design and application in ultraviolet spectroscopy. *Opt. Lett.* **2017**, *42*, 4323–4326. [CrossRef] [PubMed]

128. Wilkes, T.C.; Pering, T.D.; McGonigle, A.J.S.; Tamburello, G.; Willmott, J.R. A low cost smartphone sensor-based UV camera for volcanic SO_2 emission measurements. *Remote Sens.* **2017**, *9*, 27. [CrossRef]

129. Kern, C.; Werner, C.; Elias, T.; Sutton, A.J.; Lübcke, P. Applying UV cameras for SO_2 detection to distant or optically thick plumes. *J. Volcanol. Geotherm. Res.* **2013**, *262*, 80–89. [CrossRef]

130. Wagner, T.; Dix, B.; Friedeburg, C.; Frieß, U.; Sanghavi, S.; Sinreich, R.; Platt, U. MAX-DOAS O_4 measurements: A new technique to derive information on atmospheric aerosols—Principles and Information content. *J. Geophys. Res.* **2004**, *109*, D22205. [CrossRef]

131. Frieß, U.; Monks, P.S.; Remedios, J.J.; Rozanov, A.; Sinreich, R.; Wagner, T.; Platt, U. MAX-DOAS O_4 measurements: A new technique to derive information on atmospheric aerosols. (II) Modelling studies. *J. Geophys. Res.* **2006**, *111*, D14203. [CrossRef]

geosciences

MDPI

Article

Ozone Depletion in Tropospheric Volcanic Plumes: From Halogen-Poor to Halogen-Rich Emissions

Tjarda J. Roberts

LPC2E Laboratoire de Physique et Chimie de l'Environnement et de l'Espace, UMR 7328 CNRS and Université d'Orléans, 3 Avenue de la Recherche Scientifique, 45071 Orléans, France; Tjarda.Roberts@cnrs-orleans.fr; Tel.: +33-238-255-282

Received: 15 November 2017; Accepted: 24 January 2018; Published: 10 February 2018

Abstract: Volcanic halogen emissions to the troposphere undergo a rapid plume chemistry that destroys ozone. Quantifying the impact of volcanic halogens on tropospheric ozone is challenging, only a few observations exist. This study presents measurements of ozone in volcanic plumes from Kīlauea (HI, USA), a low halogen emitter. The results are combined with published data from high halogen emitters (Mt Etna, Italy; Mt Redoubt, AK, USA) to identify controls on plume processes. Ozone was measured during periods of relatively sustained Kīlauea plume exposure, using an Aeroqual instrument deployed alongside Multi-Gas SO_2 and H_2S sensors. Interferences were accounted for in data post-processing. The volcanic H_2S/SO_2 molar ratio was quantified as 0.03. At Halema'uma'u crater-rim, ozone was close to ambient in the emission plume (at 10 ppmv SO_2). Measurements in grounding plume (at 5 ppmv SO_2) about 10 km downwind of Pu'u 'Ō'ō showed just slight ozone depletion. These Kīlauea observations contrast with substantial ozone depletion reported at Mt Etna and Mt Redoubt. Analysis of the combined data from these three volcanoes identifies the emitted Br/S as a strong but non-linear control on the rate of ozone depletion. Model simulations of the volcanic plume chemistry highlight that the proportion of HBr converted into reactive bromine is a key control on the efficiency of ozone depletion. This underlines the importance of chemistry in the very near-source plume on the fate and atmospheric impacts of volcanic emissions to the troposphere.

Keywords: BrO; reactive halogen; O_3; atmospheric chemistry; plume

1. Introduction

Volcanoes release large quantities of gases and aerosols to the atmosphere. Very large explosive eruptions inject gases directly to the stratosphere, but a significant number of smaller eruptions and continuously passive degassing volcanoes release their emissions to the troposphere; the total global SO_2 flux from passive degassing over 2004–2016 was recently estimated as 23 Tg/yr on average, exceeding volcanic eruptive emissions by about one order of magnitude [1]. To date, studies of the atmospheric chemistry and climate impacts of volcanic emissions have mostly focused on SO_2 and its oxidation to sulfate particles in both the stratosphere (e.g., [2]) and troposphere (e.g., [3]). Volcanic sulfates also catalyse gas-aerosol reactions leading to reductions in stratospheric ozone levels, e.g., [4,5]. However, the volcanic release contains a number of other gases and particles, including notably the emission of volcanic halogens such as HBr and HCl (e.g., [6]). These were initially assumed to be simply washed out of the plume in the troposphere and deposited. Volcanic halogens can occasionally be injected into the stratosphere as evidenced by recent observations of HCl, OClO, BrO, and IO by satellite [7–10] and of HCl and HF by an instrumented aircraft that transected a high-altitude volcanic cloud [11]. Volcanic halogens that reach high altitudes may cause reductions in stratospheric ozone levels. This has been both observed [11] and simulated by numerical models (e.g., [12,13]) using atmospheric chemistry schemes that were originally developed to study impacts from anthropogenic sources of halogens (chlorofluorocarbons, CFCs) on stratospheric ozone.

The discovery of BrO in a volcanic plume in the troposphere through ground-based remote sensing by Borowski et al. [14] showed that volcanic HBr emissions can become transformed into reactive bromine over very short time-scales (minutes). Volcanic BrO has subsequently been observed in many tropospheric volcano plumes globally, e.g., [15,16] and references therein. A global analysis of satellite observations of volcanic BrO is provided by Hörmann et al. [17]. Volcanic OClO has also been reported in some tropospheric plumes [18–21]. The mechanism for reactive halogen formation in the troposphere at such very fast rates was not well understood. It was proposed that the formation of BrO in volcanic plumes could occur via a volcanic version of the autocatalytic "bromine explosion" (similar to that first identified in the polar boundary layer) [14,22]. This mechanism would lead to depletion of tropospheric ozone immediately downwind from the volcano, as has been confirmed by observations [23,24]. To better understand the fate of volcanic halogen emissions in the troposphere and their atmospheric chemistry impacts, two 1D or box models of the volcanic plume halogen chemistry were developed, [18,25,26], and the plume halogen chemistry was recently incorporated into a regional model [27]. The tropospheric chemistry model mechanisms are briefly outlined below.

Bromine is the main halogen species responsible for ozone depletion in volcanic plumes in the troposphere. Volcanic HBr emissions are converted into reactive bromine via an autocatalytic "bromine explosion" mechanism, involving gas-phase, photolytic, and heterogeneous (gas-aerosol) reactions; for a review, see Von Glasow et al. [28]. Cycles involving the formation and reactive uptake of HOBr and $BrONO_2$ act to convert volcanic HBr into the reactive form Br_2 that photolyses and can then deplete ozone via the reaction of Br with O_3. A rapid interconversion cycle between BrO (self-reaction) and Br (that reacts with ozone to reform BrO) is suggested to be a major cause of ozone loss at high halogen concentrations (e.g., in the near-source plume). Upon depletion of plume HBr the reactive uptake of HOBr can form BrCl, promoting the formation of reactive chlorine through a non-autocatalytic cycle that can also contribute to ozone loss. Key controls on the plume chemistry are described by Roberts et al. [29] and include the following: the halogen emission flux, the primary volcanic aerosol emission that catalyses key heterogeneous (gas-aerosol) reactions of reactive halogens, and the rate of plume-air mixing that entrains oxidants including ozone into the plume. A high-temperature near-vent chemistry generates radicals that act to "kick-start" the downwind plume chemistry, thereby accelerating the (low-temperature) formation of BrO.

Comparison of the box/1D model simulations to plume observations enables testing of the underlying model mechanisms and is an important first step towards assessing regional-to-global-scale tropospheric impacts from volcanic halogen emissions. In such comparisons, the data is typically scaled to SO_2 that acts as a quasi-plume-tracer over short time-scales. This enables to distinguish between plume chemistry and dilution effects. The model simulations were able to reproduce the observed magnitude and trends in volcanic BrO/SO_2 downwind from Mt Etna [25,26,29], and to broadly capture the magnitude of reported $OClO/SO_2$ [30]. Recent studies have also demonstrated a modelling capability to reproduce ozone depletion observed in the Mt Redoubt (AK, USA) plume [23] and at Mt Etna (Italy) [24]. Other model-predicted impacts of volcanic plume reactive halogen chemistry in the troposphere include depletion of HO_x and NO_x (the latter converted into nitric acid) [25] and conversion of inert mercury to a more reactive and easily deposited form [26]. A regional modelling study by Jourdain et al. [27] also highlighted the potential for volcanic plume reactive halogens to undergo secondary transport to the stratosphere (e.g., via convective events) for further impacts. However, uncertainties remain in our understanding of the very complex volcanic plume chemistry given the model parameter-space is vast and is constrained by rather few observations, particularly regarding the impacts of volcanic plume halogens on tropospheric ozone. This study presents field-measurements of ozone in the Halema'uma'u and Pu'u 'Ō'ō plumes of Kīlauea (HI, USA) on 3 September 2007 and 19 July 2008, respectively, and compares them to reported observations at Mt Etna and Mt Redoubt. As a low halogen emitter, Kīlauea provides a useful "end-member" to complement the existing case-studies of the higher halogen emitters Mt Etna and Mt Redoubt, [23,24].

The goals of this study are: (i) to quantify the H$_2$S emission alongside SO$_2$ (i.e., H$_2$S/SO$_2$ ratio) enabling the prediction of cross-sensitivity on the Aeroqual ozone measurement; (ii) to measure ozone in the crater-rim plume; (iii) to measure ozone in the chemically more evolved downwind plume; and (iv) to interpret the ozone observations in the context of volcanic plume halogen chemistry and tropospheric ozone depletion reported in volcanic plumes globally.

Section 2 below outlines the challenges to measuring ozone in volcanic plumes; Section 3 describes the instruments and field-deployment; Section 4 presents the ozone measurements made at Kīlauea; and Section 5 discusses these observations in a wider volcanic plume chemistry context.

2. Challenges to In-Situ Measurement of Ozone in the Near-Source Volcano Plume

2.1. Interferences of Volcanic Gases on the Ozone Measurement

Ozone is present in the background troposphere at mixing ratios of tens of nmol/mol (or ppbv). Two main in-situ approaches to measuring atmospheric ozone are ultra-violet (UV) spectroscopy on ground-based or aircraft platforms (with instruments often operating at 254 nm, within the Hartley ozone absorption band) and the ozone (electrochemical) balloon sonde. A challenge to measuring ozone in volcanic plumes is the presence of other interfering gases at much higher abundances (e.g., μmol/mol or ppmv) than typically occur in the background atmosphere. These can induce positive or negative interferences to yield an erroneously high or low ozone measurement, respectively.

Specifically, volcanic SO$_2$ induces a negative interference on electrochemical cell measurements of ozone. For example, ozone-sondes launched into the Eyjafjallajökull 2010 eruption plume over Europe detected severely disturbed profiles [31], but the low ozone signal could not be quantifiably attributed to volcanic plume chemistry due to the interference from SO$_2$. Volcanic SO$_2$ induces a positive interference on UV spectroscopic measurements of ozone due to its absorption at 254 nm. This interference can be automatically corrected in dual channel instruments that contain a second (ozone scrubbed) channel, or may be subtracted in data post-processing using co-measured SO$_2$, provided that the plume is sufficiently dilute [23]. Several airborne measurements of ozone in volcanic plumes have been reported; ozone concentrations were observed below ambient levels in a study of volcanic plumes predominantly from Alaska [32]. Ozone was depleted by at least a third compared to background, reaching up to 90% depletion in the 1980 Mount St. Helens eruption plume, during both ash-rich and ash-poor (passive degassing following the eruption) conditions [33,34]. Oppenheimer et al. [35] reports ozone depletion up to 35% with respect to background in the Mt Erebus (Antarctica) plume, alongside measurements of volcanic sulfur and nitrogen species, but also with some evidence for a very rapid (and unexplained) near-source reduction in SO$_2$. Instrumented aircraft campaigns by Vance et al. [36] and Schumann et al. [37] measured ozone depletion alongside volcanic SO$_2$ and other gases and particles in the 2010 Eyjafjallajokull eruption plume dispersed over Europe. Volcanic BrO was also observed in the downwind plume. However, the total volcanic bromine emission from Eyjafjallajokull was not well monitored. Finally, an instrumented aircraft campaign spatially mapped ozone depletion alongside SO$_2$ in the 2010 Mt Redoubt eruption plume over 2–20 km downwind [27]. In this study, Kelly et al. also quantified the volcanic bromine emission by filterpack (HBr/SO$_2$ = 4.1 × 10^{-3} mol/mol), enabling interpretation of the observations and a comparison to an atmospheric box model. Their study captured a rapid decrease in ozone in the near-downwind plume reaching up to tens of ppbv O$_3$ loss, followed by a slow (partial) recovery in the dispersing plume. This demonstrated the need to characterise plume ozone-depleting chemistry over short spatial-timescales very near to the source. To do so, diffusion tubes were installed on Mt Etna flanks by Vance et al. [36], who reported an ozone depletion signature that was anti-correlated with SO$_2$. However, corrosion problems prevented a measurement very close to the crater-rim. Some initial measurements were also made using a UV ozone instrument with CrO$_3$ scrubber [36]. This approach was significantly furthered by Surl et al. [24], whose novel observations quantified the rate of ozone depletion through measurements on Mt Etna's flanks. Their approach involved a portable single-channel UV spectroscopic instrument combined in series with CrO_3 scrubbers to remove SO_2 from the air inlet, with the ozone losses on the scrubber quantified and treated in the data post-processing. This enabled—for the first time—quantification of ozone depletion in very near-to-source plumes containing tens of ppmv of volcanic SO_2 that was co-measured by small electrochemical sensor (Multi-Gas instrument). The volcanic bromine emission was also characterised during the field-campaign (HBr/SO_2 = 6.13 × 10^{-4} mol/mol). Through observations made at up to several hundred meters distance from the summit emission source (equivalent to up to a few minutes travel time), Surl et al. [24] derived a linear equation for ozone loss in Mt Etna's near-source plume as a function of travel time downwind from the summit craters, finding a gradient of $\Delta O_3/\Delta SO_2$ = $(-1.02 \pm 0.07) \times 10^{-5}$ s^{-1} and intercept of $(-6.2 \pm 0.05) \times 10^{-4}$ mol/mol. The non-zero intercept suggests that some depletion of ozone had already occurred before the gases reached the crater-rim. The ozone depletion rate (gradient) has been compared to atmospheric 1D and box model simulations by Surl et al. [24] and Roberts et al. [30].

Here, we used an Aeroqual instrument (WO_3 sensor; Aeroqual Limited, Auckland, New Zealand) to measure ozone in the Kīlauea volcanic plumes (Halemaʻumaʻu and Puʻu ʻŌʻō, HI, USA). The instrument is much less sensitive to interference from volcanic SO_2 than other methods, see Table 1. However, the sensor exhibits other cross-sensitivities as listed in Table 1. The most important cross-sensitivity for volcanic plumes is that of H_2S at −2.5% that needs to be considered in data post-processing. The importance of this interference depends on the magmatic conditions specific to each volcano. Fumarolic emissions are H_2S-rich (e.g., Vulcano, with H_2S/SO_2 molar ratios of ~1 [38]). Oxidized magma emissions are H_2S-poor and dominated by SO_2 (e.g., Masaya, Mt Etna, Kīlauea with H_2S/SO_2 molar ratios of just a few percent [39]). The H_2S-poor nature of the Kīlauea volcanic emission is further verified by our in-situ real-time sensing of H_2S and SO_2 alongside ozone (see Results). Alongside our field-measurements of ozone, the bromine content of Kīlauea emissions was characterized by Mather et al. [40], finding HBr/SO_2 = 2.9 × 10^{-5} mol/mol.

Table 1. Cross-sensitivities of Aeroqual ozone instrument (tungsten oxide sensor). Source: www.Aeroqual.com [41].

Test Gas	Mixing Ratio (ppmv)	Sensor Reading (ppmv)	Sensitivity (%)
Ammonia (NH_3)	25	−0.02	−0.08
Butane	1000	0	0
Carbon monoxide (CO)	100	0	0
Carbon dioxide (CO_2)	1000	0	0
Chlorine (Cl_2)	0.5	0.2	−40
Ethanol	20	−0.02	−0.1
Ethyl acetate	100	−0.02	−0.02
Heptane	100	0	0
Hydrogen sulfide (H_2S)	4	−0.1	−2.5
Isopropanol	20	0	0
Methane (CH_4)	5000	0	0
Nitrogen dioxide (NO_2)	0.5	0.1	20
Ozone (O_3)	0.3	0.3	100
Perchloroethylene	50	0	0
Propane	5000	0	0
Sulfur dioxide (SO_2)	10	0	0
Toluene	50	0	0

2.2. Intermittent Plume Exposure and Sensor Response Times

In-situ measurements of SO_2 at the volcano crater-rim typically exhibit a high temporal variability. Rapid variations in local gas concentrations are often observed near to the volcanic source due to the complex local wind fields that advect the volcanic gases towards and away from the sensors. To characterize the emissions, the correlation in the time-series from two sensors (e.g., SO_2 and H_2S) is analyzed to derive a gas ratio, e.g., H_2S/SO_2. However, the response times of Multi-Gas and other sensors are not all identical, which can introduce uncertainty in the volcanic gas measurement [42]. Rapid variations in gas concentrations are also a challenge to the Aeroqual instrument that makes measurements of ozone at one-minute resolution. As mentioned in Section 2.1 above, the instrument sensor is sensitive to other gases, including (volcanic) H_2S. These cross-sensitivities have been individually quantified by Aeroqual in laboratory experiments at constant gas abundance, Table 1. However, field-deployment of the Aeroqual instrument at volcanoes can expose it to time-varying abundances of volcanic gases. Very large and rapid temporal variations in interfering gas abundances may lead to anomalous values in the ozone measurement that are difficult to correct in data post-processing (see Section 3, Materials and Methods). A potential solution is to attempt to perform measurements under conditions of more sustained exposure to the volcanic plume. Such conditions are rather uncommon due to the local wind-field conditions near-to-source and because buoyant volcanic plumes tend to become elevated above ground. Here, we present in-situ observations of SO_2, H_2S, and ozone in the near-source and near-downwind Kīlauea (Halema'uma'u and Pu'u 'Ō'ō, HI, USA) plumes obtained during two rare periods of relatively sustained plume exposure (>10 min at "plume strengths" of 5–10 ppmv SO_2) on 3 September 2007 and 19 July 2008.

3. Materials and Methods

The Aeroqual portable ozone monitor is a hand-held instrument for measuring ozone at low mixing ratios at one-minute resolution using a tungsten oxide (WO_3) sensor. The range is 0 to 150 ppbv and a reported resolution of 1 ppbv, accuracy $<\pm5$ ppbv (datasheets: www.Aeroqual.com [41]). The WO_3 sensor is highly sensitive to ozone, but a major challenge to its use for atmospheric ozone measurements is sensitivity drift. The Aeroqual instrument uses propriety software to correct for WO_3 sensitivity drift that involves different phases of sensor operation during the one-minute measurement period. Cross-sensitivities to the ozone measurement are reported by Aeroqual for fixed gas concentrations (such as for H_2S, Table 1), however, the interferences caused by time-varying gas exposure under time-varying sensor operation are more complex to characterize or quantitatively account for in data post-processing. Therefore, this study focuses on Aeroqual measurements during some (rare) periods of relatively sustained plume exposure, and relies on co-measurements of SO_2 and H_2S to aid interpretation of the Aeroqual ozone observations.

Volcanic gases were measured at 1 Hz resolution using the pumped Multi-Gas system described by Roberts et al. [43] that includes Alphasense Ltd (Essex, UK) electrochemical sensors SO_2-AF and H_2S-A1 to measure SO_2 and H_2S respectively. Reported resolution of the sensors is <0.1 ppmv (SO_2), <0.05 ppmv (H_2S), and response time is <35 s (www.Alphasense.com [44]). The sensor current output was converted into ppmv mixing ratio time-series using sensitivities and cross-sensitivities determined from laboratory calibrations prior to the fieldwork. The H_2S-A1 sensor exhibits a positive interference to SO_2 that was subtracted in the data post-processing (using SO_2 simultaneously measured by the SO_2-AF sensor and calibration (cross)-sensitivities). Further details on the instrument and data analysis are provided by Roberts et al. [43].

Kīlauea volcano was emitting three plumes during the fieldwork: an emission from the Halema'uma'u summit crater (measured prior to the 2008 appearance of a lava-lake), the emission from Pu'u 'Ō'ō (east rift zone vents), and an emission resulting from lava contact with sea-water (not sampled). Field-measurements were made at relatively sustained plume exposure on 3 September 2007 to sample Halema'uma'u crater-rim emissions and on 19 July 2008 sampling grounding plume downwind from Pu'u 'Ō'ō, on chain of craters road. This latter site is about 10 km distance from the Pu'u 'Ō'ō vent emission source. Linear regressions were performed to determine H_2S/SO_2 and $\Delta O_3/\Delta SO_2$ using robust-fit algorithms that yield similar results to least-squares but are less affected by outliers. The reported linear model trends exhibit p-values < 0.05 that indicate significant explanatory power. The R^2 coefficient of determination (ratio of the explained variation to the total variation) was calculated using coefficients of correlation and is also provided in the Results.

4. Results

4.1. Volcanic SO₂, H₂S, and Ozone Measured in the Halema'uma'u Crater Rim Emission Plume

Time-series of volcanic SO_2 and H_2S gas abundances measured at Halema'uma'u crater-rim are well-correlated (p-value < 0.05 and R-squared $= 0.52$), and linear-regression finds a H_2S/SO_2 molar ratio of 0.030, see Figure 1. This result confirms the low H_2S content of the Kīlauea plume, although the H_2S/SO_2 molar ratio is somewhat higher than has been reported previously [39]. This might reflect changing magmatic conditions over time. The scatter in the data is most likely due to instrument response times [41]. Nevertheless, a H_2S measurement is still possible, due to the relatively sustained exposure (e.g., around 10 ppmv SO_2 for tens of minutes). The statistical uncertainty in the gas ratio is very low ($<10^{-3}$) due to the very large number of data points. The actual measurement uncertainty is probably somewhat higher as discussed in Section 4.2.

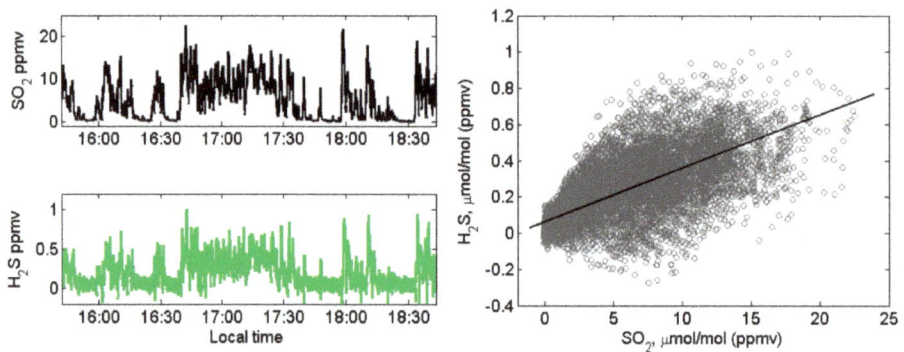

Figure 1. Volcanic plume SO_2 and H_2S abundances measured at 1 Hz resolution in the crater-rim emissions from Halema'uma'u on 3 September 2007. These two volcanic gas time-series are well-correlated; a scatter plot with linear regression yields the H_2S/SO_2 molar ratio of 0.030 mol/mol. p-value < 0.05 and R-squared $= 0.52$.

The volcanic SO_2 time-series was filtered to make one-minute averaged data and these are compared to the ozone time-series (1 min resolution) in Figure 2. The measured ozone abundance fluctuates around 20 to 40 ppbv with a few (anomalous) zero data points. No clear trend is visible in the time-series although a scatter plot of the SO_2, and ozone data show a weak anti-correlation (p-value < 0.05 and R-squared $= 0.13$). The gradient of the linear regression is -0.72 (± 0.13) $\times 10^{-3}$ mol/mol. This measured $\Delta O_3/\Delta SO_2$ closely matches the predicted Aeroqual instrument response to a 2.5% H_2S interference for volcanic plume with $H_2S/SO_2 = 0.03$ mol/mol ($-2.5/100 \times 0.03 = -0.75 \times 10^{-3}$ mol/mol). Accounting for the H_2S interference leads to the deduction in that actual in-plume ozone concentrations were very close to ambient levels in the emission plume at Halema'uma'u crater rim.

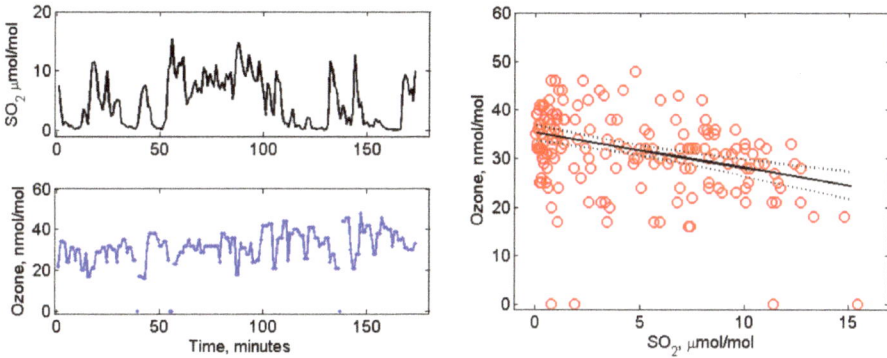

Figure 2. Volcanic plume SO_2 and ozone abundances measured at 1 min resolution in the crater-rim emissions from Halemaʻumaʻu on 3 September 2007. The data show a degree of scatter (with some zero ozone observations that are likely instrument anomalies) and weak anti-correlation between measured ozone and SO_2. The linear regression has gradient $\Delta O_3/\Delta SO_2 = -0.72\ (\pm0.13) \times 10^{-3}$ mol/mol with intercept corresponding to 35 nmol/mol ambient ozone. p-value < 0.05 and R-squared $= 0.13$. Dashed lines denote regression trend with coefficients $\pm 2 \times$ standard error (95% CI).

4.2. Volcanic SO_2, H_2S, and Ozone Measured in the Near-Downwind Plume from Puʻu ʻŌʻō

Time-series of volcanic SO_2 and H_2S gas abundances measured in the plume ~10 km downwind from Puʻu ʻŌʻō are also well-correlated (p-value < 0.05 and R-squared $= 0.50$), and linear-regression finds a H_2S/SO_2 molar ratio of 0.034, Figure 3. Visually, the data appear less scattered compared to those in Figure 1, reflecting improved instrument performance when exposed to more slowly fluctuating gas concentrations. However, instrument noise becomes important at these low H_2S abundances in relatively dilute plume, leading to similar R^2 as for Figure 1. The gas ratio is slightly higher than found for the crater-rim emissions from the Halemaʻumaʻu ($H_2S/SO_2 = 0.030$ mol/mol) which could be due to (i) slight differences in the H_2S/SO_2 emitted from Puʻu ʻŌʻō and Halemaʻumaʻu despite their common magma source; (ii) partial atmospheric oxidation of SO_2 in the Puʻu ʻŌʻō plume by up to 14% (this would indicate a relatively fast in-plume SO_2 oxidation); or (iii) uncertainty in the Multi-Gas measurement.

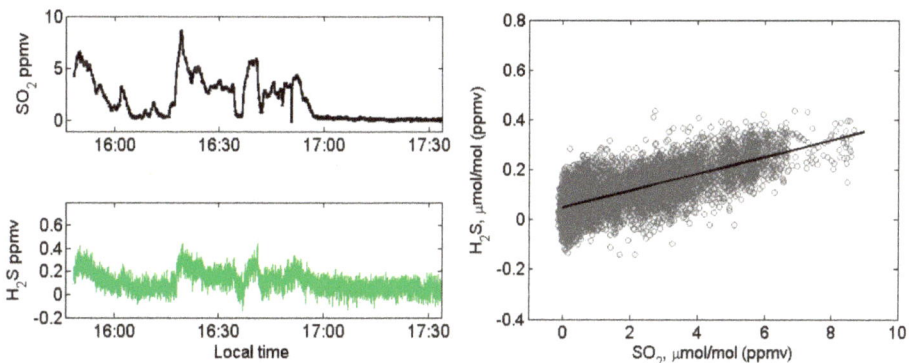

Figure 3. Volcanic plume SO_2 and H_2S abundances measured at 1 Hz resolution in the grounding plume ~10 km from the Puʻu ʻŌʻō emission source on 19 July 2008. These two volcanic gas time-series are well-correlated; a scatter plot with linear regression yields an H_2S/SO_2 molar ratio of 0.0335 mol/mol. p-value < 0.05 and R-squared $= 0.50$.

The one-minute averaged SO_2 time-series is compared to the ozone measurement (at 1 min resolution) in Figure 4. The measured ozone abundance fluctuates between 30 and 50 ppbv. The last 20 min of the time-series appear mostly plume-free (low SO_2) so show the natural variability in ozone abundance. There were no (anomalous) zero data points. A clear tendency (seen particularly in the first 50 min) is that ozone decreases in conjunction with maximum peaks in SO_2. A scatter plot of the SO_2 and ozone data confirms this anti-correlation (p-value < 0.05 and R-squared $= 0.30$). The R^2 is higher than found at the crater-rim (Figure 2), i.e., the anti-correlation of ozone to SO_2 explains a greater ratio of the variation in ozone compared to total variation for the Pu'u 'Ō'ō downwind plume measurements than for that of the Halema'uma'u crater-rim measurements. The gradient of the linear regression is $-1.15 \, (\pm 0.18) \times 10^{-3}$ mol/mol. This measured $\Delta O_3/\Delta SO_2$ exceeds the Aeroqual instrument response to volcanic H_2S calculated as -0.84×10^{-3} mol/mol for a plume with H_2S/SO_2 molar ratio of 0.034. Accounting for the negative H_2S interference on measured ozone at $H_2S/SO_2 = 0.034$ suggests an actual in-plume $\Delta O_3/\Delta SO_2$ of $-0.31 \, (\pm 0.18) \times 10^{-3}$ mol/mol.

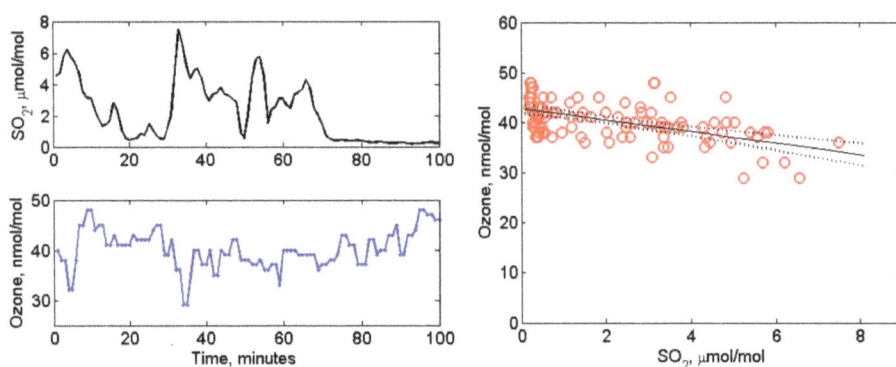

Figure 4. Volcanic plume SO_2 and ozone abundances measured at 1 min resolution in grounding plume on a chain of craters road on 19 July 2008. This location is approximately 10 km downwind of the Pu'u 'Ō'ō volcanic emission source. Peak maxima in the volcanic SO_2 time-series correspond to clearly defined minima in the ozone time-series. The anti-correlation between ozone and SO_2 is confirmed by scatter-plot; the linear regression has gradient $\Delta O_3/\Delta SO_2 = -1.15 \, (\pm 0.18) \times 10^{-3}$ mol/mol with intercept corresponding to 43 nmol/mol ambient ozone. Dashed lines denote regression trend with coefficients $\pm 2 \times$ standard error (95% CI). p-value < 0.05 and R-squared $= 0.30$. This measured $\Delta O_3/\Delta SO_2$ exceeds the Aeroqual instrument response to volcanic H_2S, therefore, it indicates an ozone depletion. See text for further details and quantification.

A source of uncertainty in the ozone depletion measurement is variation in background ozone, which appears to descend then ascend during the time-period, and also exhibits short-term variability. Additional uncertainties can arise from uncertainty in the interference from H_2S, for example, an error of ± 0.005 in the measured H_2S/SO_2 molar ratio would induce an uncertainty in the detected $\Delta O_3/\Delta SO_2$ of $\pm 0.125 \times 10^{-3}$ mol/mol, whilst an error of ± 0.01 would induce an uncertainty in $\Delta O_3/\Delta SO_2$ of $\pm 0.25 \times 10^{-3}$ mol/mol. Nevertheless, results from the crater-rim observations suggested that H_2S/SO_2 measured by the Multi-Gas sensors is fully consistent with the expected interference. In summary, the data show that ozone in the downwind Pu'u 'Ō'ō plume was likely slightly depleted below ambient levels but only by a small magnitude.

5. Discussion and Conclusion: Ozone Depletion from Halogen-Poor to Halogen-Rich Volcanic Plumes

Measurements of ozone in volcanic plumes are challenging. This study demonstrates the need to consider the interference of H_2S on Aeroqual measurements of ozone, the need for sustained plume exposure (as 1 min measurements can be affected by rapid changes in gas concentration), and also illustrates natural variability in background ozone that can act to mask an observable ozone depletion signature. Nevertheless, the Aeroqual measurements provide a useful constraint on the magnitude of ozone depletion in the Kīlauea volcanic plumes at short distances from the source. In particular, the observations provide a valuable end-member example that quantifies ozone in the plume of a low halogen emitter. This contrasts to studies to date that mostly focused on volcanic plumes with higher halogen contents. To interpret the field-data, ozone depletion is scaled relative to co-measured SO_2 in order to distinguish between chemistry and plume dilution effects, and it is interpreted in terms of distance or travel time downwind and the bromine emission.

For the Kīlauea plumes, ozone was shown to be close to ambient levels during plume exposures of up to 10 ppmv SO_2 at the Halema'uma'u crater-rim, and they were only slightly depleted in the grounding plume at around 10 km downwind from Pu'u 'Ō'ō (e.g., by up to 2 ppbv measured at 5–8 ppmv SO_2). The measurements indicate that the volcano emission has only a limited impact on tropospheric ozone, as might be expected for a low halogen emitter. This contrasts with findings for higher halogen emitters. A study of Mt Etna plume ozone by Surl et al. [24], for example, found 11 ppbv ozone depletion measured in plume exposure of 9 ppmv SO_2 at just 300 m downwind. Surl et al. [24] also suggest that ozone in the Mt Etna crater-rim plume was depleted below background, finding from their near-source field-campaign that $\Delta O_3/\Delta SO_2 = (-1.02 \pm 0.07) \times 10^{-5}$ s^{-1} with an intercept of $(-6.2 \pm 0.05) \times 10^{-4}$ mol/mol. A much greater depletion of tropospheric ozone (tens of ppbv ozone loss in plume of 1 ppmv SO_2) was observed by Kelly et al. [23] through spatial mapping of the Mt Redoubt 2009 eruption plume. The $\Delta O_3/\Delta SO_2$ data from Kelly et al. [23] is shown in Figure 5 as a function of distance downwind for two flight campaigns. A smaller ozone depletion was observed in August compared to June 2009. We focus our comparison on the June 2009 data for which the volcanic halogen emission was also quantified (just one day prior to the aircraft campaign) that yields a linear regression in $\Delta O_3/\Delta SO_2$ of $(-7.2 \pm 1.3) \times 10^{-3}$ mol/mol km^{-1}, with intercept close to zero (within statistical uncertainty). Table 2 summarizes these field measurements and interconverts $\Delta O_3/\Delta SO_2$ per unit travel time or per distance downwind using available or estimated wind-speeds. Not shown in Table 2 are aircraft studies of the 2010 Eyjafjallajökull eruption plume over Europe [36,37] that found large ozone depletions in the far-field dispersed plume but which are difficult to interpret because the halogen emission was poorly constrained and the assumption of SO_2 as a quasi-plume-tracer (in $\Delta O_3/\Delta SO_2$) may not be valid.

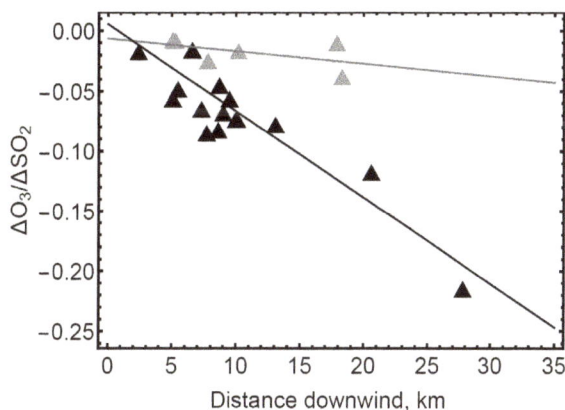

Figure 5. Ozone depletion (scaled to co-measured SO_2) in the Mt Redoubt 2009 eruption plume as a function of distance downwind during the aircraft campaigns on 21 June (black triangles) and 19 August (grey triangles) 2009. Data from Kelly et al. [23]. The gradient of the linear regression for 21 June is $(-7.2 \pm 1.3) \times 10^{-3}$ mol/mol km^{-1}.

Table 2. Studies of ozone depletion in volcanic plumes where the halogen emission was also reported: Br/S in the emission and $\Delta Ozone/\Delta SO_2$ per km downwind or per second of travel time.

Volcano	Br/S (mmol/mol)	Distance (km)	Wind Speed * m/s	Time (min)	$\Delta O_3/\Delta SO_2$ Per km $(km^{-1}) \times 10^{-3}$	$\Delta O_3/\Delta SO_2$ Per Second $(s^{-1}) \times 10^{-5}$	Reference
Mt Redoubt 2009 eruption	4.1	2.4–27.8	4.4	9–105	-7.2 ± 1.3	-3.2 ± 0.6	[23] (Figure 5 this study)
Mt Etna passive degassing 2012	0.613	0.15–0.40	~2.2	1–4	(-4.4 ± 0.3)	-1.02 ± 0.07	[24]
Kīlauea plumes 2007–2008	0.029 ± 0.025	~10	~5	~30	Up to -0.03	(Up to 0.02)	This study $(\Delta O_3/\Delta SO_2)$ [40] (2012) (Br/S)

* Wind-speed reported directly by Kelly et al. [23] for the study of Mt Redoubt, calculated from the reported 380 m distance downwind equivalent to about 175 s travel time in the study of Mt Etna by Surl et al. [24], and estimated from Hilo radiosonde data for the study of Kīlauea. Values of ozone depletion rate in brackets were obtained by interconverting time and distance data using the wind-speeds of Table 2.

Figure 6 presents the $\Delta O_3/\Delta SO_2$ scaled to distance or travel time downwind (i.e., rate of ozone depletion) as a function of Br/S in the volcano emission for the three case studies of Table 2 (Kīlauea, Mt Etna, and Mt Redoubt). The data show that ozone depletion is a non-linear function of the bromine emission; greater ozone loss occurs for a higher Br/S emission, as expected. However, the rate of ozone depletion for the volcano with the highest Br/S emission (Mt Redoubt) is disproportionately slow, i.e., the Mt Redoubt plume is found to be less efficient at destroying tropospheric ozone than would be expected. This non-linearity is initially a surprising result given the role of volcanic bromine in the depletion of tropospheric ozone through a "bromine explosion" that is autocatalytic (i.e., self-enhancing). A potential reason could be the complete titration of plume ozone (thereby preventing any further ozone depletion), but this is ruled out based on the measurements of Kelly et al. [23]; ozone in the Mt Redoubt plume was reduced to less than ambient levels but was not fully depleted. Instead, an explanation is provided by atmospheric box modelling of the near-source plume chemistry. Model studies point to non-linearities in the conversion of emitted HBr into reactive bromine species, depending on the bromine content of the emissions [24,29]. For high Br/S emissions, not all of the emitted HBr may become converted into reactive bromine. This was shown in a model sensitivity study by Roberts et al. [29] for Mt Etna that predicts HBr is rapidly and fully converted into reactive forms over tens of minutes for an emission with HBr/SO_2 molar ratio of 7.4×10^{-4}, but HBr is more slowly and only partially (~50% after one hour) converted into reactive bromine for an emission with a higher HBr/SO_2 molar ratio of 2.4×10^{-3}. Furthermore, the model simulation of the Mt Redoubt 2009 eruption plume by Kelly et al. [23] predicts that only 30% of emitted HBr was converted into reactive forms, for an emission with even higher HBr/SO_2 molar ratio of 4.1×10^{-3}. However, another difference between the plumes is the emission flux, with 4.3 kg/s SO_2 flux for Mt Redoubt and ~20 kg/s SO_2 flux for the Mt Etna simulation. The net conversion of emitted volcanic HBr into reactive bromine reflects the balance of plume chemistry processes (that form reactive halogens from HBr and that can re-form HBr from reactive halogens) as a function of plume properties such as gas flux, aerosol, plume-air mixing, etc., and is also a function of time. For plumes younger than one minute, recent observations by Rüdiger et al. [45] suggest reactive bromine accounts for less than 44% of total bromine, consistent with the models. The modeled plume Br-speciation predicted by Roberts et al. [23] and [29] are shown in Figure 7, illustrating the proportion of emitted HBr converted into reactive bromine during the first hours of plume chemical evolution. These model outputs are used to re-evaluate the non-linear trend in Figure 6. When the Br/S content of the emissions is adjusted to reflect the modeled plume "reactive bromine"/S (e.g., reduced to 30% of the total emitted Br/S for Mt Redoubt: open triangle, Figure 6), there is more linearity (broadly proportional trend) between the volcanic (reactive) bromine/S and the rate of plume ozone depletion. It is nevertheless expected that other variables, e.g., aerosol and plume-air mixing rate, may affect $\Delta O_3/\Delta SO_2$ depending on the volcanic-meteorological setting.

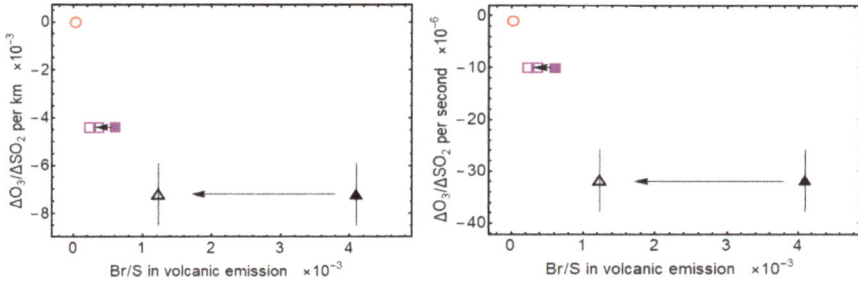

Figure 6. Depletion in ozone scaled to volcanic SO_2 (as a quasi-plume-tracer) and scaled with (i) km distance downwind or (ii) seconds travel time downwind, plotted as a function of the bromine/SO_2 molar ratio in the emission. Data are available for three volcanic systems: Kīlauea (red circle, this study), Mt Etna (purple square, [24]), and Mt Redoubt (black triangle, [23]), in ascending order of Br/S in their emissions. Arrow and open triangle denote adjustments in the Br/S content that accounts for partial rather than full conversion of emitted HBr into reactive bromine. See text for details.

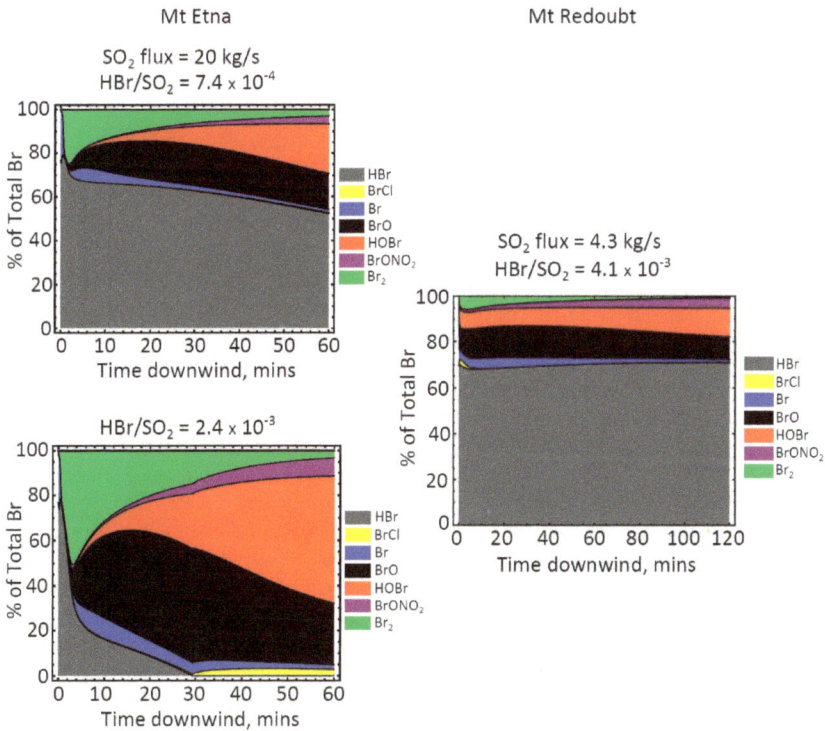

Figure 7. Model simulated reactive bromine speciation in the plumes of Mt Etna and Mt Redoubt, from Roberts et al. [29], and Kelly et al. [23] illustrating varying degrees (complete, partial) of conversion of HBr into reactive bromine. Reproduced with permission from Kelly et al. [23], JVGR; published by Elsevier.

In conclusion, chemistry in the near-source volcanic plume exerts an important influence on the fate of volcanic halogens entering the troposphere. Plume chemistry converts the volcanic HBr emission into reactive bromine, causing the depletion of tropospheric ozone. Measurements of ozone in the near-source volcanic plume are challenging to make but they can provide useful observational constraints on the volcanic plume halogen chemistry. This study presents observations of ozone in the volcanic plumes from Kīlauea, a low-halogen emitter. Ozone was close to ambient levels in crater-rim Halemaʻumaʻu emissions and only slightly depleted in the plume 10 km downwind from Puʻu ʻŌʻō. This contrasts with observations of much larger ozone depletion (tens of ppbv) in the near-source and near-downwind plumes of two higher halogen emitters: Mt Etna and Mt Redoubt [23,24]. The available observations combined with numerical modelling of the plume chemistry suggest that the volcanic Br/S emission and its (partial or complete) conversion into reactive bromine are key controls in the depletion of tropospheric ozone downwind from the volcano. Characterizing the near-source volcanic plume chemistry is thus an essential step to quantifying the fate and downwind impacts of volcanic emissions to the atmosphere.

Acknowledgments: This work was funded by the Orleans Labex VOLTAIRE (VOLatils-Terre Atmosphère Interactions—Ressources et Environnement) ANR-10-LABX-100-01, and a UK NERC studentship, and acknowledges MDPI for the invitation to cover publication fees. I would like to thank the following people for very useful discussions on volcano science, sensors, and ozone observations: Clive Oppenheimer, Rod Jones, John Saffell (Alphasense Ltd., Essex, UK), Peter Kelly, Luke Surl; field-collaborators Tamsin Mather and Mel Witt; and the Hawaiʻi Volcano Observatory. I also thank the editor P. Sellitto and two anonymous reviewers.

Conflicts of Interest: The authors declare no conflict of interest. The founding sponsors had no role in the design of the study; in the collection, analyses, or interpretation of data; in the writing of the manuscript, and in the decision to publish the results.

References

1. Carn, S.A.; Fioletov, V.E.; McLinden, C.A.; Li, C.; Krotkov, N.A. A decade of global volcanic SO_2 emissions measured from space. *Sci. Rep.* **2017**, 44095. [CrossRef] [PubMed]
2. Stoffell, M.; Khodri, M.; Corona, C.; Guillet, S.; Poulain, V.; Bekki, S.; Guiot, J.; Luckman, B.H.; Oppenheimer, C.; Lebas, N.; et al. Estimates of volcanic-induced cooling in the Northern Hemisphere over the past 1500 years. *Nat. Geosci.* **2015**, *8*, 784–788. [CrossRef]
3. Sellitto, P.; Zanetel, C.; di Sarra, A.; Salerno, G.; Tapparo, A.; Meloni, D.; Pace, G.; Caltabiano, T.; Briole, P.; Legras, B. The impact of Mount Etna sulfur emissions on the atmospheric composition and aerosol properties in the central Mediterranean: A statistical analysis over the period 2000–2013 based on observations and Lagrangian modelling. *Atmos. Environ.* **2017**, *148*, 77–88. [CrossRef]
4. Berthet, G.; Jégou, F.; Catoire, V.; Krysztofiak, G.; Renard, J.B.; Bourassa, A.E.; Degenstein, D.A.; Brogniez, C.; Dorf, M.; Kreycy, S.; et al. Impact of a moderate volcanic eruption on chemistry in the lower stratosphere: Balloon-borne observations and model calculations. *Atmos. Chem. Phys.* **2017**, *17*, 2229–2253. [CrossRef]
5. Ivy, D.J.; Solomon, S.; Kinnison, D.; Mills, M.J.; Schmidt, A.; Neely, R.R., III. The influence of the Calbuco eruption on the 2015 Antarctic ozone hole in a fully coupled chemistry-climate model. *Geophys. Res. Lett.* **2017**, *44*, 2556–2561. [CrossRef]
6. Aiuppa, A.; Baker, D.R.; Webster, J.D. Halogens in volcanic systems. *Chem. Geol.* **2009**, *263*, 1–18. [CrossRef]
7. Carn, S.; Clarisse, L.; Prata, A.J. Multi-decadal satellite measurements of global volcanic degassing. *J. Volcanol. Geotherm. Res.* **2016**, *311*, 99–134. [CrossRef]
8. Theys, N.; Van Roozendael, M.; Dils, B.; Hendrick, F.; Hao, N.; De Mazière, M. First satellite detection of volcanic bromine monoxide emission after the Kasatochi eruption. *Geophys. Res. Lett.* **2009**, *36*. [CrossRef]
9. Theys, N.; De Smedt, I.; Van Roozendael, M.; Froidevaux, L.; Clarisse, L.; Hendrick, F. First satellite detection of volcanic OClO after the eruption of Puyehue-Cordón Caulle. *Geophys. Res. Lett.* **2014**, *41*, 667–672. [CrossRef]
10. Schönhardt, A.; Richter, A.; Theys, N.; Burrows, J.P. Space-based observation of volcanic iodine monoxide. *Atmos. Chem. Phys.* **2017**, *17*, 4857–4870. [CrossRef]

11. Rose, W.I.; Millard, G.A.; Mather, T.A.; Hunton, D.E.; Anderson, B.; Oppenheimer, C.; Thornton, B.F.; Gerlach, T.; Viggiano, A.A.; Kondo, Y.; et al. Atmospheric chemistry of a 33–34 hour old volcanic cloud from Hekla Volcano (Iceland): Insights from direct sampling and the application of chemical box modeling. *J. Geophys. Res.* **2006**, *111*, D20206. [CrossRef]

12. Millard, G.A.; Mather, T.A.; Pyle, D.M.; Rose, W.I.; Thornton, B. Halogen emissions from a small volcanic eruption: Modeling the peak concentrations, dispersion, and volcanically induced ozone loss in the stratosphere. *Geophys. Res. Lett.* **2006**, *33*, L19815. [CrossRef]

13. Lurton, T.; Jégou, F.; Berthet, G.; Renard, J.-B.; Clarisse, L.; Schmidt, A.; Brogniez, C.; Roberts, T. Model simulations of the chemical and aerosol microphysical evolution of the Sarychev Peak 2009 eruption cloud compared to in-situ and satellite observations. *Atmos. Chem. Phys. Discuss.* **2017**. [CrossRef]

14. Bobrowski, N.; Honninger, G.; Galle, B.; Plat, U. Detection of bromine monoxide in a volcanic plume. *Nature* **2003**, *423*, 273–276. [CrossRef] [PubMed]

15. Boichu, M.; Oppenheimer, C.; Roberts, T.J.; Tsanev, V.; Kyle, P. On bromine, nitrogen oxides and ozone depletion in the tropospheric plume of Erebus volcano (Antarctica). *Atmos. Environ.* **2011**, *45*, 3856–3866. [CrossRef]

16. Lübcke, P.; Bobrowski, N.; Arellano, S.; Galle, B.; Garzón, G.; Vogel, L.; Platt, U. BrO/SO$_2$ molar ratios from scanning DOAS measurements in the NOVAC network. *Solid Earth* **2014**, *5*, 409–424. [CrossRef]

17. Hörmann, C.; Sihler, H.; Bobrowski, N.; Beirle, S.; de Vries, M.P.; Platt, U.; Wagner, T. Systematic investigation of bromine monoxide in volcanic plumes from space by using the GOME-2 instrument. *Atmos. Chem. Phys.* **2013**, *13*, 4749–4781. [CrossRef]

18. Bobrowski, N.; von Glasow, R.; Aiuppa, A.; Inguaggiato, S.; Louban, I.; Ibrahim, O.W.; Platt, U. Reactive halogen chemistry in volcanic plumes. *J. Geophys. Res.* **2007**, *112*, D06311. [CrossRef]

19. General, S.; Bobrowski, N.; Pöhler, D.; Weber, K.; Fischer, C.; Platt, U. Airborne I-DOAS measurements at Mt. Etna: BrO and OClO evolution in the plume. *J. Volcanol. Geotherm. Res.* **2015**, *300*, 175–186. [CrossRef]

20. Gliß, J.; Bobrowski, N.; Vogel, L.; Pöhler, D.; Platt, U. OClO and BrO observations in the volcanic plume of Mt. Etna—Implications on the chemistry of chlorine and bromine species in volcanic plumes. *Atmos. Chem. Phys.* **2015**, *15*, 5659–5681. [CrossRef]

21. Donovan, A.; Tsanev, V.; Oppenheimer, C.; Edmonds, M. Reactive halogens (BrO and OClO) detected in the plume of Soufrière Hills Volcano during an eruption hiatus. *Geochem. Geophys. Geosyst.* **2014**, *15*, 3346–3363. [CrossRef]

22. Oppenheimer, C.; Tsanev, V.I.; Braban, C.F.; Cox, R.A.; Adams, J.W.; Aiuppa, A.; Bobrowski, N.; Delmelle, P.; Barclay, J.; McGonigle, A.J. BrO fomation in volcanic plumes. *Geochim. Cosmochim. Acta* **2006**, *70*, 2935–2941. [CrossRef]

23. Kelly, P.J.; Kern, C.; Roberts, T.J.; Lopez, T.; Werner, C.; Aiuppa, A. Rapid chemical evolution of tropospheric volcanic emissions from Redoubt Volcano, Alaska, based on observations of ozone and halogen-containing gases. *J. Volcanol. Geotherm. Res.* **2013**, *259*, 317–333. [CrossRef]

24. Surl, L.; Donohoue, D.; Aiuppa, A.; Bobrowski, N.; von Glasow, R. Quantification of the depletion of ozone in the plume of Mount Etna. *Atmos. Chem. Phys.* **2015**, *15*, 2613–2628. [CrossRef]

25. Roberts, T.J.; Braban, C.F.; Martin, R.S.; Oppenheimer, C.; Adams, J.W.; Cox, R.A.; Jones, R.L.; Griffiths, P.T. Modelling reactive halogen formation and ozone depletion in volcanic plumes. *Chem. Geol.* **2009**, *263*, 151–163. [CrossRef]

26. Von Glasow, R. Atmospheric chemistry in volcanic plumes. *Proc. Natl. Acad. Sci. USA* **2010**, *107*, 6594–6599. [CrossRef] [PubMed]

27. Jourdain, L.; Roberts, T.J.; Pirre, M.; Josse, B. Modeling the reactive halogen plume from Ambrym volcano and its impact on the troposphere with the CCATT-BRAMS mesoscale model. *Atmos. Chem. Phys.* **2016**, *16*, 12099–12125. [CrossRef]

28. Von Glasow, R.; Bobrowski, N.; Kern, C. The effects of volcanic eruptions on atmospheric chemistry. *Chem. Geol.* **2009**, *263*, 131–142. [CrossRef]

29. Roberts, T.J.; Martin, R.S.; Jourdain, L. Reactive halogen chemistry in Mt Etna's volcanic plume: The influence of total Br, high temperature processing, aerosol loading and plume-air mixing (volcanic emissions flux). *Atmos. Chem. Phys.* **2014**, *14*, 11201–11219. [CrossRef]

30. Roberts, T.J.; Vignelles, D.; Liuzzo, M.; Giudice, G.; Aiuppa, A.; Coltelli, M.; Salerno, G.; Chartier, M.; Couté, B.; Berthet, G.; et al. The primary volcanic aerosol emission from Mt Etna: Size-resolved particles with SO_2 and role in plume reactive halogen chemistry. *Geochim. Cosmochim. Acta* **2018**, *222*, 74–93. [CrossRef]

31. Flentje, H.; Claude, H.; Elste, T.; Gilge, S.; Köhler, U.; Plass-Dülmer, C.; Steinbrecht, W.; Thomas, W.; Werner, A.; Fricke, W. The Eyjafjallajökull eruption in April 2010—Detection of volcanic plume using in-situ measurements, ozone sondes and lidar-ceilometer profiles. *Atmos. Chem. Phys.* **2010**, *10*, 10085–10092. [CrossRef]

32. Stith, J.L.; Hobbs, P.V.; Radke, L.F. Airborne particle and gas measurements in the emissions from six volcanoes. *J. Geophys. Res.* **1978**, *83*, 4009–4017. [CrossRef]

33. Fruchter, J.S.; Robertson, D.E.; Evans, J.C.; Oslen, K.B.; Lepel, E.A.; Laul, J.C.; Abel, K.H.; Sanders, R.W.; Jackson, P.O.; Wogman, N.S.; et al. Mount St. Helens ash from the 18 May 1980 eruption: Chemical, physical, mineralogical and biological properties. *Science* **1980**, *209*, 1116–1125. [CrossRef] [PubMed]

34. Hobbs, P.V.; Tuell, J.P.; Hegg, D.A.; Radke, L.F.; Eltgroth, M.W. Particles and gases in the emissions from the 1980–1981 volcanic eruptions of Mt. St. Helens. *J. Geophys. Res.* **1982**, *87*, 11062–11086. [CrossRef]

35. Vance, A.; McGonigle, A.J.S.; Aiuppa, A.; Stith, J.L.; Turnbull, K.; von Glasow, R. Ozone depletion in tropospheric volcanic plumes. *Geophys. Res. Lett.* **2010**, *37*, L22802. [CrossRef]

36. Schumann, U.; Weinzierl, B.; Reitebuch, O.; Schlager, H.; Miniki, A.; Forster, C.; Baumann, R.; Saile, T.; Graf, K.; Mannstein, H.; et al. Airborne observations of the Eyjafjalla volcano ash cloud over Europe during air space closure in April and May 2010. *Atmos. Chem. Phys.* **2011**, *11*, 2245–2279. [CrossRef]

37. Oppenheimer, C.; Kyle, P.; Eisele, F.; Crawford, J.; Huey, G.; Tanner, D.; Saewung, K.; Mauldin, L.; Blake, D.; Beyersdorf, A.; et al. Atmospheric chemistry of an Antarctic volcanic plume. *J. Geophys. Res.* **2010**, *115*, D04303. [CrossRef]

38. Aiuppa, A.; Federico, C.; Giudice, G.; Gurrieri, S. Chemical mapping of a fumarolic field: La Fossa Crater, Vulcano Island (Aeolian Islands, Italy). *Geophys. Res. Lett.* **2005**, *32*, L13309. [CrossRef]

39. Gerlach, T.M. Volcanic sources of tropospheric ozone-depleting trace gases. *Geochem. Geophys. Geosyst.* **2004**, *5*, Q09007. [CrossRef]

40. Mather, T.A.; Witt, M.L.I.; Pyle, D.M.; Quayle, B.M.; Aiuppa, A.; Bagnato, E.; Martin, R.S.; Sims, K.W.W.; Edmonds, M.; Sutton, A.J.; et al. Halogens and trace metal emissions from the ongoing 2008 summit eruption of Kīlauea volcano, Hawai'I. *Geochim. Cosmochim. Acta* **2012**, *83*, 292–323. [CrossRef]

41. Aeroqual Limited. Datasheets for Aeroqual Series 500 Portable Ozone Instrument. Available online: https://www.aeroqual.com/product/series-500-portable-ozone-gas-monitor (accessed on 24 January 2018).

42. Roberts, T.J.; Saffell, J.R.; Dawson, D.H.; Oppenheimer, C.; Lurton, T. Electrochemical sensors applied to pollution monitoring: Measurement error and gas ratio bias—A volcano plume case study. *J. Volcanol. Geotherm. Res.* **2014**, *281*, 85–96. [CrossRef]

43. Roberts, T.J.; Braban, C.; Oppenheimer, C.; Martin, R.S.; Saffell, J.R.; Dawson, D.; Freshwater, R.A.; Griffiths, P.T.; Jones, R.L. Electrochemical sensing of volcanic plumes. *Chem. Geol.* **2012**, *332–333*, 74–91. [CrossRef]

44. Alphasense Sensor Technology Company. Datasheets for Alphasense SO_2-AF and H_2S-A1 Sensors. Available online: http://www.alphasense.com/index.php/safety/downloads/ (accessed on 24 January 2018).

45. Rüdiger, J.; Bobrowski, N.; Liotta, M.; Hoffman, T. Development and application of a sampling method for the determination of reactive halogen species in volcanic gas emissions. *Anal. Bioanal. Chem.* **2017**, *409*, 5975–5985. [CrossRef] [PubMed]

geosciences

MDPI

Review

Pyplis—A Python Software Toolbox for the Analysis of SO₂ Camera Images for Emission Rate Retrievals from Point Sources

Jonas Gliß [1,2,3,*], **Kerstin Stebel** [1], **Arve Kylling** [1], **Anna Solvejg Dinger** [1,4], **Holger Sihler** [5] **and Aasmund Sudbø** [3]

[1] NILU—Norwegian Institute for Air Research, NO-2007 Kjeller, Norway; Kerstin.Stebel@nilu.no (K.S.); Arve.Kylling@nilu.no (A.K.); anna.solvejg.dinger@nilu.no (A.S.D.)
[2] Department of Physics, University of Oslo (UiO), NO-0371 Oslo, Norway
[3] Department of Technology Systems, University of Oslo (UiO), NO-2007 Kjeller, Norway; aasmund.sudbo@its.uio.no
[4] Institute of Environmental Physics (IUP), University of Heidelberg, D-69120 Heidelberg, Germany
[5] Max Planck Institute for Chemistry (MPIC), D-55128 Mainz, Germany; holger.sihler@mpic.de
* Correspondence: jg@nilu.no; Tel.: +47-94885617

Received: 11 October 2017; Accepted: 11 December 2017; Published: 15 December 2017

Abstract: Ultraviolet (UV) SO₂ cameras have become a common tool to measure and monitor SO₂ emission rates, mostly from volcanoes but also from anthropogenic sources (e.g., power plants or ships). Over the past decade, the analysis of UV SO₂ camera data has seen many improvements. As a result, for many of the required analysis steps, several alternatives exist today (e.g., cell vs. DOAS based camera calibration; optical flow vs. cross-correlation based gas-velocity retrieval). This inspired the development of *Pyplis* (**Py**thon **pl**ume **i**maging **s**oftware), an open-source software toolbox written in Python 2.7, which unifies the most prevalent methods from literature within a single, cross-platform analysis framework. *Pyplis* comprises a vast collection of algorithms relevant for the analysis of UV SO₂ camera data. These include several routines to retrieve plume background radiances as well as routines for cell and DOAS based camera calibration. The latter includes two independent methods to identify the DOAS field-of-view (FOV) within the camera images (based on (1) Pearson correlation and (2) IFR inversion method). Plume velocities can be retrieved using an optical flow algorithm as well as signal cross-correlation. Furthermore, *Pyplis* includes a routine to perform a first order correction of the signal dilution effect (also referred to as light dilution). All required geometrical calculations are performed within a 3D model environment allowing for distance retrievals to plume and local terrain features on a pixel basis. SO₂ emission rates can be retrieved simultaneously for an arbitrary number of plume intersections. Hence, *Pyplis* provides a state-of-the-art framework for more efficient and flexible analyses of UV SO₂ camera data and, therefore, marks an important step forward towards more transparency, reliability and inter-comparability of the results. *Pyplis* has been extensively and successfully tested using data from several field campaigns. Here, the main features are introduced using a dataset obtained at Mt. Etna, Italy on 16 September 2015.

Keywords: volcanic gases; SO₂; remote sensing; UV cameras; image processing; analysis software; Python 2.7

1. Introduction

Sulfur dioxide (SO₂) is a toxic gas emitted by anthropogenic and natural sources (e.g., power plants, ships, volcanoes). The pollutant has impacts on the atmosphere both on local and global scales (e.g., particle formation, radiation budget, e.g., [1,2]). Furthermore, the monitoring of SO_2 emissions from active volcanoes can provide insight into the magmatic degassing behaviour and hence plays an important role for the development of new risk assessment approaches (e.g., [3–6], and references therein).

The gas composition of the emission plumes can, for instance, be studied using ground-based passive remote sensing techniques. The column-densities (CDs) of the gases in the plumes are quantified based on the absorption signature carried by scattered sunlight that has penetrated the plume. SO_2-CDs, for instance, can be retrieved at ultraviolet (UV) wavelengths (i.e., around 310 nm) where it exhibits distinct absorption bands. Prominent examples for passive remote sensing instrumentation are the correlation spectrometer (COSPEC, [7]), or instruments based on the technique of Differential Optical Absorption Spectroscopy (DOAS, [8], e.g., [9,10]). Over the past decade, the comparatively young technique of UV SO_2 cameras has gained in importance, since it enables the study of volcanic SO_2 emissions at unprecedented spatial and temporal resolution (e.g., [11–15]). This is particularly helpful to study multiple sources independently (e.g., [16]) or to investigate volcanic degassing characteristics by studying periodicities in the SO_2 emission rates (e.g., [17–19] and references therein). The technique of UV SO_2 cameras has seen remarkable improvements over the past decade (e.g., [20–24]) and can now be considered one of the standard methods for ground-based remote sensing of SO_2 plumes. A drawback, however, is the low spectral resolution, restricting the technique to a single species and furthermore requiring external calibration.

The retrieval of SO_2 emission rates from plume imagery comprises several analysis steps that are summarised in Table 1 and are illustrated in the flowchart shown in Appendix Figure A1. Thanks to ongoing developments, today, researchers can choose between several methods for nearly all of the required steps (e.g., cell vs. DOAS calibration, velocity retrieval using optical flow vs. cross-correlation method).

Available software solutions include Vulcamera [25], the IDL® source code provided by [22] and Plumetrack [26]. The first two programs include routines for cell calibration and cross-correlation based plume velocity retrievals. The IDL® program also includes a routine to perform a first order correction for the signal dilution effect (commonly referred to as light dilution, e.g., [22]). The software Plumetrack provides an interface to calculate gas velocities using an optical flow algorithm and can be applied to pre-calibrated SO_2-CD images in order to retrieve SO_2 emission rates.

Table 1. Analysis blocks for emission rate retrievals (for details see Section 2).

Analysis Block	Quantities	Analysis Options	Section
Geometrical Calculations	Δs		3.1
Plume Background Analysis	$\tau_{on}, \tau_{off}, \tau_{AA}$		3.3
Camera Calibration	S_{SO2}	Cell, DOAS	3.4
Plume Velocity Retrieval	$\langle \vec{v} \cdot \hat{n} \rangle$	Optical flow, cross-correlation	3.5
Emission rate	Φ_{SO2}	Signal dilution correction	3.6, 3.7

Pyplis ([27,28]) is a cross-platform, open-source software toolbox for the analysis of UV SO_2 camera data. The code is written in Python 2.7 and emerged from the idea of a common software platform incorporating the most relevant analysis methods, including recent developments. The most important features of *Pyplis* 1.0.0 are (details follow in Section 3):

1. 3D distance retrievals to plume and local terrain features at pixel-level,
2. several methods to retrieve plume background radiances,
3. cell and DOAS based camera calibration including two independent DOAS FOV search routines,
4. cross-correlation and optical flow based plume velocity retrievals,
5. histogram based correction for ill-posed optical flow vectors in low-contrast image regions,
6. image based correction for the signal dilution effect,

7.   automated emission rate retrievals along linear plume intersections.

Pyplis comes with numerous example scripts providing an easy and comprehensive introduction into the software. The following Section 2 introduces the technique of UV SO_2 cameras and the required analysis steps for SO_2 emission rate retrievals. The implementation of the individual analysis methods is discussed in Section 3.

2. Methodology

2.1. UV SO_2 Cameras

UV SO_2 cameras analyse scattered sunlight that has penetrated a plume containing SO_2 gas. Plume optical densities (ODs) τ are retrieved in two wavelength windows of approximately 10 nm width, using optical bandpass filters. One filter is centred around 310–315 nm, where SO_2 has strong absorption bands (referred to as SO_2 on-band). A second filter is situated around 330 nm, where SO_2 absorption is weak (SO_2 off-band). The latter is used to correct for additional broadband light extinction, for instance, resulting from aerosols or water droplets in the plume. From the retrieved ODs in both channels, the apparent absorbance (AA) of SO_2 (τ_{AA}, e.g., [20]) can be retrieved as

$$\tau_{AA} = \tau_{on} - \tau_{off} = \ln\left(\frac{I_0}{I}\right)_{on} - \ln\left(\frac{I_0}{I}\right)_{off}, \tag{1}$$

where I and I_0 denote the plume and plume background radiances, respectively, in both channels (on, off). Note that the method requires all additional optical densities in the on and off-band regime to be of broadband nature.

2.2. Image Analysis—Retrieval of S_{SO2} Images

AA images are determined from a set of pre-processed (e.g., dark / offset corrected) plume and background images using Equation (1). Next, the AA images are converted into SO_2 column-density (CD) images, where

$$S_{SO2}(i,j) = \int_{\mathcal{C}_{ij}} c(x,y,z)ds \tag{2}$$

denotes the SO_2-CD in the viewing direction $\mathcal{C}_{ij}$ of the image pixel i, j. $c(x,y,z)$ is the concentration distribution of SO_2 in real world coordinates x,y,z (cf. Figure 1a) and $ds = \sqrt{dx^2 + dy^2 + dz^2}$ is the integration differential.

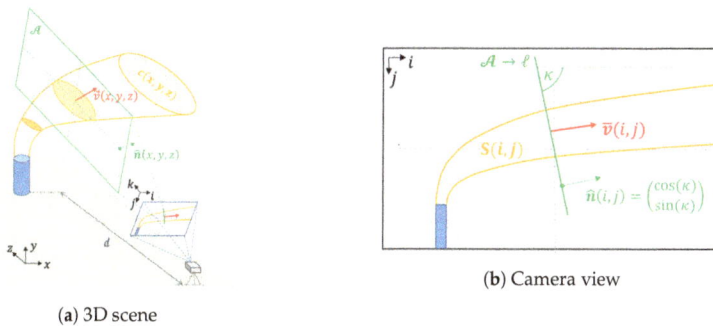

(a) 3D scene

(b) Camera view

Figure 1. Measurement geometry-sketched (**a**) in three dimensions and (**b**) as the camera sees it. The emission plume is indicated in yellow colours, gas velocities in red. The orientation of the plume cross-section (PCS) $\mathcal{A}$ (green colours) is aligned with the camera optical axis k.

The camera calibration (i.e., the conversion of AA values into SO_2-CDs) can be performed either using gas cells (e.g., [11]) or a co-located DOAS spectrometer (e.g., [21]). The latter is more accurate in case aerosols are present in the plume [21,29]. The position and shape of the DOAS FOV within the camera images are required in order to perform the DOAS calibration. The DOAS FOV can either be measured experimentally (e.g., in the lab) or can be retrieved directly from the field data [21,30].

The camera calibration curve depends on the pixel-position in the images (e.g., [21]). This is due to shifts in the filter transmission curves for non-perpendicular illumination (i.e., off-axis image pixels), resulting in an increased SO_2 sensitivity towards the image edges (e.g., [31]). A first order correction for this effect can be performed using a normalised sensitivity correction mask retrieved from cell calibration data, as described in [21]. The impact of this effect can be dramatically reduced by placing the transmission filters between lens and detector rather than in front of the lens [31]. Please note that this effect of an increased SO_2 sensitivity towards the image edges is not to be confused with lens-vignetting effects, as discussed by [31].

2.3. Emission Rate Retrieval

SO_2 emission rates Φ are retrieved from SO_2-CD images by performing a discrete path integration along a suitable plume cross section (PCS) projected into the image plane, for instance a straight line ℓ (illustrated in Figure 1b). Then,

$$\Phi(\ell) = f^{-1} \sum_{m=1}^{M} S_{SO2}(m) \cdot \langle \bar{v}(m) \cdot \hat{n}(m) \rangle \cdot d_{pl}(m) \cdot \Delta s(m) \tag{3}$$

corresponds to the SO_2 emission rate through ℓ, where m denotes one of a total of M sample positions along ℓ in the image plane and Δs is the integration step length, measured in physical distances on the detector. f is the focal length of the camera, d_{pl} is the distance between camera and plume and $\hat{n}$ is the normal of ℓ. $\bar{v}$ is a 2-vector containing projected plume velocities averaged in the viewing direction.

The plume distances d_{pl} can be estimated from the measurement geometry and require information about the camera position and viewing direction as well as the source position and the meteorological wind direction. The gas velocities are usually retrieved from the images directly either using cross-correlation based methods (e.g., [32]) or optical flow algorithms (e.g., [23]).

Equation (3) is equivalent to commonly used retrieval methods (e.g., [11,29]) and is based on the assumption that over or underestimations of the measured quantities S_{SO2} and $\bar{v}$ (e.g., due to non-perpendicular plume transects) cancel out in the emission rate retrieval. This is a valid approximation for typical measurement conditions (i.e., plume nearly perpendicular to the optical axis, and plume extent small compared to plume distance). However, care has to be taken for unfavourable geometries (e.g., tilted or overhead plume; retrieval close to the image edges), which may require additional corrections (e.g., [33]). The software Plumetrack [26] includes an alternative "2D" method to compute the emission rates, which considers all image pixels that are crossing the PCS line between consecutive frames. This method is not included in the current version of *Pyplis* but may be a valuable extension in a future release of the software.

If a locally uniform gas velocity can be assumed (i.e., $\bar{v}(i, j) \rightarrow \bar{v}$) and a planar PCS is used for the retrieval (i.e., $\hat{n}(i, j) \rightarrow \hat{n}$), then, Equation (3) can be further approximated as $\Phi(\ell) \approx \langle \bar{v} \cdot \hat{n} \rangle \cdot \chi(\ell)$, where

$$\chi(\ell) = f^{-1} \sum_{m=1}^{M} S_{SO2}(m) \cdot d_{pl}(m) \cdot \Delta s(m) \tag{4}$$

denotes the integrated column amount (ICA) along ℓ.

2.4. Radiative Transfer Corrections

Radiative transfer effects may introduce systematic errors in the retrieved emission rates (see e.g., [34,35]). The signal dilution effect describes a decrease in the measured CDs due to scattering between instrument and plume. The magnitude of this effect primarily depends on the local visibility (i.e., amount of molecules and particles in the ambient atmosphere) and on the distance between camera and plume. A first order correction can be performed using the atmospheric scattering extinction coefficients in the viewing direction between camera and source (e.g., [12,36]). The latter can be retrieved, for instance, by studying brightness variations of topographic terrain features in the images as a function of their distance to the instrument [22]. More complicated radiative transfer issues (e.g., optically thick plumes) require corrections using radiative transfer models (e.g., [29]). *Pyplis* can correct for the signal dilution effect based on the method suggested by [22].

3. Implementation

In the following, the main features and modules of *Pyplis* are presented. The application programming interface (API) is designed to be modular using an object oriented architecture. It can therefore be used in parts or as a whole. Figure 2 illustrates the *Pyplis* API for SO_2 emission rate retrievals (see also Appendix E). *Pyplis* includes 19 example scripts, which provide an introduction to the main features of the software (summarised in Table A2). The scripts are based on an example dataset recorded at Mt. Etna in Italy on 16 September 2015 between 6:40 a.m. and 7:30 a.m. UTC. These data are freely available and can be downloaded from the website (for details, see Appendix F.1).

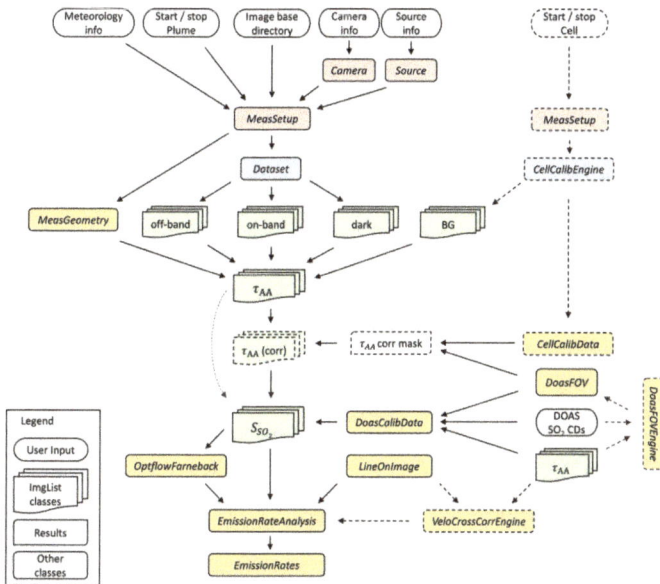

Figure 2. *Pyplis* API flowchart illustrating the central analysis steps and the corresponding classes of the *Pyplis* API for SO_2 emission rate retrievals. *Italic* denotations correspond to class names in *Pyplis*. Optional/alternative analysis procedures are indicated by dashed boxes. Setup classes (red) include relevant meta information and can be used to create Dataset objects (blue). The latter perform file separation by image type and create ImgList objects (green) for each type (e.g., on, off, dark). Further analysis classes are indicated in yellow. Note that the routine for signal dilution correction is not shown here (cf. Figure A1).

3.1. Geometrical Calculations

Geometrical calculations are performed within an instance of the `MeasGeometry` class (part of the *geometry.py* module). The plume distances d_{pl} (cf. Equation (3)) can be retrieved per pixel-column i using intersections of the respective viewing azimuth with the plume azimuth (see Figure 3). This requires specification of the camera position, viewing direction and optics (e.g., detector specifications, focal length), source coordinates and meteorological wind direction. The distances are determined based on the horizontal plume distance and the altitude difference between source and camera (i.e., assuming a horizontally aligned plume). It is pointed out that this approach may be inapplicable for complicated measurement geometries (e.g., overhead plumes), which would require a more detailed knowledge of the plume shape and altitude.

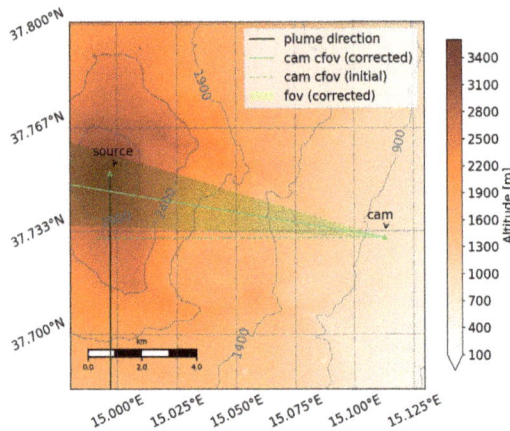

Figure 3. Measurement geometry 2D overview map of the measurement setup at Mt. Etna from the *Pyplis* example data. The map includes plume orientation (dark green line), camera azimuth retrieved using a compass (dashed light green line), the corrected camera azimuth (light green line) and the corresponding FOV (semi-transparent green), retrieved automatically using the pixel-position i, j of the Etna south-east crater within the images and the corresponding coordinates of the crater (longitude, latitude, altitude). The map was generated using an instance of the `MeasGeometry` class (see Section 3.1).

Further features of the `MeasGeometry` class include a routine to retrieve the camera viewing direction based on the position of distinct objects in the images (e.g., summit of volcano, cf. Figure 3), or the calculation of distances to topographic features in the images (cf. Figure 10). The `MeasGeometry` class makes use of the Geonum library [37], which is briefly introduced in Appendix A, and which provides automatic online access to topographic data from the NASA shuttle radar topography mission (SRTM, [38]) and also supports the ETOPO1 dataset [39]. Furthermore, 2D and 3D topographic overview maps of the measurement setup can be created automatically (as shown in Figure 3, see also Figure 10c).

3.2. Image Representation and Pre-Processing Routines

Images are represented by the `Img` class (*image.py* module). The `Img` class includes reading routines for all image formats supported by the Python Imaging Library (PIL, e.g., png, tiff, jpeg, bmp) as well as the FITS format (Flexible Image Transport System). It further allows to store relevant meta information (e.g., exposure and acquisition time, focal length, f-number) and includes several basic processing methods (e.g., smoothing, cropping or resizing using a Gaussian pyramid approach). The image data itself is loaded and stored as a 2D-NumPy array within an `Img` object. Automatic dark and offset correction can be performed as described in Appendix C. `Img` objects can be saved as FITS files at any stage of the analysis (e.g., dark corrected image, τ_{AA} image, calibrated SO_2-CD image).

3.3. Retrieval of Plume Background Radiances

The calculation of the OD images in both wavelength channels requires the retrieval of the sky-background intensities I_0 behind the plume (cf. Equation (1)). The *plumebackground.py* module provides several alternatives to retrieve the background intensities, either from the plume images directly or using an additional sky reference image (I_0-image). For the latter, several methods are available to correct for offsets and non-uniformities in the sky-background between the plume and I_0-image. The corrections are based on suitable sky-reference-areas in the plume image (rectangles or profile-lines, cf. Table 2) and use polynomials of a suitable order (e.g., linear or quadratic) to model the I_0-image such that the corresponding OD image satisfies $\tau = 0$ within the specified sky reference areas.

If no I_0-image is available, the plume background radiances can also be estimated from the plume images directly using a masked 2D polynomial surface fit. The required mask specifies clear-sky image pixels that are considered during the fit. The mask can either be provided by the user or can be retrieved automatically using the function `find_sky_background` of the *plumebackground.py* module.

The `PlumeBackgroundModel` class (part of *plumebackground.py*) provides high level access to eight default methods for the background retrieval (in the code denoted with *mode*). These different retrieval methods are summarised in Table 2. Figure 4 shows four example plume on-band OD images calculated using the background modelling methods 0, 1, 4 and 6 (cf. Table 2). It is not intended to give a recommendation for a "best" method here, as this strongly depends on the data (e.g., availability of suitable reference areas; acquisition time and relative viewing direction of I_0-image; solar position).

Table 2. Available plume background modelling methods of the `PlumeBackgroundModel` class (cf. names of sky reference areas in Figure 4).

Method	I_0-img	Corrections		
		Scaling	Vertical	Horizontal
1	yes	Scaling in *scale_rect*	None	None
2	yes	See 1	Linear correction using *scale_rect* and *ygrad_rect*	None
3	yes	See 1	Curvature correction by fitting polynomial of *n-th* order using sky reference pixels along *vertical* profile line (default: $n = 2$, i.e., quadratic polynomial)	None
4	yes	See 1	Linear correction (see 2)	Linear correction using *scale_rect* and *xgrad_rect*
5	yes	See 1	Curvature correction (see 3)	Linear correction (see 4)
6	yes	See 1	Curvature correction (see 3)	Curvature correction by fitting polynomial of *n-th* order using sky reference pixels along *horizontal* profile line (default: $n = 2$, i.e., quadratic polynomial)
0	no	Masked 2D polynomial surface fit		
99	yes	None (use I_0-img as is)		

(a) Method 0

(b) Method 1

(c) Method 4

(d) Method 6

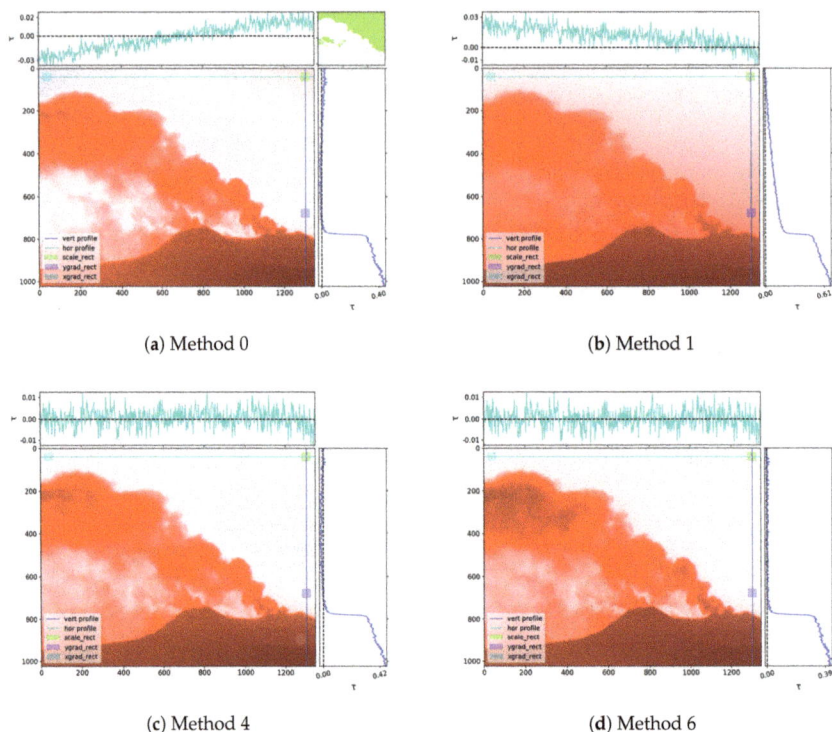

Figure 4. Plume background modelling. Four examples of on-band OD images (τ_{on}) determined using background modelling methods 0 (polynomial surface fit) and 1, 4 and 6 (based on I_0-image, cf. Table 2). Horizontal and vertical profiles are plotted on the top and on the right, respectively. The top-left plot (Method 0) further includes the mask specifying sky-reference pixels (green area) that was used for the polynomial surface fit. The plume and background images used for the displayed examples were recorded at 7:06 a.m. UTC and at 7:02 a.m. UTC, respectively.

3.4. Camera Calibration

The camera calibration can either be performed using data based on cuvettes (gas cells) filled with a known amount of SO_2-gas, or using plume SO_2-CDs retrieved from a co-located DOAS spectrometer (cf. Table 1).

3.4.1. Calibration Using SO_2 Cells

Cell calibration can be performed using the `CellCalibEngine` class (*cellcalib.py* module) and requires a dataset containing both cell and sky background images. It is assumed that the camera is pointed into a gas and cloud free area of the sky and that the cells (containing different SO_2-CDs) are consecutively placed in front of the lens, such that they cover the whole FOV of the camera. Figure 5a shows a time-series of the image mean intensities retrieved from such a dataset. The individual cells can be identified by abrupt intensity drops in the time-series while the ambient background only changes gradually. The corresponding time-intervals for cell and background images can be specified by the user or, alternatively, detected automatically within an instance of the `CellCalibEngine` class (shown in Figure 5a, see also example scripts 0.7 and 5).

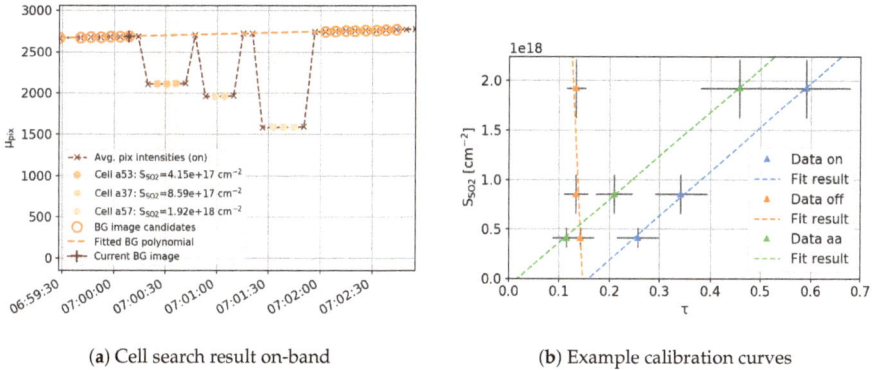

(a) Cell search result on-band

(b) Example calibration curves

Figure 5. Cell calibration — (**a**) shows the output of the automatic cell search routine (included in the `CellCalibEngine` class); (**b**) shows example calibration curves for τ_{on}, τ_{off} and τ_{AA} images for the image center pixel from the dataset shown in (**a**). The error bars in (**b**) correspond to uncertainties in the cell SO_2-CDs from the DOAS analysis (*y*-axis errors) and the *x*-errorbars correspond to the min / max-range of the corresponding τ-values over the entire image (arising from effects due to non-perpendicular illumination, see Section 2.2 for details).

Cell OD images in both channels (τ_{on}, τ_{off}) can be determined using a suitable background image. Care has to be taken for measurements performed at large solar zenith angles (early morning, late afternoon) due to rapid changes of the ambient sky intensity (cf. Figure 5a). In this case, the background image needs to be scaled to the sky intensity present at the acquisition time of each cell in order to calculate the OD images. This correction was performed for the data shown in Figure 5a (i.e., dashed line "Fitted BG polynomial"). It requires at least two background images, one recorded before, and a second one after the cells were put in front of the lens.

The calibration results (e.g., one AA image for each cell) are stored within `CellCalibData` objects together with the corresponding cell SO_2-CDs (which need to be provided by the user). Calibration curves can then be retrieved per image pixel or within a certain pixel neighbourhood. Figure 5b shows example calibration curves for τ_{on}, τ_{off} and τ_{AA} images.

3.4.2. Calibration Using DOAS Data

The DOAS calibration is performed using a set of plume optical density images (usually AA images) and a corresponding time-series of SO_2-CDs retrieved from a DOAS spectrometer. In a first step, the AA images are stacked into a 3D-NumPy array and merged in time with the DOAS data. The latter can be performed in three different ways (for details see Appendix D.1). The calibration data (i.e., merged time-series of SO_2-CDs and camera AA values within the DOAS FOV) can be retrieved by convolving the AA stack with a mask specifying position and shape of the DOAS field-of-view (FOV) within the image plane. The calibration data is stored within instances of the `DoasCalibData` class, which is also used to retrieve the calibration curve. The latter is done by fitting a polynomial of appropriate order to the calibration data. Optionally, the fit can be performed using a weighted regression, to account for statistical uncertainties in the DOAS SO_2-CDs (e.g., fit-errors, cf. Figure 6). `DoasCalibData` objects can be stored using the FITS standard (including the FOV mask).

Note that the DOAS calibration curve is only valid within the image pixel area covered by the DOAS FOV. This is due to cross-detector variations in the SO_2 sensitivity (see Section 2.2) and can be corrected for using a mask retrieved from a cell calibration dataset. The mask is determined by fitting a 2D polynomial to a cell AA image (see prev. Section 3.4.1), which is then normalised to the centre-position of the DOAS FOV (illustrated in example script 7). Please also note that *Pyplis* cannot perform the DOAS analysis. Thus, the SO_2-CDs need to be retrieved using a suitable 3rd party software (e.g., DOASIS, [40]).

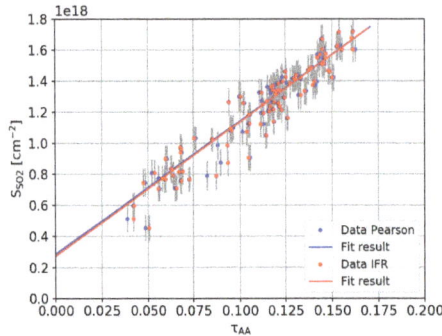

Figure 6. DOAS calibration curves. Plot showing example DOAS calibration data corresponding to the two different FOV parametrisations shown in Figure 7. The fit was performed using a first-order weighted polynomial fit. The weights were calculated using the relative errors ΔS_{SO2} / S_{SO2} of the DOAS SO_2-CDs. The y-axis offset is likely due to uncertainties in the DOAS retrieval (e.g., due to O_3 interference) and is compensated by the calibration.

3.4.3. DOAS FOV Search

Pyplis includes two routines to retrieve the DOAS FOV mask (included in the `DoasFOVEngine` class) based on a stack of AA images and a DOAS data vector:

1. Pearson routine: this method loops over all image pixels in the AA stack and determines the Pearson correlation coefficient between the corresponding AA time-series ($\tau_{AA}(t)$) and the DOAS SO_2-CD vector ($S_{SO2}(t)$). The method yields a correlation image as shown in Figure 7a, from which the pixel coordinate with highest correlation (i_0, j_0) is extracted (see also [21]). Assuming a circular FOV shape, the pixel extent of the FOV is estimated around i_0, j_0, by iteratively searching the disk radius with highest correlation between the AA and the DOAS time-series.
2. In-operation field-of-view retrieval (IFR) routine: this method is based on [30] and uses an inversion algorithm to retrieve the FOV. Position and shape of the FOV is parametrised by fitting a 2D Super-Gaussian to the retrieved FOV distribution (shown in Figure 7b, see Appendix D.2 for details).

The retrieved FOV information (position, shape, convolution mask) is represented by the `DoasFOV` class and can be stored as a FITS file.

3.5. Plume Velocity Analysis

Plume velocities can be retrieved either using the ICA cross-correlation method or using an optical flow algorithm. Both methods yield displacement estimates in units of pixels / time. These are converted into plume gas velocities based on the measurement geometry (`MeasGeometry`, see Section 3.1). The relevant code is implemented in the *plumespeed.py* module.

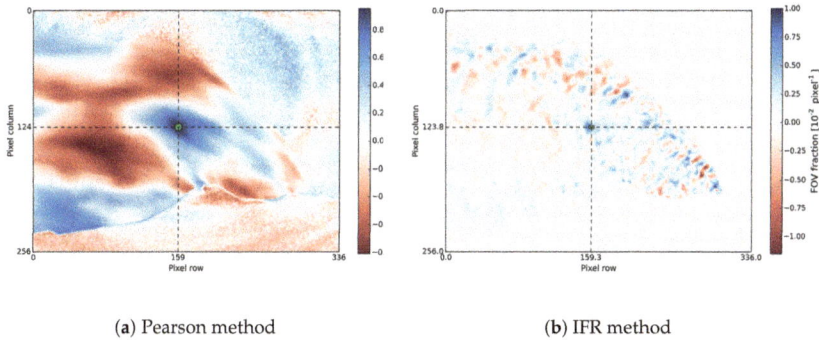

(a) Pearson method   (b) IFR method

Figure 7. DOAS FOV search. DOAS FOV search results using (**a**), the *Pearson* and (**b**), the *IFR* method. The Pearson method (a) yields an FOV centered at $i = 159$, $j = 124$ and a disk radius of four pixels. For the IFR retrieval (b), a tolerance factor of $\lambda = 2 \times 10^{-3}$ was chosen and a Super-Gaussian (without tilt) was fitted to the correlation image yielding an FOV centered at $i = 159.3, j = 123.8$, $\sigma = 7.1$, asymmetry parameter $a = 1.9$ and a shape parameter of $b = 0.3$ (for details see Equation (A2)). The retrieved FOV positions show good agreement. Note that the FOV was retrieved from downscaled images (Gauss pyramid level 2).

3.5.1. Velocity Retrieval Using the ICA Cross-Correlation Method

For the cross-correlation method, ICA time-series (see Equation (4)) are determined using two PCS lines located at different positions downwind the emission source. Ideally, the PCS lines should be parallel and should cover an entire plume cross-section (indicated in Figure 8, left).

In a first step, the two time-series are re-sampled onto a regular grid (default is 1 s resolution). In a second step, a correlation analysis is applied to the re-sampled data vectors in order to find the time lag corresponding to the highest correlation between both signals. The method yields one average velocity vector, which is representative for the used image region and time-series. Note that the method intrinsically assumes that the average plume propagation direction $\bar{\varphi}$ in the i, j-plane is aligned with the PCS normal (i.e., $\bar{\varphi} \parallel \hat{n}$) and furthermore that SO_2 is conserved between the two PCS lines used. Figure 8 shows results from an example cross-correlation analysis, resulting in a plume velocity of 3.5 m/s.

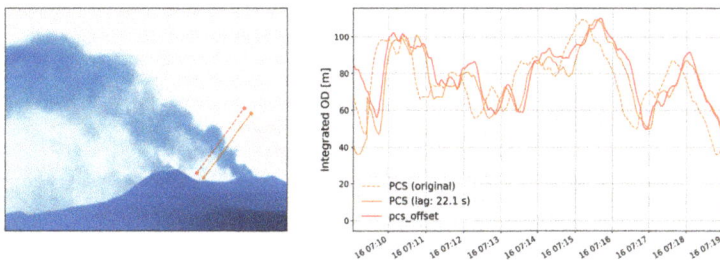

Figure 8. Plume velocity retrieval using the cross-correlation method applied to a time-series of on-band OD (τ_{On}) images. Left: example plume image including the two used PCS lines. Right: Time-series of the integrated optical densities along both lines (orange dashed and cyan line) in addition to the PCS signal shifted using the retrieved correlation lag of 22.1 s (orange profile). The analysis yields an average gas velocity of 3.5 m/s.

3.5.2. Optical Flow Based Velocity Retrievals

Optical flow velocity retrievals are performed using the Farnebäck algorithm [41], which is implemented in the OpenCV library [42]. The algorithm can estimate motion at the pixel-level by tracking local contrast features between consecutive frames. Note, however, that optical flow algorithms may fail to detect motion in extended homogeneous image areas. In this case, appropriate corrections may be required (e.g., [43]).

All relevant calculations for optical flow based velocity retrievals are performed within the `OptflowFarneback` class. The class includes the Farnebäck algorithm itself as well as several pre- and post-analysis routines. The latter includes a routine that can detect and correct for unphysical motion estimates in homogeneous image regions based on the method proposed by [43]. The correction identifies the local average velocity vector using peaks in histograms of an optical flow motion field.

The Farnebäck algorithm itself requires specification of several input parameters (see e.g., [23], or OpenCV documentation). The *Pyplis* default settings follow the suggestions of [23]. An example flow field is shown in Figure 9 including results from the histogram post analysis within a narrow region-of-interest (ROI) around an example retrieval line. The latter yields an average velocity magnitude of 4.2 (± 0.4) m/s and a predominant movement direction of $\varphi = -65\,(\pm 14)\,^\circ$ within the image plane. Optical flow plume velocity retrievals, including the histogram post analysis method, are introduced in example script 9 (see Table A2).

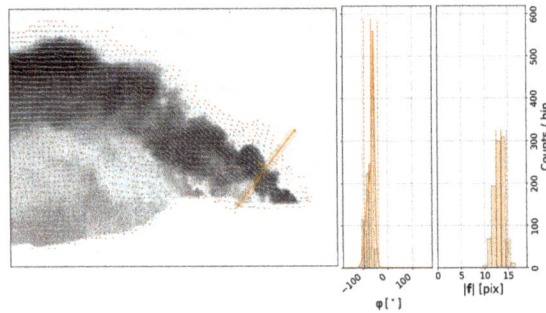

Figure 9. Plume velocity optical flow. Left: example optical flow vector field (blue lines with red dots depicting the movement direction) calculated using the Farnebäck algorithm. Middle, right: histograms of the flow vector orientation angles φ and magnitudes $|f|$, respectively, corresponding to the ROI shown in the left image (semi-transparent rectangular area around the displayed PCS line). From the latter, a plume velocity of 4.2 (± 0.4) m/s and a predominant movement direction of $\theta = -65\,(\pm 14)\,^\circ$ was retrieved using first and second moments of the displayed histogram distributions.

3.6. Image Based Signal Dilution Correction

A correction for the signal dilution effect (see Section 2.4) can be performed using the `DilutionCorr` class. The method is based on [22] and uses the model function

$$I_{\text{meas}}(\lambda) = I_0(\lambda)e^{-\epsilon(\lambda)d} + I_A(\lambda)\left(1 - e^{-\epsilon(\lambda)d}\right) \tag{5}$$

to retrieve atmospheric extinction coefficients ϵ in the on and off-band regime. Here, λ denotes the wavelength, I_{meas} are measured intensities of dark terrain features within the images, d is the distance to these features and I_0 their initial intensity. The ambient intensity I_A can be approximated using gas free sky areas in the plume images [22]. For the retrieval, a set of measured intensities I_{meas} is extracted from vignetting corrected plume images using suitable terrain features (e.g., pixels along a volcanic flank). These intensities are fitted to Equation (5) (as a function of their distance d) in order to retrieve the extinction coefficients in both wavelength regimes. The required distances d to the features can be retrieved automatically, based on intersections of individual pixel viewing directions with the local topography (using the Geonum library; for details, see Appendix A).

The plume images can then be corrected for the signal dilution effect using the retrieved extinction coefficients (see Equation (4) in [22]). The correction is only applied to plume pixels (e.g., using a τ_{AA} threshold mask). The required plume distances can be retrieved from the measurement geometry (for details, see Section 3.1).

Figure 10 shows results of an example dilution correction using an on and off-band image from the Etna example data, recorded at 6:45 a.m. UTC. Extinction coefficients of $\epsilon_{on} = 0.0743\,\text{km}^{-1}$ (Figure 10a) and $\epsilon_{off} = 0.0654\,\text{km}^{-1}$ (Figure 10b) were retrieved using the two topographic profile lines shown in Figure 10c,d. The correction yields an emission rate of 4.8 (±1.2) kg/s using the PCS line shown in Figure 10d. The emission rate of the uncorrected image is 1.6 (±0.8) kg/s, corresponding to an underestimation of approximately 67 % at an average plume distance of 10.4 (±0.1) km.

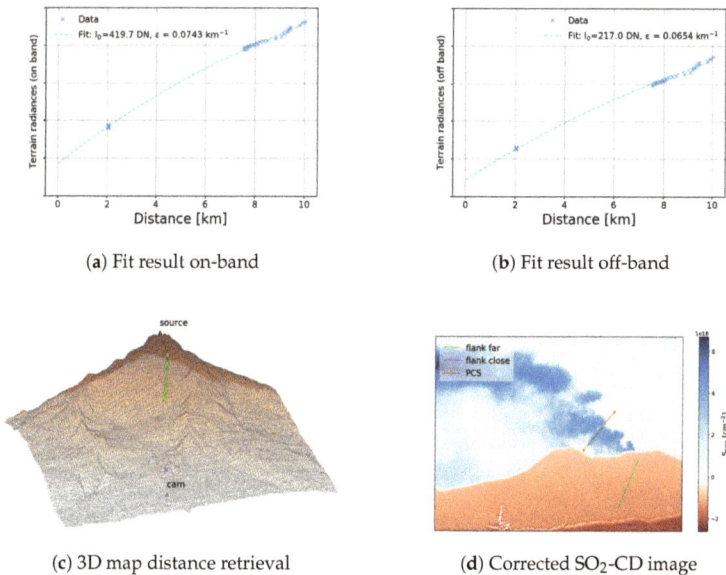

(a) Fit result on-band

(b) Fit result off-band

(c) 3D map distance retrieval

(d) Corrected SO_2-CD image

Figure 10. Signal dilution correction. (a) and (b) show the fit results of an example dilution analysis for the on and off-band regime, respectively; (c) shows a 3D map of the camera scene and (d) a dilution corrected SO_2-CD image. The terrain features used for the dilution analysis are indicated in (c) and (d) (blue and lime coloured lines); (d) further includes the image region used to estimate the ambient intensity I_A (blue rectangle) and an example PCS line used to compare emission rates before and after the correction.

3.7. Emission Rate Retrieval

Emission rates can be determined using the `EmissionRateAnalysis` class. The analysis is performed based on an image list containing calibrated SO_2-CD images (see also Appendix E.2) and a specification of one or more retrieval PCS lines (`LineOnImage` objects, cf. Figure 2). Plume velocities can be provided by the user (e.g., using the result of a cross-correlation analysis, see Section 3.5.1) or can be retrieved automatically during the emission rate analysis using the Farnebäck optical flow routine (see Section 3.5.2). For the latter, three options are available:

1. *flow_raw* → the raw output of the Farnebäck algorithm is used. This should only be done if it can be assured that the algorithm yields reliable output in the considered plume area (i.e., ROI around the PCS line) and for all images of the time-series.
2. *flow_histo* → performs the histogram post-analysis proposed by [43] (cf. Section 3.5.2). The retrieved local average velocity vector for each PCS line is then used as a velocity estimate for the corresponding retrieval line.
3. *flow_hybrid* → reliable motion vectors from the flow field are used while unreliable ones are identified and replaced based on the histogram post-analysis (see previous point).

Figure 11 shows results from an emission rate analysis for the Etna test data (upper panel) including plume velocities retrieved using the *flow_hybrid* method (lower panel). The AA images were calculated from dilution corrected on and off-band plume images using background modelling method 6 (cf. Section 3.3) and were corrected for cross-detector sensitivity variations using a mask retrieved from cell "*a53*" (cf. Figure 5a). The AA images were calibrated using the *Pearson* DOAS calibration curve shown in Figure 6.

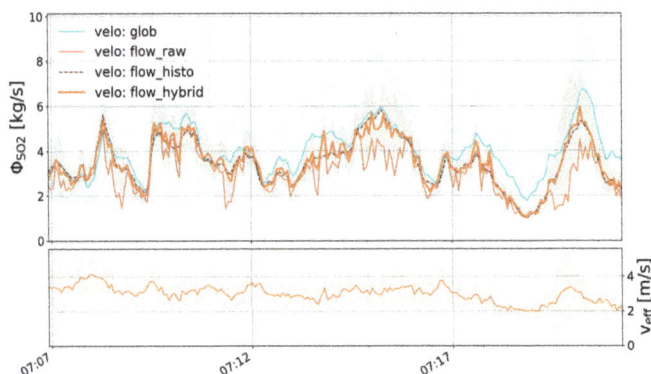

Figure 11. Emission rate retrieval. Emission rates (top) and plume velocities (bottom) of the Etna example dataset on 16 September 2015 between 7:06 a.m. and 7:22 a.m. UTC using the retrieval line "PCS" shown in Figure 8. The analysis was performed using (1) a global velocity of 3.5 m/s (cyan, from cross-correlation analysis); (2) the raw output of the Farnebäck algorithm (thin orange); (3) using the *flow_histo* method (thin dashed brown) and (4) using the *flow_hybrid* method (bold orange). The retrieved effective velocities are plotted in the lower panel and correspond to the average velocities along the PCS line using the *flow_hybrid* method. Uncertainties are displayed as shaded areas.

Remark on Performance

The emission rate analysis (cf. Figure 11) was performed using a Gaussian pyramid level of 1 (i.e., reduction of image size by a factor of 2). For these settings, the typical computation time to calculate a calibrated SO_2-CD image from the raw data amounts to approximately 0.2 s using an Intel(R) Core(TM) i7-6500U CPU (2.50 GHz, 8 GB RAM) (Santa Clara, CA, USA). Approximately the same time of 0.2 s is required to compute the optical flow between two frames at this pyramid level (cf. Appendix Table A1). For the discussed dataset of 209 on-band plume images, this results in a total computation time of about two minutes, including the four different velocity retrieval methods introduced above and shown in Figure 11. Note that the latter does not include the time to compute the DOAS FOV and calibration data, nor the time required for the cross-correlation velocity retrieval, since these analyses were performed beforehand (for details, see Appendix B and example script 12).

3.8. Remark on Uncertainties

The uncertainties in the emission rates (cf. shaded areas in Figure 11) were calculated using Gaussian error propagation (see method `det_emission_rate` of the *fluxcalc.py* module) considering (1) uncertainties in the meteorological wind direction and camera viewing direction (accessible using the method `plume_dist_err` of the `MeasGeometry` class); (2) uncertainties in the calibration curve (based on slope error retrieved from covariance matrix of fit result, accessible using the attribute `slope_err` of either the `DoasCalibData` or the `CellCalibData` class); and (3) the uncertainties in the velocity retrieval. For optical flow based velocity analyses, the uncertainties are calculated as discussed in Section 2.4.1 in [43], based on the retrieved effective velocities $v_{eff} = \langle \bar{v} \cdot \hat{n} \rangle$ along ℓ (cf. Equation (3)). Note that *Pyplis* does not provide a method to estimate the uncertainties of cross-correlation based velocity retrievals. These need to be provided by the user (cf. example script 12). The errors in the effective velocities can be accessed via the `velo_eff_err` attribute of the `EmissionRates` class. Note that all uncertainties computed by *Pyplis* are assumed to be of statistical nature. Hence, they do not account for any potential systematic errors, for instance arising from ill-constrained optical flow vectors when using the *flow_raw* retrieval method (see discussion in Section 2.4.1 in [43]).

4. Conclusions

In this paper, the software package *Pyplis* was introduced. *Pyplis* contains an extensive collection of relevant algorithms for the analysis of UV SO_2 camera data, particularly for the retrieval of emission rates from SO_2-emitters (e.g., volcanoes, power plants, ships).

Apart from established analysis methods, such as cross-correlation based velocity retrievals (e.g., [32]) or cell and DOAS calibration (e.g., [21]), *Pyplis* incorporates more recent developments. These include an implementation of the DOAS FOV retrieval algorithm proposed by [30], a routine to correct for the signal dilution effect based on [22], or pixel based gas velocity retrievals using an optical flow algorithm (e.g., [23]). The latter incorporates a method to detect and correct for ill constrained optical flow motion-vectors based on [43]. Furthermore, *Pyplis* includes a framework to perform a detailed 3D-analysis of the observed camera scene including automatic online access to high resolution topography data from the SRTM dataset. This enables, for example, to retrieve the camera viewing direction based on distinct topographic features in the images (e.g., volcano summit), or to calculate distances between the camera and the local topography at a pixel-level. The latter is of particular relevance for the image based correction of the signal dilution effect.

Due to this extensive collection of algorithms, *Pyplis* provides flexibility with regard to the analysis strategy and is highly adaptable to different data situations. The object oriented architecture of the API gives intuitive access to the main features and makes it easy to compare individual methods (e.g., different plume velocity retrievals, as illustrated in this paper). *Pyplis* is open-source and can be operated both on Windows and Unix machines. Thus, it is well suited for inter-comparison studies, the exchange of analysis results or for the development and verification of new methods.

The *Pyplis* installation includes numerous example scripts that were used in this paper to introduce the main features of the software. The examples are based on a 15 min dataset recorded at Mt. Etna, Italy in September 2015, which is freely available and can be downloaded from the website.

Finally, the authors wish to point out that *Pyplis* may also be used for other applications based on the same measurement principle (e.g., NO_2 cameras) and that parts of it can also be useful for other remote sensing applications (e.g., the engine for geometrical calculations). The *Pyplis* software is hosted on GitHub ([27,28]). The code documentation and further information (e.g., installation instructions) can be found on the documentation website (see [27,28]).

Acknowledgments: The authors like to thank the *Atmosphere and Remote Sensing* group from the Institute of Environmental Physics in Heidelberg, Germany for providing image data for test purposes and for collaboration during the Etna field campaign in 2015. The work of K.S., A.K. and A.S.D. was supported by the European Research Council (ERC) under the European Union's Horizon 2020 research and innovation programme under grant agreement No 670462 (COMTESSA). J.G. wishes to thank Torbjørn Skauli, Matthias Vogt, and Axel Donath for helpful discussions related to the development of the software and the drafting of the manuscript. Furthermore, the authors wish to thank the Editor Cecile Zheng for support during the review process and four anonymous referees for very helpful comments and suggestions.

Author Contributions: J.G. designed, developed and tested the software including the example scripts and wrote the manuscript. K.S., A.S.D., A.K., H.S. and A.S. contributed to the API design (selection of implemented methods/algorithms) and helped with testing and debugging the code. H.S. provided the IFR algorithm and wrote the corresponding part in the paper.

Conflicts of Interest: The authors declare no conflict of interest.

Abbreviations

The following abbreviations are used in this manuscript:

UV	Ultraviolet
CD	Column density
DOAS	Differential optical absorption spectroscopy
FOV	Field of view
OD	Optical density
AA	SO_2 apparent absorbance
PCS	Plume cross section
ICA	Integrated column amount
API	Application programming interface
IFR	In-operation field-of-view retrieval

Appendix A. The Geonum Python library

The *Pyplis* class `MeasGeometry` (see Section 3.1) makes use of the Python library Geonum [37]. Geonum was developed by Jonas Gliß in parallel to *Pyplis* and features vector based geographical calculations in three dimensions as well as access and handling of high resolution topographic data. It supports topographic data based on the Etopo1 global relief model [39] and from the NASA shuttle radar topography mission (SRTM, [38]). The latter can be accessed and downloaded automatically within Geonum from the SRTM online database (for details, see information on the Geonum [37] website).

Appendix A.1. Pixel Based Retrieval of Distances to Local Terrain Features

Distances to topographic features can be retrieved at the pixel-level based on the camera position and viewing direction by calculating the intersections of individual pixel viewing directions (i.e., azimuth and elevation angle) with the local topography. This is particularly helpful for the image based correction of the signal dilution effect (see Section 3.6 and Figure 10).

Appendix B. Performance of Typical Analysis Chain

The time required to perform a full emission rate analysis with *Pyplis* (cf. Section 3.7) depends on many factors, most importantly on the performance of the computer, but also largely on the chosen analysis settings and the methods used. Below, we provide information about the typical computation time required to calculate a calibrated SO_2-CD image from the raw data and to compute the optical flow, using an Intel(R) Core(TM) i7-6500U CPU (2.50 GHz, 8 GB RAM) and the Envicam-2 image type (i.e., 12 bit, 1344 × 1024 pixels). The main analysis steps per image comprise:

1. Image import and dark and offset correction (on and off-band).
2. Further image preparation operations (e.g., noise reduction using Gaussian blurring filter, size reduction using Gaussian pyramid).
3. Plume background modelling (on, off) and calculation of τ_{AA}-image.
4. Image calibration (i.e., requires availability of calibration curve).
5. Optional: computation of optical flow field.

Table A1 summarises typical computation times for these steps that are performed for each image during the emission rate analysis. Further settings for this performance analysis (e.g., background modelling method) correspond to the settings provided in Section 3.7. Note that here we omit the discussion of potential additional analysis procedures that may be required and that are typically performed prior to the emission rate analysis (e.g., dilution correction, DOAS and / or cell calibration, cross-correlation velocity retrieval, cf. Figure 2). These highly depend on the chosen settings and often include looping over a number of images and pre-computation of OD images (e.g., DOAS FOV search, cross-correlation velocity analysis) and we encourage the reader to use the corresponding example scripts for individualised performance tests of these additional analysis procedures.

Table A1. Performance analysis of the five image preparation steps listed above, dependent on Gaussian pyramid level (*pyrlevel*) and Gaussian blurring (*blur*) including the relative percentage impact of the optical flow computation.

Blur	Pyrlevel	Computation Time [s]		
		Image Preparation (Steps 1–4)	Optical Flow (Step 5)	Total
10	0	0.350	0.823 (70 %)	1.173
0	0	0.188	0.813 (81 %)	1.001
0	1	0.205	0.202 (50 %)	0.407
0	2	0.203	0.103 (34 %)	0.306

Appendix C. Dark and Offset Correction

Pyplis includes two options to perform dark and offset corrections for image data. Both methods require access to the exposure time of the images (e.g., from the image file names, see also Appendix F).

1. **Option 1: Modelling of Dark/Offset Image**
   The correction is performed based on two dark images, one being recorded at short(est) exposure time (offset signal O) and the second one at long(est) exposure time (dark current + offset signal, D). A dark image is then calculated based on the exposure time t_{exp} of the input image I using the following formula:

$$D_{mod} = O + \frac{(D - O) * t_{exp,I}}{(t_{exp,D} - t_{exp,O})}. \tag{A1}$$

   This mode is, for instance, used for the Envicam-2 camera type (NILU, Norway, for details see [24]). The corresponding method `model_dark_image` is part of the *processing.py* module.

2. **Option 2: Subtraction of Dark Image**
   Dark and offset correction is performed by subtracting a single dark image D (containing dark and offset), which, thus, needs to be recorded at the same camera exposure time. This mode is, for instance, used for the HD-Custom camera (Heidelberg, Germany, for details see [24]).

The detector dark current depends on the temperature. In case of temperature variations, it is therefore recommended to use dark images recorded close in time to a given plume image.

Appendix D. Spectrometer FOV Search: Additional Information

Appendix D.1. Temporal Merging of Image and DOAS Data

The `ImgStack` class includes three methods to merge a set of camera images (stacked within such an `ImgStack`) and a DOAS time series vector, both sampled on arbitrary irregular grids.

- **First Method: Averaging of Camera Images**
  This method averages all images in the stack based on start/stop time stamps of the spectrometer data (i.e., the image sampling rate should be larger than the spectrum sampling rate). Spectra for which no image data could be found are removed.

- **Second Method: Vice Versa Interpolation of Both Grids**
  This method uses the unified sampling grid (all time stamps from both datasets) and performs interpolation of the DOAS data vector (at image acquisition time stamps) and vice versa. The method is slow compared to method 1 since each image pixel of the stack is interpolated. However, it results in more data points, which can be an advantage especially for short time series. This method can be significantly accelerated by reducing the image size or by only performing the analysis within a certain image region (c.f. example script no. 6, Table A2, script option: DO_FINE_SEARCH). The time series interpolation is done using the *pandas* Python library.
- **Third Method: Nearest Data Point**
  This method loops over all spectra and for each spectrum, finds the image which is nearest in time. This method is for instance used, if only the acquisition time stamps are provided and not the start / stop stamps of each exposure (which is required for the first method).

Appendix D.2. FOV Determination Applying the IFR Method

The In-operation Field of view Retrieval (IFR) method is an implementation of the method proposed by [30]. IFR applies a linear camera model to invert the FOV of a low-resolution instrument (in this case, a DOAS spectrometer) from imager data without a priori information (e.g., FOV position, size and shape). The inversion problem is typically under-determined for SO_2 camera applications. Therefore, the iterative LSMR method [44] is applied to retrieve the FOV coefficients depending on the regularisation parameter λ.

The choice of λ is somewhat arbitrary but may influence the IFR results depending on the input data quality. The default value is $\lambda = 10^{-6}$. However, in case only a small sample size is available, λ may need to be increased (e.g., $\lambda = 10^{-3}$) in order to produce meaningful results. A side effect of increasing λ is a spatial smoothing of the results potentially leading to unrealistic large FOVs. Figure 7b shows a sample FOV distribution retrieved from the Etna test data (88 images) using $\lambda = 2 \times 10^{-3}$.

In order to reach the final goal of gaining a calibration curve from the image stack containing AA images, individual images need to be convolved with the FOV of the low-resolution instrument. Therefore, a parametrised FOV is fitted to the IFR results, which is more applicable to ground-based instruments than the parametrisation proposed by [30]. We propose the following elliptical Super-Gaussian FOV parametrisation g of the IFR result depending on pixel coordinates i, j in horizontal and vertical direction, respectively:

$$g(i,j) = C + Ae^{-\left(\left[\frac{i-i_m}{\sigma}\right]^2 + \left[\frac{(j-j_m)a}{\sigma}\right]^2\right)^b},$$ (A2)

where C is a constant offset, A the amplitude, i_m and j_m define the centre position, σ measures the width in i-direction, the asymmetry parameter a measures the ratio of the semi-axes (e.g., $a = 1$ yields a circular FOV), and b is the shape parameter of the Super Gaussian (e.g., $b = 1$ yields a Gaussian FOV, $b = 10$ approximates a flat-disk FOV).

If an additional tilt of the FOV is required in case of an elliptical FOV, the above fit may be performed in a transformed coordinate system and

$$\begin{pmatrix} i' \\ j' \end{pmatrix} = \begin{pmatrix} \cos\theta & -\sin\theta \\ \sin\theta & \cos\theta \end{pmatrix} \begin{pmatrix} i - i_m \\ j - j_m \end{pmatrix}$$ (A3)

defines the transformation into tilted coordinates i' and j'. Equation (A2) is then replaced by

$$g = C + A\exp\left[-\left(\left[\frac{i'}{\sigma}\right]^2 + \left[\frac{j'a}{\sigma}\right]^2\right)^b\right].$$ (A4)

Appendix E. Basic Data Structure

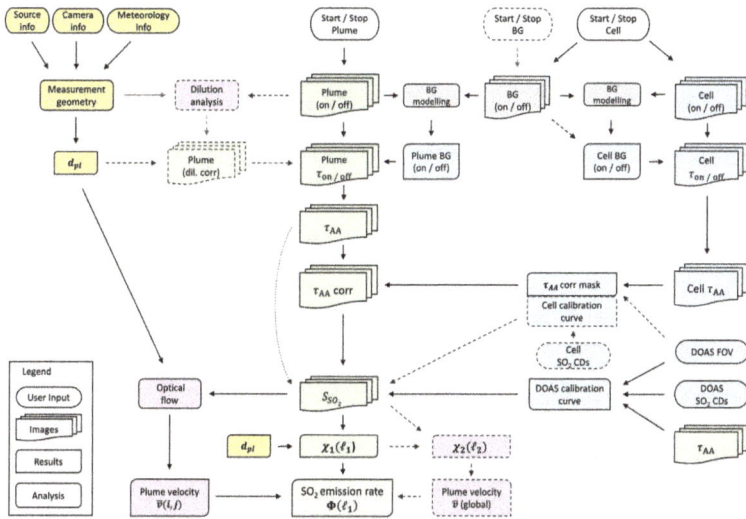

Figure A1. Main analysis steps. Flowchart showing the main analysis steps for emission rate retrievals (cf. Table 1). The colours indicate geometrical calculations (yellow), background modelling (light gray), camera calibration (light blue), plume velocity retrieval (light purple) and the central processing steps for the emission rate retrieval (light green). Shaded and dashed symbols indicate optional or alternative analysis methods. Note that the colour scheme is not related to the scheme used in Figure 2.

The *Pyplis* code hierarchy for the emission rate analysis is shown in Figure 2. The structure is based on the work flow shown in Figure A1 and includes most of the relevant classes required for the emission rate analysis.

Appendix E.1. Setup and Dataset Classes

The most important classes related to data import and image management are:

- Setup classes (e.g., `Camera`, `Source`, `MeasSetup`), which can be used to specify all relevant meta information.
- Dataset classes (`Dataset`, `CellCalibEngine`), which can be used for automatic image separation, for instance by image type (e.g., on, off, dark, offset) or acquisition time.

Appendix E.2. ImgList classes

Image list classes are central for the data analysis. They can be found in the *imagelists.py* module (e.g., `ImgLisxt`, `CellImgList`). An `ImgList` typically contains images of one specific type (e.g., on-band) corresponding to a certain time window. In order to avoid potential memory overflows, images are loaded, processed and unloaded successively within `ImgList` objects. The most important features are described in the following.

Appendix E.2.1. Linking of `ImgList` Objects

`ImgList` can be linked to each other (e.g., off to on-band list). This means that, whenever the list index (i.e., the current image) is changed in `ImgList` *A* (e.g., the on-band images), it is also changed in `ImgList` *B* (if *B* is linked to *A*), such that the current image in *B* is closest in time to the one in *A*.

Appendix E.2.2. Image Preparation and Processing Modes

Image lists include several image preparation options (e.g., size reduction, cropping, blurring). Furthermore, if certain requirements are fulfilled, additional preparation options and routines can be activated:

- *darkcorr_mode* → images are automatically corrected for dark and offset and requires a dark image (or an `ImgList` containing dark images) to be available in the list. For dark correction mode 1 (see appendix C), an offset image (or list) must also be available.
- *tau_mode* → if active, images are converted into τ images on image load (using the `PlumeBackgroundModel` class to retrieve the plume background intensities) and requires availability of a sky reference image in the list (only for background modelling methods 1–6, see Section 3.3).
- *aa_mode* → if active, images are converted into τ_{AA} images on image load and requires an off-band image list to be linked to the list and availability of a sky reference image in both lists (only for background modelling methods 1–6, see Section 3.3).
- *dilcorr_mode* → if active, images are loaded as dilution corrected images (cf. Section 3.6) and requires extinction coefficients to be available in the list (list attribute `ext_coeffs`, cf. example script 11). Furthermore, availability of plume distances (list attribute `plume_dists`) and pre-computation of a τ-image (see two points above) is required to retrieve a boolean mask specifying plume-pixels (identified from the τ-image using a provided τ threshold).
- *sensitivity_corr_mode* → if active, images will be corrected for sensitivity variations due to shifts in the filter transmission windows (see Section 2.2) and requires a corresponding correction mask to be available in the list. The latter can, for instance, be retrieved from cell calibration data (see Section 3.4.1).
- *calib_mode* → if active, images are loaded as calibrated SO_2-CD images and requires the list to be in *aa_mode* and calibration data to be available in the list. The latter can be of type `CellCalibData` or `DoasCalibData` (see Figure 2), and warns if `sensitivity_corr_mode` is inactive.
- *optflow_mode* → if active, the Farnebäck optical flow will be calculated between current and the next list image (using the `OptflowFarneback` class, see Section 3.5.2).
- *vigncorr_mode* → if active, images will be corrected for vignetting and requires availability of a vignetting mask in the list or a sky reference image from which the mask is determined.

All active image preparation options are applied on image load (i.e., whenever the current image is changed in the `ImgList`).

Appendix E.3. Processing Classes

Most of the relevant processing classes are shown in Figure 2. They include:

- `MeasGeometry` (*geometry.py*) → all relevant geometrical calculations (Section 3.1).
- `OptflowFarneback` (*plumespeed.py*) → calculation and post analysis of optical flow field between two images (Section 3.5.2).
- `CellCalibData` (*cellcalib.py*) → pixel based retrieval of cell calibration polynomial (based on a set of cell τ images) and retrieval of sensitivity correction mask (Section 3.4.1).
- `DoasFovEngine`(*doascalib.py*) → performs FOV search of DOAS spectrometer within camera images (Section 3.4.2 and Appendix D).
- `DoasFov` (*doascalib.py*) → DOAS FOV information such as position, shape, convolution mask (Section 3.4.2 and Appendix D), can be saved as FITS file.
- `DoasCalibData` (*doascalib.py*) → DOAS calibration data, i.e., vector of τ and SO_2-CD values for fitting of calibration polynomial (Section 3.4.2), can be saved as FITS file.
- `LineOnImage` (*processing.py*) → data extraction (interpolation) along a line on a discrete 2D image grid (e.g., SO_2-CDs from calibrated images or displacement vectors from optical flow field, Section 3.7).

- `EmissionRateAnalysis` (*fluxcalc.py*) → Performs emission rate analysis based on an `ImgList` containing calibrated images. Emission rates can be retrieved along one (or more) plume cross section lines (`LineOnImage` objects) and has several options related to the plume velocity retrieval (Section 3.7).
- `EmissionRates` (*fluxcalc.py*) → Contains results (time series) of an emission rate analysis (i.e., including plume velocity data), specific for one PCS line and one velocity retrieval (e.g., the analysis shown in Figure 11 creates three `EmissionRates` objects for each of the three different velocity retrievals, Section 3.7).

Further important classes (not shown in Figure 2) are:

- `PlumeBackgroundModel` (*plumebackground.py*) → performs τ image modelling using either of the available modelling methods (Section 3.3).
- `VeloCrossCorrEngine` (*plumespeed.py*) → high level class to calculate the plume velocity using the cross-correlation method (Section 3.5.1).
- `DilutionCorr` (*dilutioncorr.py*) → engine to perform signal dilution correction (Section 3.6).
- `ImgStack` (*processing.py*) → contains a series of images (stored as 3D numpy array) as well as supplementary data (e.g., acq. time stamps, exposure times of all stacked images) and basic processing operations (time merging with other data, up / downscaling), can be saved as FITS file.

Appendix F. Supplementary Information and Test Data

Appendix F.1. Example Dataset and Example Scripts

Most of the example and introduction scripts provided with *Pyplis* are based on a short example dataset recorded at Mt. Etna, Italy on 16 September 2015, using a type Envicam-2 camera. It includes ~15 min of plume data (between 7:06 a.m. and 7:22 a.m. UTC, see e.g., Figure 11.) as well as cell calibration data including suitable background images (between 6:59 a.m. and 7:04 a.m. UTC, see Figure 5). These data are used for demonstration purposes in the provided example scripts, which are summarised in Table A2. The data is not part of the *Pyplis* installation and can be downloaded from the website.

Table A2. *Pyplis* example scripts, sub-categorised into introductory scripts (0.1–0.7) and scripts related to the emission rate analysis of the Etna test data (1–12).

No.	Name	Description	Section
0.1	ex0_1_img_handling.py	The `Img` class - Image import and dark correction	3.2
0.2	ex0_2_camera_setup.py	The `Camera` class - Definition of camera specifications and image file name convention	E
0.3	ex0_3_imglists_manually.py	Introduction into `ImgList` objects	F.2
0.4	ex0_4_imglists_auto.py	Automatic creation of `ImgList` objects using the ECII default `Camera` type	F.2
0.5	ex0_5_optflow_livecam.py	Interactive optical flow using web cam	3.5.2
0.6	ex0_6_pcs_lines.py	Plume cross section lines (creation and orientation of `LineOnImage` objects)	3.7
0.7	ex0_7_cellcalib_manual.py	Introduction into cell calibration and the `CellCalibData` object (manually)	3.4.1
1	ex01_analysis_setup.py	Create `MeasSetup` class and initiate analysis `Dataset` object from that (see Figure 2)	3.4.1
2	ex02_meas_geometry.py	Introduction into the `MeasGeometry` class	3.1
3	ex03_plume_background.py	The `PlumeBackgroundModel` class - background modelling and τ image retrieval	3.3
4	ex04_prep_aa_imglist.py	Preparation of image list containing AA images	F.2
5	ex05_cell_calib_auto.py	Automatic cell calibration using the `CellCalibEngine` class	3.4.1
6	ex06_doas_calib.py	DOAS calibration and FOV search	3.4.2
7	ex07_doas_cell_calib.py	Retrieval of AA sensitivity correction mask	3.4
8	ex08_velo_crosscorr.py	Plume velocity retrieval using cross-correlation	3.5.1
9	ex09_velo_optflow.py	Plume velocity retrieval using Farnebäck optical flow algorithm using `OptflowFarneback` class	3.5.2
10	ex10_bg_imglists.py	Retrieval of background image lists (on, off) using `Dataset` class	E
11	ex11_signal_dilution.py	Correction for signal dilution and the `DilutionCorr` class	3.6
12	ex12_emission_rate.py	Emission rate retrieval for the test dataset	3.7
	SETTINGS.py	Global settings for example scripts	

Appendix F.2. Camera Specifications

In order to use all features of *Pyplis* (e.g., automatic file separation, automatic dark and offset correction, geometrical calculations), certain camera characteristics need to be provided by the user. This information is typically specified within a `Camera` class (*setupclasses.py* module). The required information includes specifications about the image sensor (e.g., pixel geometry) and optics (e.g., focal length) as well as file naming conventions (e.g., how to retrieve the filter type or the image acquisition time from file names). *Pyplis* provides the possibility to define new default camera types that store all relevant camera information to the *Pyplis* data file *cam_info.txt*, which can be found in the *data* directory of the installation (see example script 0.2 for details).

Appendix F.3. Source Specifications

Default source information (e.g., longitude, latitude, altitude) can be specified in the file *my_sources.txt* in the installation *data* directory.

References

1. Robock, A. Volcanic eruptions and climate. *Rev. Geophys.* **2000**, *38*, 191–219.
2. IPCC. *Climate Change 2013: The Physical Science Basis. Contribution of Working Group I to the Fifth Assessment Report of the Intergovernmental Panel on Climate Change*; Cambridge University Press: Cambridge, UK, 2013; p. 1535.
3. Carroll, M.R.; Holloway, J.R. *Volatiles in Magmas*; Mineralogical Society of America: Chantilly, VA, USA, 1994.
4. Oppenheimer, C.; Fischer, T.; Scaillet, B. 4.4 - Volcanic Degassing: Process and Impact. In *Treatise on Geochemistry, 2nd ed.*; Elsevier: Amsterdam, The Netherlands, 2014; pp. 111–179.
5. Lübcke, P.; Bobrowski, N.; Arellano, S.; Galle, B.; Garzón, G.; Vogel, L.; Platt, U. BrO / SO_2 molar ratios from scanning DOAS measurements in the NOVAC network. *Solid Earth* **2014**, *5*, 409–424.
6. Bobrowski, N.; von Glasow, R.; Giuffrida, G.B.; Tedesco, D.; Aiuppa, A.; Yalire, M.; Arellano, S.; Johansson, M.; Galle, B. Gas emission strength and evolution of the molar ratio of BrO/SO_2 in the plume of Nyiragongo in comparison to Etna. *J. Geophys. Res. Atmos.* **2015**, *120*, 277–291.
7. Moffat, A.J.; Millan, M.M. The applications of optical correlation techniques to the remote sensing of SO_2 plumes using sky light. *Atmos. Environ. (1967)* **1971**, *5*, 677–690.
8. Platt, U.; Stutz, J. *Differential Optical Absorption Spectroscopy: Principles and Application*; Springer: New York, NY, USA, 2008.
9. Platt, U.; Perner, D. Direct measurements of atmospheric CH_2O, HNO_2, O_3, NO_2, and SO_2 by differential optical absorption in the near UV. *J. Geophys. Res.* **1980**, *85*, 7453–7458.
10. Galle, B.; Johansson, M.; Rivera, C.; Zhang, Y.; Kihlman, M.; Kern, C.; Lehmann, T.; Platt, U.; Arellano, S.; Hidalgo, S. Network for Observation of Volcanic and Atmospheric Change (NOVAC)—A global network for volcanic gas monitoring: Network layout and instrument description. *J. Geophys. Res.* **2010**, *115*, doi:10.1029/2009JD011823.
11. Mori, T.; Burton, M. The SO_2 camera: A simple, fast and cheap method for ground-based imaging of SO_2 in volcanic plumes. *Geophys. Res. Lett.* **2006**, *33*, doi:10.1029/2006GL027916.
12. Bluth, G.; Shannon, J.; Watson, I.; Prata, A.; Realmuto, V. Development of an ultra-violet digital camera for volcanic SO_2 imaging. *J. Volcanol. Geotherm. Res.* **2007**, *161*, 47–56.
13. Kantzas, E.P.; McGonigle, A.J.S.; Tamburello, G.; Aiuppa, A.; Bryant, R.G. Protocols for UV camera volcanic SO_2 measurements. *J. Volcanol. Geotherm. Res.* **2010**, *194*, 55–60.
14. Stebel, K.; Amigo, A.; Thomas, H.; Prata, A. First estimates of fumarolic SO_2 fluxes from Putana volcano, Chile, using an ultraviolet imaging camera. *J. Volcanol. Geotherm. Res.* **2015**, *300*, 112–120.
15. McGonigle, A.J.S.; Pering, T.D.; Wilkes, T.C.; Tamburello, G.; D'Aleo, R.; Bitetto, M.; Aiuppa, A.; Willmott, J.R. Ultraviolet Imaging of Volcanic Plumes: A New Paradigm in Volcanology. *Geosciences* **2017**, *7*, 68.
16. D'Aleo, R.; Bitetto, M.; Delle Donne, D.; Tamburello, G.; Battaglia, A.; Coltelli, M.; Patanè, D.; Prestifilippo, M.; Sciotto, M.; Aiuppa, A. Spatially resolved SO_2 flux emissions from Mt Etna. *Geophys. Res. Lett.* **2016**, *43*, 7511–7519.

17. Sweeney, D.; Kyle, P.R.; Oppenheimer, C. Sulfur dioxide emissions and degassing behavior of Erebus volcano, Antarctica. *J. Volcanol. Geotherm. Res.* **2008**, *177*, 725–733.

18. Nicholson, E.; Mather, T.; Pyle, D.; Odbert, H.; Christopher, T. Cyclical patterns in volcanic degassing revealed by SO$_2$ flux timeseries analysis: An application to Soufrière Hills Volcano, Montserrat. *Earth Planet. Sci. Lett.* **2013**, *375*, 209–221.

19. Tamburello, G.; Aiuppa, A.; McGonigle, A.J.S.; Allard, P.; Cannata, A.; Giudice, G.; Kantzas, E.P.; Pering, T.D. Periodic volcanic degassing behavior: The Mount Etna example. *Geophys. Res. Lett.* **2013**, *40*, 4818–4822.

20. Kern, C.; Kick, F.; Lübcke, P.; Vogel, L.; Wöhrbach, M.; Platt, U. Theoretical description of functionality, applications, and limitations of SO$_2$ cameras for the remote sensing of volcanic plumes. *Atmos. Meas. Tech.* **2010**, *3*, 733–749.

21. Lübcke, P.; Bobrowski, N.; Illing, S.; Kern, C.; Alvarez Nieves, J.M.; Vogel, L.; Zielcke, J.; Delgado Granados, H.; Platt, U. On the absolute calibration of SO$_2$ cameras. *Atmos. Meas. Tech.* **2013**, *6*, 677–696.

22. Campion, R.; Delgado-Granados, H.; Mori, T. Image-based correction of the light dilution effect for SO$_2$ camera measurements. *J. Volcanol. Geotherm. Res.* **2015**, *300*, 48–57.

23. Peters, N.; Hoffmann, A.; Barnie, T.; Herzog, M.; Oppenheimer, C. Use of motion estimation algorithms for improved flux measurements using SO$_2$ cameras. *J. Volcanol. Geotherm. Res.* **2015**, *300*, 58–69.

24. Kern, C.; Lübcke, P.; Bobrowski, N.; Campion, R.; Mori, T.; Smekens, J.F.; Stebel, K.; Tamburello, G.; Burton, M.; Platt, U.; et al. Intercomparison of SO$_2$ camera systems for imaging volcanic gas plumes. *J. Volcanol. Geotherm. Res.* **2015**, *300*, 22–36.

25. Tamburello, G.; Kantzas, E.; McGonigle, A.J.S.; Aiuppa, A. Vulcamera: A program for measuring volcanic SO$_2$ using UV cameras. *Ann. Geophys.* **2011**, *54*, 219–221.

26. Peters, N. Plumetrack SO$_2$ Flux Calculator. 2014. Available online: https://ccpforge.cse.rl.ac.uk/gf/project/plumetrack/ (accessed on 11 September 2017).

27. Gliß, J.; Stebel, K.; Kylling, A.; Dinger, A.S.; Sihler, H.; Sudbø, A. Pyplis Website. 2017. GitHub. Available online: https://github.com/jgliss/pyplis (accessed on 13 December 2017).

28. Gliß, J.; Stebel, K.; Kylling, A.; Dinger, A.S.; Sihler, H.; Sudbø, A. Pyplis Website. 2017. Documentation Website. Available online: https://pyplis.readthedocs.io (accessed on 13 December 2017).

29. Kern, C.; Sutton, J.; Elias, T.; Lee, L.; Kamibayashi, K.; Antolik, L.; Werner, C. An automated SO$_2$ camera system for continuous, real-time monitoring of gas emissions from Kīlauea Volcano's summit Overlook Crater. *J. Volcanol. Geotherm. Res.* **2015**, *300*, 81–94.

30. Sihler, H.; Lübcke, P.; Lang, R.; Beirle, S.; de Graaf, M.; Hörmann, C.; Lampel, J.; Penning de Vries, M.; Remmers, J.; Trollope, E.; et al. In-operation field-of-view retrieval (IFR) for satellite and ground-based DOAS-type instruments applying coincident high-resolution imager data. *Atmos. Meas. Tech.* **2017**, *10*, 881–903.

31. Kern, C.; Werner, C.; Elias, T.; Sutton, A.J.; Lübcke, P. Applying UV cameras for SO$_2$ detection to distant or optically thick volcanic plumes. *J. Volcanol. Geotherm. Res.* **2013**, *262*, 80–89.

32. McGonigle, A.J.S.; Hilton, D.R.; Fischer, T.P.; Oppenheimer, C. Plume velocity determination for volcanic SO$_2$ flux measurements. *Geophys. Res. Lett.* **2005**, *32*. L11302.

33. Klein, A.; Lübcke, P.; Bobrowski, N.; Kuhn, J.; Platt, U. Plume propagation direction determination with SO$_2$ cameras. *Atmos. Meas. Tech.* **2017**, *10*, 979–987.

34. Kern, C.; Deutschmann, T.; Vogel, L.; Wöhrbach, M.; Wagner, T.; Platt, U. Radiative transfer corrections for accurate spectroscopic measurements of volcanic gas emissions. *Bull. Volcanol.* **2010**, *72*, 233–247.

35. Kern, C.; Deutschmann, T.; Werner, C.; Sutton, A.J.; Elias, T.; Kelly, P.J. Improving the accuracy of SO$_2$ column densities and emission rates obtained from upward-looking UV-spectroscopic measurements of volcanic plumes by taking realistic radiative transfer into account. *J. Geophys. Res. Atmos.* **2012**, *117*, doi:10.1029/2012JD017936.

36. Vogel, L.; Galle, B.; Kern, C.; Delgado Granados, H.; Conde, V.; Norman, P.; Arellano, S.; Landgren, O.; Lübcke, P.; Alvarez Nieves, J.M.; et al. Early in-flight detection of SO$_2$ via Differential Optical Absorption Spectroscopy: A feasible aviation safety measure to prevent potential encounters with volcanic plumes. *Atmos. Meas. Tech.* **2011**, *4*, 1785–1804.

37. Gliß, J. Geonum. 2016. Available online: https://github.com/jgliss/geonum (accessed on 11 September 2017).

38. Farr, T.G.; Rosen, P.A.; Caro, E.; Crippen, R.; Duren, R.; Hensley, S.; Kobrick, M.; Paller, M.; Rodriguez, E.; Roth, L.; et al. The Shuttle Radar Topography Mission. *Rev. Geophys.* **2007**, *45*, doi:10.1029/2005RG000183.
39. Amante, C.; Eakins, B.W. ETOPO1 Global Relief Model converted to PanMap layer format. *NOAA-Natl. Geophys. Data Center PANGAEA* **2009**, doi:10.1594/PANGAEA.769615.
40. Kraus, S. DOASIS: A framework design for DOAS. Ph.D. Thesis, Universitat Mannheim, Mannheim, Germany, 2006.
41. Farnebäck, G., Two-Frame Motion Estimation Based on Polynomial Expansion. In Proceedings of the 13th Scandinavian Conference on Image Analysis, (SCIA 2003), Halmstad, Sweden, 29 June–2 July 2003; Springer: Berlin, Germany, 2003; pp. 363–370.
42. Bradski, G. The OpenCV Library. *Dr. Dobb's J. Softw. Tools* **2000**, *25*, 120–123.
43. Gliß, J.; Stebel, K.; Kylling, A.; Sudbø, A. Optical flow gas velocity analysis in plumes using UV cameras—Implications for SO_2-emission rate retrievals investigated at Mt. Etna, Italy, and Guallatiri, Chile. *Atmos. Meas. Tech. Discuss.* **2017**, *2017*, 1–30.
44. Fong, D.C.L.; Saunders, M. LSMR: An Iterative Algorithm for Sparse Least-Squares Problems. *SIAM J. Sci. Comput.* **2011**, *33*, 2950–2971.

Sample Availability: The *Pyplis* software is freely available, including the example data and scripts. For more information see http://pyplis.readthedocs.io.

geosciences

MDPI

Article

Volcanic Plume CO$_2$ Flux Measurements at Mount Etna by Mobile Differential Absorption Lidar

Simone Santoro [1,2], Stefano Parracino [2,3], Luca Fiorani [2], Roberto D'Aleo [1], Enzo Di Ferdinando [2], Gaetano Giudice [4], Giovanni Maio [5,†], Marcello Nuvoli [2] and Alessandro Aiuppa [1,4,*]

[1] Dipartimento di Scienze della Terra e del Mare, Università di Palermo, 90123 Palermo, Italy; simone.santoro@unipa.it (S.S.); roberto.daleo01@unipa.it (R.D.)
[2] Nuclear Fusion and Safety Technologies Department, ENEA (Italian National Agency for New Technologies, Energy and Sustainable Economic Development), 00044 Frascati, Italy; stefano.parracino@uniroma2.it (S.P.); luca.fiorani@enea.it (L.F.); enzo.diferdinando@enea.it (E.D.F.); marcello.nuvoli@enea.it (M.N.)
[3] Department of Industrial Engineering, University of Rome "Tor Vergata", 00173 Rome, Italy
[4] Istituto Nazionale di Geofisica e Vulcanologia, 90146 Palermo, Italy; gaetano.giudice@ingv.it
[5] ARES Consortium, 00100 Rome, Italy; prof.maio@gmail.com
*   Correspondence: alessandro.aiuppa@unipa.it
†   Current address: Vitrociset SpA, 00156 Rome, Italy.

Academic Editors: Andrew McGonigle, Giuseppe Salerno and Jesús Martínez-Frías
Received: 23 January 2017; Accepted: 27 February 2017; Published: 3 March 2017

Abstract: Volcanic eruptions are often preceded by precursory increases in the volcanic carbon dioxide (CO$_2$) flux. Unfortunately, the traditional techniques used to measure volcanic CO$_2$ require near-vent, in situ plume measurements that are potentially hazardous for operators and expose instruments to extreme conditions. To overcome these limitations, the project BRIDGE (BRIDging the gap between Gas Emissions and geophysical observations at active volcanoes) received funding from the European Research Council, with the objective to develop a new generation of volcanic gas sensing instruments, including a novel DIAL-Lidar (Differential Absorption Light Detection and Ranging) for remote (e.g., distal) CO$_2$ observations. Here we report on the results of a field campaign carried out at Mt. Etna from 28 July 2016 to 1 August 2016, during which we used this novel DIAL-Lidar to retrieve spatially and temporally resolved profiles of excess CO$_2$ concentrations inside the volcanic plume. By vertically scanning the volcanic plume at different elevation angles and distances, an excess CO$_2$ concentration of tens of ppm (up to 30% above the atmospheric background of 400 ppm) was resolved from up to a 4 km distance from the plume itself. From this, the first remotely sensed volcanic CO$_2$ flux estimation from Etna's northeast crater was derived at ≈2850–3900 tons/day. This Lidar-based CO$_2$ flux is in fair agreement with that (≈2750 tons/day) obtained using conventional techniques requiring the in situ measurement of volcanic gas composition.

Keywords: volcanic plumes; volcanic CO$_2$ flux; remote sensing; Differential Absorption Lidar (DIAL)

1. Introduction

In the last two decades, there have been major advances in the instrumental monitoring of volcanic gas plume composition and fluxes [1]. These have included the first instrumental networks of scanning Differential Optical Absorption Spectrometers (DOAS) for volcanic SO$_2$ flux monitoring, the implementation of satellite-based volcanic gas observations, and the advent of sensor units for in situ gas monitoring [1–6]. Owing to this technical progress, volcanic gas plume composition and fluxes have increasingly been used to extract information on degassing mechanisms/processes [4], and to derive constraints on shallow volcano plumbing systems [5]. However, work still needs to be done to increase the number of volcanic gas species that can be detected in plumes, which remain few if compared to the countless number of chemicals quantified from fumarole direct sampling [1].

Studying volcanic gas plumes has additionally contributed to monitoring, and eventually allowed the prediction of volcano behavior [6]. In particular, it has been shown that, at open-vent persistently degassing volcanoes, volcanic eruptions are often preceded by anomalous increases of the volcanic CO_2 flux [7]. These initial observations have motivated attempts to systematically monitor the volcanic CO_2 flux, and to identify novel measurement strategies [8]. Until recently, however, attempts to remotely sense the volcanic CO_2 flux from distal locations have been limited in number [9,10], while the majority of the observations have involved in situ measurements in the proximity of hazardous active volcanic vents [3]. On Mt. Etna, for example, one of the largest volcanic CO_2 point sources on Earth [11], the volcanic CO_2 flux has systematically been measured since the mid-2000s by combining in situ measurement of the volcanic CO_2/SO_2 ratio (with portable or permanent Multi-Component Gas Analyzer Systems, Multi-GAS; [12–15]) with remotely sensed SO_2 fluxes [16–18]. No successful report exists, at least to the best of our knowledge, of spectroscopy-based detection of Etna's volcanic CO_2 flux from a remote (distal) location.

Within the context of the ERC (European Research Council) starting the grant project BRIDGE (BRIDging the gap between Gas Emissions and geophysical observations at active volcanoes), we designed a new DIAL (Differential Absorption Lidar) [19], with the specific objective to remotely sense the volcanic CO_2 flux. Lidars have only recently been introduced in volcanic gas studies. A CO_2 laser–based lidar was used at Mt. Etna in 2008 [20] and at Stromboli Volcano in 2009 [21] to measure the volcanic plume water vapor flux. More recently, lidars were first been used to target volcanic CO_2 [9,10,22–24]. Our lidar BILLI (BrIdge voLcanic LIdar) [22], for example, has recently been used to successfully retrieve three-dimensional tomographies of volcanic CO_2 in the plumes of Pozzuoli, Solfatara in 2014 [9] and Stromboli volcano in 2015 [23,24]. As such, although gas-sensing lidars remain far less exploited in volcanology than those targeting volcanic ash/particles [25,26], this novel application field may expand rapidly in the near future.

Here, we report on the first successful use of BILLI at Mt. Etna. We show that, in our July–August 2016 Etna experiment, the lidar successfully resolved a volcanic CO_2 signal of a few tens of ppm (in excess to the background air) from more than 4 km of distance, and with good spatial (5 m) and temporal (10 s) resolution. These results are used to derive the first "remote" assessment of Etna's volcanic CO_2 flux. Our observations open new perspectives for routine volcanic CO_2 flux monitoring via lidars.

2. Materials and Methods

2.1. Field Set-Up on Mt. Etna

Observations on Mt. Etna were conducted from 28 July to 1 August 2016, including an initial phase of instrumental setup (28–29 July). Successful CO_2 flux detections were obtained on 31 July, when optimal viewing conditions persisted over the day. The DIAL was mounted in a trailer loaded on a truck, parked at the INGV (Istituto Nazionale di Geofisica e Vulcanologia) observatory "Pizzi Deneri" (Figure 1). The observatory is located at 2823 m a.s.l., northeast of the summit crater of Mt. Etna (3329 m a.s.l.), and at about 3 km from the main degassing vents (Figure 1).

The lidar was used to scan the volcanic plume vertically, keeping a constant azimuth angle (230°) and varying the elevation angle from 7° to 14° (Figure 1). A full 7° to 14° vertical scan was completed in ~15 min, and one atmospheric profile every 10 s was recorded throughout. With this instrumental set-up, the volcanic gas plume of the Etna's northeast crater was investigated (Figure 1), plumes from other craters being either too dilute (southeast crater) or only partially visible (central craters).

At our measurement conditions, two rock surfaces, corresponding to the eastern, outer flanks of the central crater, were intercepted by the laser beam at distances of 1.6 and 2.1 km, and at elevation angles from 7° to 9°. These rock surfaces retro-reflected the laser beam, yielding strong return signals (see below, Figure 2). The volcanic plume, e.g., high in-plume excess CO_2 concentrations, was detected in between the two above rock surfaces, and in the 2.2–4.2 km distance range.

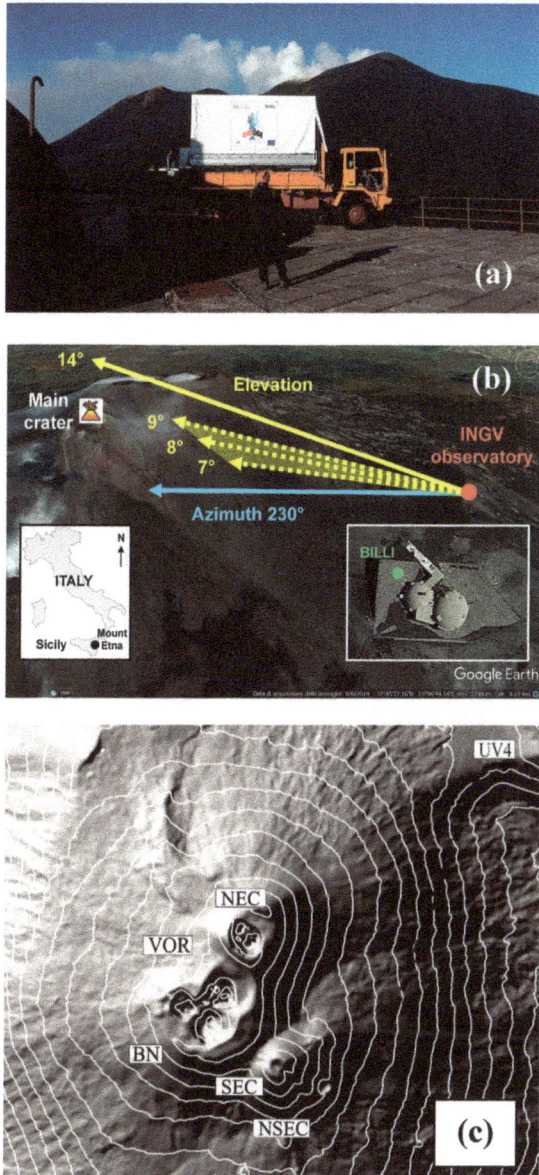

Figure 1. (**a**) The BILLI DIAL (BrIdge voLcanic Lidar, Differential Absorption Lidar) system mounted in a trailer (white) loaded on a truck (orange) at the INGV (Istituto Nazionale di Geofisica e Vulcanologia) observatory "Pizzi Deneri" (the volcanic plume of Mt. Etna is clearly visible); (**b**) Location of Mt. Etna in Sicily, southern Italy (left inset); the truck was parked at the INGV observatory "Pizzi Deneri" (right inset); the laser was fired at constant azimuth and different elevations. The volcanic plume of the northeast crater has been crossed by the laser beam. From 7° to 9° of elevation, rock faces were encountered; (**c**) A map of the summit area showing the active craters and the UV camera site (UV4). NEC: northeast crater; VOR: Voragine; BN: Bocca Nuova (VOR and BN are part of the Etna's central craters); SEC: southeast crater; NSEC: new southeast crater.

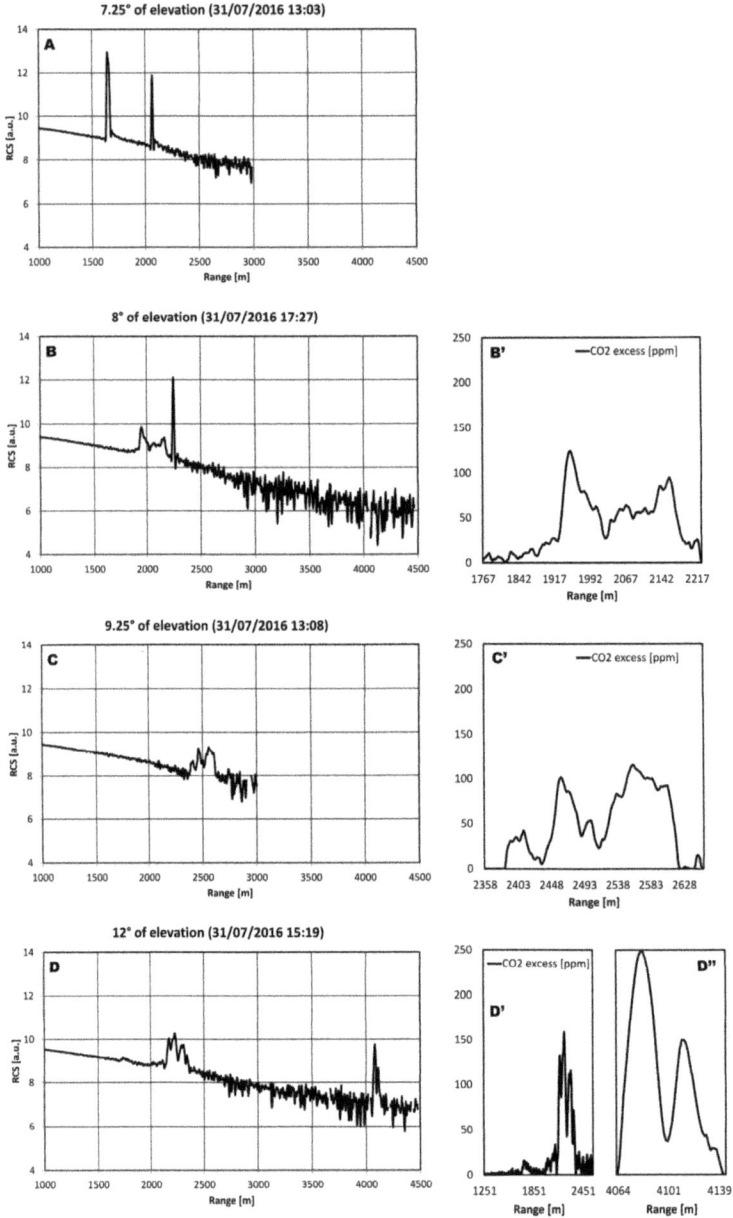

Figure 2. (**A**) Lidar return at 7.25° of elevation: two narrow and defined peaks due to beam backscattering from rock faces are clearly visible (beyond 3000 m only noise was recorded and the corresponding signal is not shown); (**B**) lidar return at 8° of elevation: a wide and jagged peak from the volcanic plume and a narrow and defined peak from a rock face are clearly visible; the CO_2 profile inside the volcanic plume is shown in (**B′**); (**C**) lidar return at 9.25° of elevation: a wide and jagged peak from the volcanic plume is clearly visible; the CO_2 profile inside the volcanic plume is shown in (**C′**) (beyond 3000 m only noise was recorded and the corresponding signal is not shown); (**D**) lidar return at 12° of elevation: two wide and jagged peaks from the volcanic plume are clearly visible; the CO_2 profiles inside the volcanic plume are shown in (**D′,D″**).

2.2. DIAL

The main components of a lidar are the transmitter (laser) and the receiver (telescope). A lidar is merely an optical radar [19]: a laser pulse is transmitted to the atmosphere, and some of its photons are backscattered to the telescope by air molecules and aerosols (droplets, particles etc.). The optical power corresponding to this photon flux is transformed into an electronic signal by photodetector and preamplifier, and converted in digital signal by an ADC (analog-to-digital converter).

The chemico-physical properties of the atmosphere along the laser beam, at distance R (range) from the lidar, can be inferred from analysis of the detected signal as a function of t, the time interval between emission and detection. R and t are linked by the relation $R = ct/2$, where c is the speed of light. The returned signal to the lidars' telescope, as a function of R (or t), then yields an atmospheric profile (Figure 2). In other words, an atmospheric profile is a range-resolved characterization of the lidar returned signal, which allows studying the air/plume optical properties along the light trajectory.

Air attenuates the laser pulse due to molecules and aerosol scattering and to the specific absorption of gases: if the laser wavelength coincides with absorption lines of a target gas, the attenuation will be stronger. A DIAL takes advantage of this effect: unlike a usual lidar, two wavelengths, ON and OFF, are transmitted, with only the former being absorbed by the target gas (Figure 3).

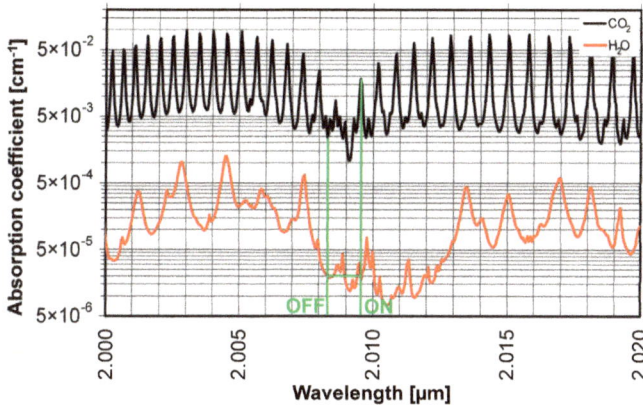

Figure 3. Carbon dioxide and water vapor absorption coefficients (from [27]) around the ON and OFF wavelengths (indicated in green).

If the absorption line is narrow, and ON and OFF wavelengths are close enough, the target gas concentration along the lidar optical path can be derived from the ratio between the OFF and ON signals. In this application, we selected the following wavelengths (Figure 3): ON, 2009.537 nm; OFF, 2008.484 nm. This selection was motivated by: (i) the CO_2 absorption is relatively low, thus allowing the system to probe far ranges (beyond 4 km). If a stronger line had been used, the ON laser pulse would have extinguished before; (ii) the beam energy (depending mainly on the dye efficiency curve) and the detector responsivity are near their maximum; and (iii) the H_2O absorption is very low (Figure 3). Moreover, ON and OFF have been chosen so that the differential absorption of H_2O is approximately zero (within the uncertainty of the spectroscopic data), thus minimizing the interference of water vapor to the carbon dioxide measurement.

In our case, the transmitter and the receiver are coaxial, and the lidar field-of-view can be aimed in the whole atmosphere thanks to a system made of two large elliptical mirrors [22]. This configuration allows the experimenters to scan the plume in both horizontal and vertical planes, thus measuring CO_2 concentrations both outside and inside the volcanic plume [9]. From this, by scanning the volcanic gas plume from different angles and viewing directions, the CO_2 distribution in a cross-section of the volcanic plume can be retrieved. This, combined with independent knowledge of plume speed and altitude, allows the CO_2 flux to be retrieved.

The reader is referred to previous work [9,22–24] for details on instrumental setup and data processing. The systematic error associated with the derived CO_2 concentrations is dominated by imprecision in wavelength setting [22]. This leads to inaccuracy in the differential absorption cross section, and thus in gas concentration. Thanks to a photo-acoustic cell filled with pure CO_2 at atmospheric pressure and temperature, the ON and OFF wavelengths were set before each atmospheric scan. The residual imprecision [23] of ± 0.02 cm^{-1} (half laser linewidth: half width at half maximum of the energy transmitted by the laser system (J) vs the wavenumber (cm^{-1})) implies a systematic error on CO_2 concentrations of 5.5% [24]. The statistical error of CO_2 measurement has been calculated by usual error propagation techniques from the standard deviation of the lidar signal. At 2.5 km, a mean range, it is about 2%, while it can exceed 5% at 4.2 km. Table 1 compares the instrumental set-up during the Mt. Etna field campaign, with those used at Solfatara [9,22], and Stromboli [23,24].

Table 1. Summary of field operational conditions at Pozzuoli Solfatara, Stromboli, and Mt. Etna (this study).

Campaign	Pozzuoli Solfatara	Stromboli Volcano	Mt. Etna
Latitude	40°49′46.28″N	38°48′06.69″N	37°45′57.28″N
Longitude	14°08′50.51″E	15°14′25.69″E	15°00′59.65″E
Period	13–17 October 2014	24–29 June 2015	28 July–1 August 2016
Azimuth scan	196°–234°	235.3°–253.6°	230°
Elevation scan	0°–18°	15.2°–27.4°	7°–14°

3. Results

Figure 2 shows examples of lidar returns obtained during our Etna campaign. Results are illustrated for four atmospheric profiles taken on 31 July (the best measurement day) at four distinct elevations, and are shown in the form of range vs. RCS (range corrected signal) scatter plots.

During its atmospheric propagation, the laser beam intensity approximately decreased:

- exponentially, due to atmospheric extinction, according to the Lambert-Beer law, and;
- as $1/R^2$, because the solid angle subtended by the receiver is A/R^2, where A is the telescope effective area.

For these reasons, it is a common practice in lidar science to express results using a RCS, this being the logarithm of the product of the signal times the square of the range. To improve the SNR (signal-to-noise ratio), the RCS was obtained by averaging 50 laser shots for each lidar return, and a 13-point Savitzky-Golay filter was applied [28].

During a vertical scan, the measured range-resolved RCS profiles varied as the laser elevation was sequentially increased. Below 7.25° elevation, the laser beam hit a first rock surface at about a 1.6 km distance. Laser beam retro-reflection at this rock surface produced, in the lidar return signal, a strong, narrow RCS peak at $R = 1.6$ km. At 7.25° elevation (Figure 2a), only part of the beam was intercepted by the $R = 1.6$ km rock surface, while the remaining part impinged on the rock surface at $R = 2.1$ km, producing a second narrow RCS peak. For geometrical reasons, an elevation increase corresponded to an increase in the range at which the rock surfaces were encountered, e.g., the second rock surface was encountered at $R = 2.1$ at 7.25° elevation, shown in Figure 2a, and at $R = 2.3$ km at 8° elevation, shown in Figure 2b. No rock surface was hit by the laser beam at elevations >9°, e.g., note the absence of narrow RCS peaks in Figure 2c,d.

Back-scattering of the laser beam by the volcanic plume produced wide and jagged RCS peaks, therefore very distinct from the narrow and defined peaks produced by beam retro-reflection at rock surfaces (compare the two peak shapes in Figure 2b).

The volcanic plume was detected at range distances in between the two rock surfaces up to a 9° elevation (e.g., Figure 2b), or beyond them at a 9° to 14° elevation (Figure 2c,d). A broad, irregular RCS peak in the lidar returns, corresponding to the volcanic plume, was resolved up to a maximum measurement range of 4.2 km (Figure 2d).

We used the procedure detailed in References [9,24] to convert the RCS profiles into range-resolved profiles of in-plume excess CO_2 concentrations (see Figure 2b′,c′,d′,d″). This procedure involves calculating the excess CO_2 concentration corresponding to each i-th ADC channel of the lidar profile from:

$$C_{CO2,i} = k\,RCS_i \tag{1}$$

$$k = \frac{\Delta C\,(R_1 - R_2)}{\Delta R \sum_i RCS_i} \tag{2}$$

where ΔR and RCS_i are, respectively, the range interval and range corrected signal corresponding to each ADC channel; R_1 and R_2 are the range distances of the two above rock surfaces; and ΔC is the average excess CO_2 concentration in the air/plume parcel between them (this is obtained from the intensity contrast of lidar returns produced by the two rock surfaces). The term "excess" implies that the reported CO_2 concentrations are after subtraction of the ambient atmospheric background, and therefore correspond to the "volcanic" CO_2 levels in the plume. The ambient atmospheric CO_2 background was obtained from the processing of lidar returns in the 0–1.6 km range distance, where no plume signal was detected (see Reference [9] for details of calculations).

At an 8° elevation, shown in Figure 2b′, the volcanic plume was evidenced by a band of excess CO_2 concentrations of ≤125 ppm. These excess CO_2 concentrations agree well with those derived by in situ in-plume measurements with conventional techniques (e.g., the Multi-GAS), from which in-plume CO_2 concentrations of tens to hundreds of ppm above ambient air are typically obtained [12]. The plume appears to be about 300 m thick; this relatively narrow plume's cross-section was probably justified by the fact that, at such an 8° elevation, the laser beam intercepted the volcanic plume at below the summit crater's rim altitude. Due to its close proximity to the crater slopes, the volcanic plume was, at least partly, protected from the local wind field, a fact that reduced its atmospheric dispersion. The volcanic plume was still relatively narrow at a ~9° elevation (Figure 2c′), where the laser beam pointed just above the summit crater's rim. At even higher elevations, the volcanic plume was wider and scattered by the wind, and the returned RCS often presented multiple peaks (Figure 2d,d′).

As explained before (Section 2), a sequence of atmospheric profiles was acquired as the lidar vertically scanned the horizon, from a 7° to 14° (max) elevation. All CO_2 profiles (e.g., Figure 2), taken at different elevations during a single lidar rotation sequence, were combined and integrated to obtain a CO_2 scan, examples of which are illustrated in Figure 4. On 31 July 2016, the most fruitful day, 19 scans were obtained. Each scan consisted of 24 profiles, all at a 230° azimuth. These profiles covered the elevation angle interval (between 7° and 13°) with an angular resolution of about 0.25°.

The results are illustrated in the form of contour maps of excess CO_2 concentrations, plotted as a function of the range and elevation. The colored spots correspond to areas of high excess CO_2 (the natural background is dark blue), and therefore illustrate the spatial distribution and temporal evolution of the volcanic plume (the yellow lines delimit the positions of the laser beam reflections off the rock surfaces). In all the maps we obtained (see examples in Figure 4), the structure of the plume was well resolved. The plume was tracked as a cluster of high CO_2 concentration spots, trending from about a 9° elevation and $R = 2.4$ km (the vent rim) to a 13° elevation and $R = 2.5–2.9$ km. As such, our CO_2 concentration maps were consistent with a gently lofting volcanic plume (Figure 4), with vertical and horizontal movements driven by thermal buoyancy and by the local wind field pattern. The maps indicate the plume was being dispersed away from the lidar during our observations, since the range of volcanic plume detection increased with the elevation in all the maps.

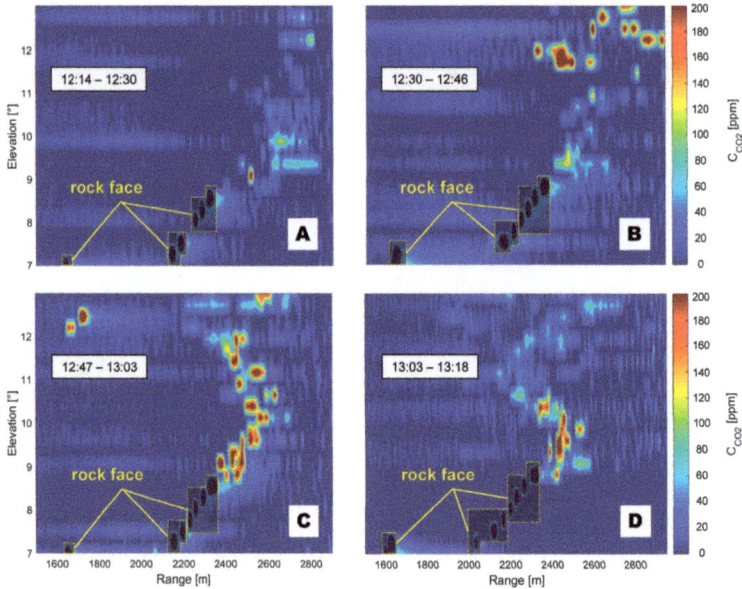

Figure 4. Vertical scans (fixed azimuth: 230°) of the volcanic plume (CO_2 excess) acquired on 31 July 2016; (**A**) from 12:14 p.m. to 12:30 p.m.; (**B**) from 12:30 p.m. to 12:46 p.m.; (**C**) from 12:47 p.m. to 1:03 p.m. and (**D**) from 1:03 p.m. to 1:18 p.m. (local civil time). At this azimuth, in the elevation interval between 7° and 9°, the laser beam is back-scattered by rock faces, thus causing signal peaks not due to the volcanic plume (rock faces are sorted out from real CO_2 by the correspondence of narrow peaks with certain range values).

The maps of Figure 4 set the basis for the calculation of the volcanic CO_2 flux. In analogy with previous work [9], we obtained the volcanic CO_2 flux by integrating the background-corrected (excess) CO_2 concentrations over a plume cross-section (from the maps of Figure 4), which allowed us to derive the plume CO_2 molecular density. This was then multiplied by the plume transport speed to obtain the CO_2 flux (ΦCO_2, in Kg·s^{-1}), as:

$$\Phi_{CO_2} = v_p \cdot \frac{PM_{CO_2}}{10^3 N_A} \cdot N_{molCO2-total} \tag{3}$$

where v_p is the plume transport speed (in m/s); $N_{molCO2-total}$ is the total-plume CO_2 molecular density (expressed in molecules·m^{-1}); and PM_{CO2} and N_A are, respectively, the CO_2 molecular weight and Avogadro's constant. The term $N_{molCO2-total}$ was obtained by integrating the effective average excess CO_2 concentrations ($\overline{C_{exc,i}}$ [ppm]) over the entire plume cross-section, according to:

$$N_{molCO2-total} = N_h \cdot 10^{-6} \cdot \sum_i \overline{C_{exc,i}} \cdot A_i \tag{4}$$

where N_h is the atmospheric number density (molecules·m^{-3}) at the crater's summit height, and A_i represents the i-th effective plume area.

The plume transport speed was inferred at 9.7 ± 0.8 m/s from the processing of plume images taken on the same day by the permanent UV camera system (UV4) in use at the Pizzi Deneri observatory since 2014; see Reference [29] for details on the instrument. The UV camera images were processed using an optical flow sub-routine using the Lukas/Kanade algorithm [30,31], integrated in the Vulcamera software [32] (same methodology as described in [29]).

Our derived CO_2 fluxes are illustrated in Figure 5. The CO_2 flux varied from 1235 to 8050 tons/day during the measurement interval, and averaged at 2850 ± 1800 tons/day.

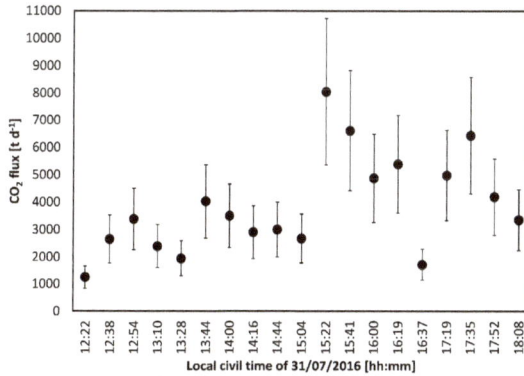

Figure 5. CO_2 flux from the northeast crater retrieved on 31 July 2016 from 12:22 p.m. to 6:08 p.m. (local civil time). The error bars indicate the inferred CO_2 flux error (±33%), as based upon the error propagation of the plume speed an in-plume integrated CO_2 amounts (procedure detailed in [24]).

4. Discussion

As long-term volcanic gas records have increased in number and quality over the last few decades [33], full empirical evidence has emerged for precursory increases of the volcanic CO_2 flux emissions prior to eruption of mafic to intermediate volcanoes [7]. However, remote direct measurements of the volcanic CO_2 flux, which are intrinsically safer for operators and more prone to provide continuous, long-term observations, have remained impossible until recently [9,10].

Our results here support the ability of the DIAL-Lidar BILLI to profile atmospheric CO_2 concentrations over large optical paths (Figures 2 and 4), and to remotely sense the CO_2 flux from distal (up to 4 km distant) sources (Figure 5). This instrument thus promises a real step ahead in the remote observation of volcanic gas emissions. Improved CO_2 flux measurements are not only vital for better gas-based volcano monitoring, but are also needed to better constrain the global volcanic CO_2 budget, which is still inaccurately known [8].

The volcanic CO_2 flux from Mt. Etna has been assessed in the past by either in-plume airborne CO_2 profiling [11], or by indirect methods involving in situ measurements of plume CO_2/SO_2 ratios, via either the Multi-GAS [3,12–15] or Fourier Transform InfraRed Spectrometers (FTIR; [16]). To the best of our knowledge, our results are the first to report a direct, remote quantification of Etna's CO_2 flux.

Our lidar results show that, in the circa 5-h-long observational widow, the CO_2 flux from Etna's northeast crater varied from 1235 to 8050 tons/day (Figure 5). The CO_2 flux was somewhat higher, typically >4000 tons/day and up to 8050 tons/day, after 3 p.m. local time, relative to the 12–3 p.m. period (<4000 tons/day) (Figure 5). No change in activity was yet observed at the northeast crater, which continued to exhibit quiescent degassing over the entire measurement interval. We therefore consider the observed variation as part of the normal fluctuations in degassing activity that occur at Etna, likely in response to temporal variations in the magma/gas transport rate in the volcano's feeding conduits [15–18]. By taking the arithmetic mean of the individual CO_2 flux measurements in Figure 5, we would obtain a time-averaged CO_2 flux of 2850 ± 1800 tons/day for 31 July 2016. In view of the non-stationary CO_2 emission behavior captured by our high-temporal resolution measurement (Figure 5), we also perform an independent exercise in which we calculated the total CO_2 output from the northeast crater by integrating (in the time domain) the available CO_2 flux measurements, each representative of 13–18 min of observation (the mean duration of scans was

15 min). From this, we obtained that ≈796 tons of CO_2 were cumulatively released during 5 h of observations, implying a time-averaged CO_2 flux of 3900 tons/day. This is about 30% higher than, but within one standard deviation of, the CO_2 flux obtained above from a simple arithmetic mean approach (2850 ± 1800 tons/day).

In the attempt to add confidence to our results, we compared our lidar-based CO_2 flux with independent estimates based upon a more conventional technique that involves a combination of SO_2 fluxes and plume CO_2/SO_2 ratios (Figure 6). Our permanent UV camera system (UV4) at Pizzi Deneri indicated, for the morning of the same 31 July, a time-averaged SO_2 flux of 645 ± 125 tons/day. This is the mean (±1 standard deviation) of 4 h of observations at a 0.5 Hz rate (Figure 6; same methodology as in [29]). Our inferred northeast crater's SO_2 flux (645 ± 125 tons/day) corresponded to about 30% of the total volcano's SO_2 emissions (≈2200 tons/day). These latter emissions were inferred using the same UV camera system, and were thus primarily determined by the central craters (not targeted by our DIAL-Lidar). The northeast crater's volcanic plume was in situ measured by a portable Multi-GAS instrument (the same as in [12,13]) two days later. These in situ observations yielded a (molar) CO_2/SO_2 ratio of ≈6, demonstrating the usual [12,16] CO_2-poor composition of the northeast crater (the simultaneously observed CO_2/SO_2 ratio of the central crater's plume was ≈16). We consider our Multi-GAS–derived composition on 2 August as still representative of the northeast crater's emissions on 31 July, since volcanic activity at that crater did not exhibit any substantial change in between the two days. By combining the two sets of data together, we converted the SO_2 flux time-series into a 4-h-long CO_2 flux time-series (Figure 6), from which an averaged (arithmetic mean) UV-Camera + MultiGAS CO_2 flux of ≈2750 tons/day was obtained. This is close to our lidar-based estimates above (≈2850–3900 tons/day). We caution that the two independent CO_2 flux time-series (from lidar and UV-Camera + MultiGAS) are not temporally overlapping, since the UV camera system ran only in the morning, when sunlight conditions were optimal [29], while our successful CO_2 flux measurement with the lidar started a few hours later in the afternoon. In addition, the UV-Camera + MultiGAS used a constant CO_2/SO_2 ratio (of six) throughout the entire UV camera temporal window, while it is valid only as a first approximation. However, the close CO_2 flux values inferred from lidar and UV-Camera + MultiGAS provide mutual validation for the two independent techniques.

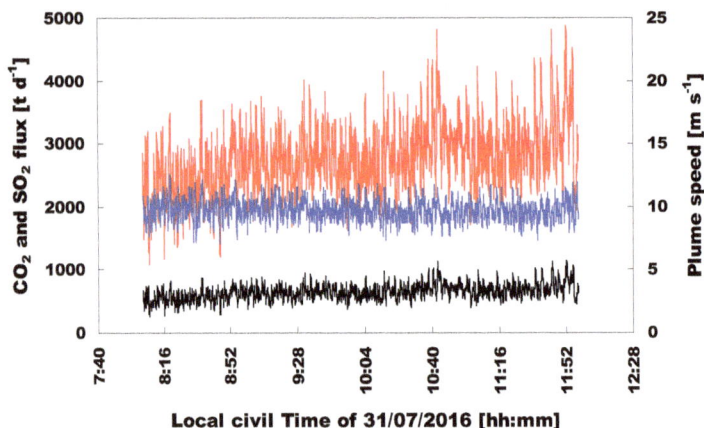

Figure 6. Time-series of CO_2 flux emissions from the northeast crater (in red) obtained from the UV-Camera + Multi-Gas technique. These were calculated by converting the SO_2 flux time-series (in black) obtained by the UV4 permanent UV camera system on 31 July 2016 (from 8 a.m. to 12 p.m., local civil time) using a CO_2/SO_2 ratio (molar) of six. The plume speed time-series calculated from the UV camera on the same 31 July is also shown (blue trend).

5. Conclusions

We have shown for the first time that the volcanic CO_2 flux can be detected with lidar from up to a 4 km distance. During our Mt. Etna field campaign, our DIAL-Lidar BILLI vertically scanned the volcanic plume while profiling CO_2 concentrations every 10 s, with a spatial resolution of 5 m. With this configuration, we successfully detected an excess volcanic CO_2 signal of a few tens of ppm, with relatively low systematic and statistical errors (5.5% and 2%, respectively). By integrating the results of the atmospheric profile taken at different heading angles, and covering a full scan of the plume, the volcanic CO_2 flux was derived (after integration, and in combination with the plume transport speed) at ≈2850–3900 tons/day. This lidar-based flux is close to that independently obtained by in situ observations of the volcanic plume (≈2750 tons/day), which combined Multi-GAS in situ sensing of the plume composition and remotely sensed (UV camera) SO_2 fluxes.

Clearly, additional field tests are required to validate our novel technique even further. Still, our results suggest BILLI is a major advance in ground-based observations of volcanic plumes. The instrument allows the remote measurements of volcanic CO_2 (and particles, if desired) from distal (safe) areas, and with unprecedented temporal resolution and high spatial coverage. Further development is now required to make this technology an operational tool for routine volcanic gas observations. Efforts are currently being undertaken to reduce the weight and power requirements (the current prototype is ~1100 kg and requires 6.5 kW), and to implement more user-friendly operational routines and software. These implementations are required to widen the application range of the lidar, and to allow its use in remote/harsh volcanic environments.

Acknowledgments: The authors are grateful to ENEA, in general, and Aldo Pizzuto, Roberta Fantoni and Antonio Palucci, in particular, for constant encouragement. They thank the staff of INGV-OE, and especially the Director Eugenio Privitera and Salvatore Consoli, for logistical support and for granting access to the INGV observatory "Pizzi Deneri". The support from the ERC project BRIDGE, n. 305377, is gratefully acknowledged. This work benefitted from the insightful comments of two anonymous reviewers.

Author Contributions: S.S., L.F and A.A. conceived and designed the experiments; S.S., L.F, R.D., E.D.F., M.N., G.G., and A.A performed the experiments; S.P., G.M., and R.D. analyzed the data; A.A. L.F. and S.S. wrote the paper with contributions from all co-authors.

Conflicts of Interest: The authors declare no conflict of interest.

References

1. Oppenheimer, C.; Fischer, T.P.; Scaillet, B. Volcanic Degassing: Process and Impact. In *Treatise on Geochemistry, The Crust*, 2nd ed.; Holland, H.D., Turekian, K.K., Eds.; Elsevier: Amsterdam, The Netherlands, 2014; Volume 4, pp. 111–179.
2. Carn, S.A.; Clarisse, L.; Prata, A.J. Multi-decadal satellite measurements of global volcanic degassing. *J. Volcanol. Geothermal Res.* **2016**, *311*, 99–134. [CrossRef]
3. Aiuppa, A. Volcanic gas monitoring. In *Volcanism and Global Environmental Change*; Schmidt, A., Fristad, K.E., Elkins-Tanton, L.T., Eds.; Cambridge University Press: Cambridge, UK, 2015; pp. 81–96.
4. Edmonds, M. New geochemical insights into volcanic degassing. *Philos. Trans. R. Soc. A Math. Phys. Eng. Sci.* **2008**, *366*, 4559–4579. [CrossRef] [PubMed]
5. Allard, P.; Carbonnelle, J.; Métrich, N.; Loyer, H.; Zettwoog, P. Sulphur output and magma degassing budget of Stromboli volcano. *Nature* **1994**, *368*, 326–330. [CrossRef]
6. Saccorotti, G.; Iguchi, M.; Aiuppa, A. In situ Volcano Monitoring: Present and Future. In *Volcanic Hazards, Risks and Disasters*; Elsevier: Amsterdam, The Netherlands, 2014; pp. 169–202.
7. Aiuppa, A.; Burton, M.; Caltabiano, T.; Giudice, G.; Guerrieri, S.; Liuzzo, M.; Murè, F.; Salerno, G. Unusually large magmatic CO_2 gas emissions prior to a basaltic paroxysm. *Geophys. Res. Lett.* **2010**, *37*, L17303. [CrossRef]
8. Burton, M.R.; Sawyer, G.M.; Granieri, D. Deep carbon emissions from volcanoes. *Rev. Mineral. Geochem.* **2013**, *75*, 323–354. [CrossRef]

9. Aiuppa, A.; Fiorani, L.; Santoro, S.; Parracino, S.; Nuvoli, M.; Chiodini, G.; Minopoli, C.; Tamburello, G. New ground-based lidar enables volcanic CO_2 flux measurements. *Sci. Rep.* **2015**, *5*, 13614. [CrossRef] [PubMed]
10. Queißer, M.; Granieri, D.; Burton, M. A new frontier in CO_2 flux measurements using a highly portable DIAL laser system. *Sci. Rep.* **2016**, *6*, 33834.
11. Allard, P.; Carbonnelle, J.; Dajlevic, D.; le Bronec, J.; Morel, P.; Robe, M.C.; Maurenas, J.M.; Faivre-Pierret, R.; Martin, D.; Sabroux, J.C.; et al. Eruptive and diffuse emissions of CO_2 from Mount Etna. *Nature* **1991**, *351*, 387–391. [CrossRef]
12. Aiuppa, A.; Federico, C.; Giudice, G.; Gurrieri, S.; Liuzzo, M.; Shinohara, H.; Favara, R.; Valenza, M. Rates of carbon dioxide plume degassing from Mount Etna volcano. *J. Geophys. Res.* **2006**, *111*, B09207. [CrossRef]
13. Aiuppa, A.; Giudice, G.; Gurrieri, S.; Liuzzo, M.; Burton, M.; Caltabiano, T.; McGonigle, A.J.S.; Salerno, G.; Shinohara, H.; Valenza, M. Total volatile flux from Mount Etna. *Geophys. Res. Lett.* **2008**, *35*, L24302. [CrossRef]
14. Patanè, D.; Aiuppa, A.; Aloisi, M.; Behncke, B.; Cannata, A.; Coltelli, M.; di Grazia, G.; Gambino, S.; Gurrieri, S.; Mattia, M.; et al. Insights into magma and fluid transfer at Mount Etna by a multiparametric approach: A model of the events leading to the 2011 eruptive cycle. *J. Geophys. Res. Solid Earth* **2013**, *118*, 3519–3539. [CrossRef]
15. Pering, T.D.; Tamburello, G.; McGonigle, A.J.S.; Aiuppa, A.; Cannata, A.; Giudice, G.; Patanè, D. High time resolution fluctuations in volcanic carbon dioxide degassing from Mount Etna. *J. Volcanol. Geotherm. Res.* **2014**, *270*, 115–121. [CrossRef]
16. La Spina, A.; Burton, M.; Salerno, G.G. Unravelling the processes controlling gas emissions from the central and northeast craters of Mt. Etna. *J. Volcanol. Geotherm. Res.* **2010**, *198*, 368–376. [CrossRef]
17. Caltabiano, T.; Burton, M.; Giammanco, S.; Allard, P.; Bruno, N.; Murè, F.; Romano, R. Volcanic Gas Emissions from the Summit Craters and Flanks of Mt. Etna, 1987–2000. In *Mt. Etna: Volcano Laboratory*; Bonaccorso, A., Calvari, S., Coltelli, M., del Negro, C., Falsaperla, S., Eds.; Geophysical Monograph Series; American Geophysical Union: Washington, DC, USA, 2004; Volume 143, pp. 111–128.
18. Allard, P.; Behncke, B.; D'Amico, S.; Neri, M.; Gambino, S. Mount Etna 1993–2005: Anatomy of an evolving eruptive cycle. *Earth-Sci. Rev.* **2006**, *78*, 85–114. [CrossRef]
19. Fiorani, L. Lidar application to litosphere, hydrosphere and atmosphere. In *Progress in Laser and Electro-Optics Research*; Koslovskiy, V.V., Ed.; Nova: New York, NY, USA, 2010; pp. 21–75.
20. Fiorani, L.; Colao, F.; Palucci, A. Measurement of Mount Etna plume by CO_2-laser-based lidar. *Opt. Lett.* **2009**, *34*, 800–802. [CrossRef] [PubMed]
21. Fiorani, L.; Colao, F.; Palucci, A.; Poreh, D.; Aiuppa, A.; Giudice, G. First-time lidar measurement of water vapor flux in a volcanic plume. *Opt. Commun.* **2011**, *284*, 1295–1298. [CrossRef]
22. Fiorani, L.; Santoro, S.; Parracino, S.; Nuvoli, M.; Minopoli, C.; Aiuppa, A. Volcanic CO_2 detection with a DFM/OPA-based lidar. *Opt. Lett.* **2015**, *40*, 1034–1036. [CrossRef] [PubMed]
23. Fiorani, L.; Santoro, S.; Parracino, S.; Maio, G.; Nuvoli, M.; Aiuppa, A. Early detection of volcanic hazard by lidar measurement of carbon dioxide. *Nat. Hazards* **2016**, *83*, S21–S29. [CrossRef]
24. Aiuppa, A.; Fiorani, L.; Santoro, S.; Parracino, S.; D'Aleo, R.; Liuzzo, M.; Maio, G.; Nuvoli, M. Advances in Dial-Lidar-based remote sensing of the volcanic CO_2 flux. *Front. Earth Sci.* **2017**, *5*, 15. [CrossRef]
25. Scollo, S.; Boselli, A.; Coltelli, M.; Leto, G.; Pisani, G.; Spinelli, N.; Wang, X. Monitoring Etna volcanic plumes using a scanning LiDAR. *Bull. Volcanol.* **2012**, *74*, 2383–2395. [CrossRef]
26. Wanga, X.; Boselli, A.; D'Avino, L.; Pisani, G.; Spinelli, N.; Amodeo, A.; Chaikovsky, A.; Wiegner, M.; Nickovic, S.; Papayannis, A.; et al. Volcanic dust characterization by EARLINET during Etna's eruptions in 2001–2002. *Atmos. Environ.* **2008**, *42*, 893–905. [CrossRef]
27. Rothman, L.S.; Gordon, I.E.; Babikov, Y.; Barbe, A.; Benner, D.C.; Bernath, P.F.; Birk, M.; Bizzocchi, L.; Boudon, V.; Brown, L.R.; et al. The HITRAN2012 molecular spectroscopic database. *J. Quant. Spectrosc. Radiat. Transf.* **2013**, *130*, 4–50. [CrossRef]
28. Schafer, R.W. What is a Savitzky-Golay Filter? *IEEE Signal Process Mag.* **2011**, *28*, 111–117. [CrossRef]
29. D'Aleo, R.; Bitetto, M.; Delle Donne, D.; Tamburello, G.; Battaglia, A.; Coltelli, M.; Patanè, D.; Prestifilippo, M.; Sciotto, M.; Aiuppa, A. Spatially resolved SO_2 flux emissions from Mt Etna. *Geophys. Res. Lett.* **2016**, *43*. [CrossRef] [PubMed]

30. Bruhn, A.; Weickert, J.; Schnörr, C. Lucas/Kanade Meets Horn/Schunck: Combining Local and Global Optic Flow Methods. *Int. J. Comput. Vis.* **2005**, *61*, 211–231. [CrossRef]
31. Peters, N.; Hoffmann, A.; Barnie, T.; Herzog, M.; Oppenheimer, C. Use of motion estimation algorithms for improved flux measurements using SO_2 cameras. *J. Volcanol. Geotherm. Res.* **2015**, *300*, 58–69. [CrossRef]
32. Tamburello, G.; Kantzas, E.P.; McGonigle, A.J.S.; Aiuppa, A. Vulcamera: A program for measuring volcanic SO_2 using UV cameras. *Ann. Geophys.* **2011**, *54*, 2. [CrossRef]
33. Fischer, T.P.; Chiodini, G. Volcanic, Magmatic and Hydrothermal Gas Discharges. In *Encyclopaedia of Volcanoes*, 2nd ed.; Elsevier: Amsterdam, The Netherlands, 2015; pp. 779–797.

MDPI

St. Alban-Anlage 66

4052 Basel

Switzerland

Tel. +41 61 683 77 34

Fax +41 61 302 89 18

www.mdpi.com

Geosciences Editorial Office

E-mail: geosciences@mdpi.com

www.mdpi.com/journal/geosciences

www.ingramcontent.com/pod-product-compliance
Lightning Source LLC
Chambersburg PA
CBHW051728210326
41597CB00032B/5644